Microbial Systematics
Biomolecules and Natural Macromolecules Applications

Editor

Bhagwan Narayan Rekadwad

Assistant Professor of Microbiology
Division of Microbiology and Biotechnology
Yenepoya Research Centre
Yenepoya (Deemed to be University)
Mangalore, Karnataka
India

CRC Press
Taylor & Francis Group
Boca Raton London New York

CRC Press is an imprint of the
Taylor & Francis Group, an **informa** business

A SCIENCE PUBLISHERS BOOK

Cover credit: The image used on the cover is prepared by the editor, Dr. Bhagwan Narayan Rekadwad.

First edition published 2023
by CRC Press
6000 Broken Sound Parkway NW, Suite 300, Boca Raton, FL 33487-2742

and by CRC Press
4 Park Square, Milton Park, Abingdon, Oxon, OX14 4RN

CRC Press is an imprint of Taylor & Francis Group, LLC

Library of Congress Cataloging-in-Publication Data (applied for)

ISBN: 978-1-032-30983-5 (hbk)
ISBN: 978-1-032-31025-1 (pbk)
ISBN: 978-1-003-30767-9 (ebk)

DOI: 10.1201/9781003307679

Typeset in Palatino
by Radiant Productions

Preface

Microbial Systematics is the science of *the characterization, classification,* and *nomenclature* of Prokaryotes. When studying microorganisms, the most important factor is a characteristic of microbial species that distinguishes them from others or makes them similar to each other. This provides a thorough understanding of their origins and products that benefit humanity. Nowadays, the world is getting more familiar with products from microbial sources from extreme environments (hot springs, hydrothermal vents, volcanoes, marine water, coastal soil, salterns, mangroves, cold-environments (glaciers, Antarctica, Arctic), ionised places, etc.). Species isolated from these habitats are always diverse, which indicates biomolecules secreted by these species might have importance. A cure for disease is a desire considering the catastrophic situations that threaten the lives of humans and animals. Microbial Systematics has always been understood as basic science by the world. Indeed, it is not! Microbial Systematics is an advanced science that uses characteristics of microorganisms to explore further applications and uses of the microbial world for the benefit of mankind.

August 2022 **Bhagwan Narayan Rekadwad**

Contents

Preface iii

1. **Current Scenario of Application of Anti-infective Compounds from Microbial Origin Perspective** 1
   Sanjana N.S., Bhagwan Narayan Rekadwad, Punchappady Devasya Rekha, Arun A. Bhagwath and *Mangesh Vasant Suryavanshi*

2. **Potential Biomolecules of Microbial Origin Against Infectious Diseases** 14
   Homem N.C., Paixão R.M., Miranda C.S., Antunes J.C., Amorim M.T.P. and *Felgueiras H.P.*

3. **Biomolecules and Natural Macromolecules Against Tuberculosis** 43
   Awalagaway Dhulappa, Shuang Wang and *Manik Prabhu Narsing Rao*

4. **Biomolecules for Treatment of Communicable Diseases** 56
   Chamma Gupta, Abhishek Byahut, Karma G. Dolma and *Mingma L. Sherpa*

5. **Marine Organisms-Derived Bioactive Molecules for Infectious Disease Treatment: Challenges and Future Prospect** 74
   Abhay B. Fulke, Atul Kotian, Santosh K. Gothwal, Parth Sharma and *Manisha D. Giripunje*

6. **Plant-Derived Microbial Bio-Similar for the Treatment of Tuberculosis** 86
   Rachana Khati, Alok K. Paul, Maria de Lourdes Pereira, Mohammed Rahmatullah, Veeranoot Nissapatorn and *Karma G. Dolma*

7. **Effects of Antimicrobial Peptides on Bacteria and Viruses** 112
   Tânia D. Tavares, Marta O. Teixeira, Marta A. Teixeira, Joana C. Antunes and *Helena P. Felgueiras*

8. **Therapeutic Applications of Cyclodextrins Against Trypanosomiases and Leishmaniasis** **154**
Zeinab Dirany, Rima El-Dirany, Gustavo González-Gaitano and *Paul A. Nguewa*

9. **Biosensors as a Diagnostic Tool for Analysis and Clinical Determination of Biomolecules** **178**
Asmaa Missoum

10. **Future Challenges in the Use of Biomolecules: Microbial and Natural Origin** **204**
Bidita Khandelwal, Arundhati Bag, Chamma Gupta and *Abhishek Byahut*

Index **223**

CHAPTER 1

Current Scenario of Application of Anti-infective Compounds from Microbial Origin Perspective

Sanjana N.S.,[1] *Bhagwan Narayan Rekadwad,*[1,*] *Punchappady Devasya Rekha,*[1] *Arun A. Bhagwath*[1,2,3] and *Mangesh Vasant Suryavanshi*[4]

Introduction

Natural and microbial sources of anti-infective/bioactive compounds from marine bacteria have been frequently reported as the best sources to treat bacterial (Liu et al. 2019), fungal (Lockhart et al. 2019), parasitic (Dixon et al. 2021) and viral (Tompa et al. 2021) infections and have acted as a lead compound for developing a new cure for communicable diseases, viz. gonorrhoea (Unemo et al. 2014), typhoid (Masuet-Aumatell et al. 2020), cholera (Pal et al. 2021) viral, dengue (Soo et al. 2016), and rhinovirus (Atkinson et al. 2016). There is a huge pool of bioactive compounds available for the treatment of the above microbial infections. Despite

[1] Division of Microbiology and Biotechnology, Yenepoya Research Centre, Yenepoya (Deemed to be University), Mangalore 575018, KN, India.
[2] Yenepoya Institute of Arts, Science, Commerce and Management, Mangalore 575002, KN, India.
[3] Dean, Faculty of Science, Yenepoya (Deemed to be University), Mangalore 575018, KN, India.
[4] Lerner Research Institute, Cleveland Clinic, 9620 Carnegie Ave N Bldg, Cleveland, OH 44106, U.S.
* Corresponding author: rekadwad@gmail.com

some of them being used as medications to treat human infections for decades, there is an untapped pool of potential biomolecules with varied structural and functional antibacterial properties that needs attention. The development of low-cost and convenient model organisms, cutting-edge molecular biology, omics technology, and machine learning is aiding the bioprospecting of new antimicrobial medications and the discovery of new therapeutic targets (Amaning Danquah et al. 2022). Microbial and natural marine products offer a wide range of chemical structures and functionalities that give microorganisms a competitive advantage and can be also used in biotechnology. Many secondary metabolites are natural compounds encoded by biosynthetic gene clusters (BGCs), which comprise a collection of unique genes.

Anti-microbial Resistance: Origins, Evolution, Resistance to Antibiotics and Bioactive Compounds

Whichever microorganisms start tolerance, it resists the further effect of the therapeutic agent at a particular concentration and the compromise response of antibiotic from time to time, which is generally termed as acquired anti-microbial resistance. This is proven in the case of therapeutic agents becoming ineffective during the treatment of bacteria, fungi, parasites and virus infections. Any inappropriate use of this wide range of biochemical may be responsible for acquired resistance, and the complex processes of transfer resistance in other microorganisms cannot be stopped in a given time. Hence, either the lack of knowledge or no further options available are some of the primary reasons for ineffective prevention and control of resistance development. The majority of international, national, and local institutions have recognized these critical issues.

Antibiotic resistance has been the subject of countless resolutions and recommendations as well as numerous regenerated reports, but this has been of no advantage since the spread of antibiotic resistance is unstoppable. Publication of antibiotic discoveries, suggestions on modes of action of antibiotics, and well-deduced mechanisms of resistance have been in the limelight and even been subjects of much interest in academia and until recently in the pharmaceutical sector. The history of antibiotics discovery and mode of action have provided important information about the effect of biochemical ligands and targets and have guided about further use of antibiotics with the emergence of phenotypic mutants (Davies and Davies 2010).

Mechanisms and pathways involved in antibiotic resistance in bacteria have been explained in the lean antibiotics research, which has been active in years from 1960 to 2015. Most of these discussions were concluded by existing acquired resistance in bacteria, such as *Acinetobacter baumannii*,

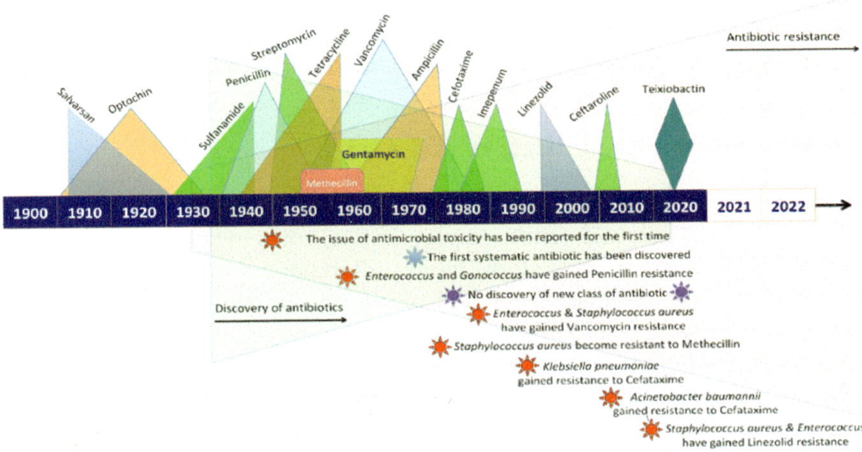

Figure 1. 100 years of antibiotic discovery vs. antibiotic resistance acquired by bacteria.

Klebsiella pneumonia, Pseudomonas aeruginosa, members of Enterobacteriaceae, Escherichia coli, Enterobacter spp., *Serratia* spp., *Proteus* spp., *Providencia* spp., *Morganella* spp. (Critical priority), *Enterococcus faecium, Staphylococcus aureus, Helicobacter pylori, Campylobacter, Salmonella* spp., *Neisseria gonorrhoeae* (High priority), *Streptococcus pneumoniae, Haemophilus influenzae, Shigella* spp. (Medium priority) listed by WHO (Asokan et al. 2019). Overlooked discussions of antibiotics discoveries were stumbled upon after the discovery of teixobactin (Ling et al. 2015). Considering the spread and continuous transmission of antibiotic resistance threat to public health, combative policy should be implemented for a comprehensive and multipronged response. Unfortunately, in a time of colossal need, we have no antibiotic in the pipeline. This is not a gradual process but more a man-made predicament overlaid on nature; there will be no clearer example for this acquired resistance other than the Darwinian theory of natural selection and survival of the fittest (Davies and Davies 2010). Perhaps, we have established assays and protocols for instigations of hundreds of secondary metabolites against different types of pathogens.

Several phytochemicals are known to have the ability to cause microbial cell death or growth inhibition *in vivo* while without impacting beneficial bacteria in the gastrointestinal system. The phytochemicals were discovered to be effective against a wide range of microbes when used in combination or conjunction with conventional antibiotics. Furthermore, secondary metabolites have been discovered to function synergistically and improve the action of less effective antibiotics against a variety of infections, including MDRs. Because this component has received little attention, it is proposed that all probable effects of secondary metabolites-drug interactions be determined. Finally, the research implies that using secondary metabolites as an alternative therapy or in conjunction with traditional antimicrobial medicines might be critical in the creation of future 21st-century treatments (Allemailem 2021).

Biosynthetic Gene Clusters (BCGs)

Terpene and bacteriocin, which were coded by BCGs, were abundant in the microbial mat. Microbial products play critical roles in both medicine and the functioning of an ecosystem. Heimdallarchaeota and Lokiarchaeota, two evolutionarily relevant archaeal phyla previously unknown to have BGCs, were discovered to also have potentially novel BGCs. This research adds to our understanding of how BGCs secondary metabolites may help various microbial mat populations adapt to severe settings (Chen et al. 2020). In terms of marine bioactive compounds, Actinobacteria are one of the most diverse and prolific taxonomic bacterial groups with the potential to be sources of antibiotics and drugs with anti-cancer and pharmacological potential (Hui et al. 2021, Zamora-Quintero et al. 2022). It has been revealed that this actinobacterial genus typically produces secondary metabolites belonging to the macrolides, aminoglycosides, and ansamycins family (Hifnawy et al. 2020). About 40% of the bioactive compounds identified were biosynthesized by species from marine habitats, whereas 43% came from terrestrial Actinomycetes and 2% came from endophytic strains. Others are of unknown provenance (Jakubiec-Krzesniak et al. 2018). Actinomycetes have been recognized as an important class of bacteria to produce a range of secondary metabolites and bioactive compounds. Among Actinomycetes, some genera (*Actinomadura*, *Micromonospora*, and *Amycolatopsis*) have been reported as potential sources of bioactive compounds (Ding et al. 2019). For instance, Rosamicin has wide clinical use and is one of the most effective broad-spectrum antibiotics that act against Gram-positive and Gram-negative bacteria and shows better activity than erythromycin (Crowe et al. 1974). It also shows considerable efficacy against MRSA (Anzai et al. 2010, Huong et al. 2014). A 24-membered macrolide with Macrolactin-S is a known macrolide isolated from the marine bacterium *Bacillus amyloliquefaciens* SCSIO 00856. Anti-microbial activity of this compound showed potent against *Escherichia coli*, *Bacillus subtilis*, and *Staphylococcus aureus* with a minimum inhibitory concentration (MIC) of 0.1 mg/ml (Gull et al. 2012). Marine bacteria and fungi are the most commonly studied as possible sources of new anti-microbial agents, e.g., cyclic peptides like Mathiapeptide A, Destotamide B, and Marfomycins A, B & E; spirotetronates polyketides like Abyssomycin C, Lobophorin F & H; and alkaloids and sesquiterpenes derivatives like Caboxamyxin and Mafuraquinocins A & D. These drugs possess significant antimicrobial activity against antibiotic-resistant strains of bacteria, such as MRSA, *Micrococcus luteus*, *Bacillus subtilis*, and *Enterococcus faecalis*. Additionally, effective biomolecules for the treatment of diseases, such as tetanus, gonorrhoea, tuberculosis, typhoid, candidiasis, aspergillosis, black fungus, malaria, common cold, dengue, rhinovirus, and H1N1 infections reported by various research groups are given in Table 1.

Current Anti-infective Compound for Upper- and Lower-Respiratory System Infections

Infectious agents causing upper respiratory infections affect the upper respiratory organs of the human respiratory system including sinuses and throat infections (Atkinson et al. 2016) and other frequently reported illnesses such as common cold, sinusitis, meningitis, sore throat, cough, pneumonia, and Covid-19. These infections are mainly caused by various pathogenic bacteria (*Streptococcus pyogenes*, *Staphylococcus aureus*, *Moraxella catarrhalis*, and *Haemophilus influenzae*) and viruses (adenovirus, enterovirus, influenza virus, respiratory syncytial virus and rhinovirus). Studies have been carried out to treat these infections using natural products isolated from marine microbial sources and plant products.

A large source of natural marine compounds with biopharmaceutical activities associated with anti-viral and immune-stimulatory properties and some specific algae-derived antiviral metabolites is the ocean. A herb, *Scutellaria baicalensis Georgi*, is a Chinese herb that has the potential effect to treat many diseases. The primary active compounds found in *S. baicalensis* are Baicalein, Baicalin, Wogonin, Wogonoside, and Oroxylin are effective against pneumonia, colitis, hepatitis, and allergy illnesses (Alam et al. 2021, Liao et al. 2021). TLR2 expression, IL-4, IL-13, and eosinophil infiltrate into the respiratory tract, which is then reduced by baicalein (Ulaganathan et al. 2020). Wogonin and wogonoside have also been found to reduce neutrophil inflammatory activity by preventing neutrophil infiltration into the respiratory system (Takagi et al. 2014). Moreover, marine bacteria, microalgae, seaweeds, and other marine organisms produce varied biometabolites that help them acclimatize and survive under harsh conditions. This reveals that these specialized natural metabolites can be a great source of anti-viral agents (Formiga et al. 2021). These bioactive agents have been evaluated and exploited as microbiota-based therapeutic agents, immunomodulators, glycan therapeutic agents, and antioxidants for their biotherapeutics in preventing and treating the SARS-CoV-2 virus. Indeed, the above-discussed literature evidence proves that marine red algae-derived metabolites have a high efficiency against the SARS-CoV-2 virus (Asif et al. 2021, Tassakka et al. 2021).

Table 1. Effective biomolecule against various diseases and their causatives.

Microbes	Disease/Causative	Effective Bioactive Compound	References
Bacteria	Tetanus *Clostridium tetani* (Chapeton-Montes et al. 2020)	Berberine, Coumarin, Catechol, and Curcumin	Skariyachan et al. 2012
	Gonorrhea *Neisseria gonorrhoeae* (Unemo et al. 2014)	Chloroform extract (*Ocimum sanctum* L. *Drynaria quercifolia* (L.) J. Sm *Annona squamosa* L.	Shokeen et al. 2005
	Tuberculosis *Mycobacterium tuberculosis* (Zhang et al. 2021)	Epigallocatechin gallate (Phenolic compound), plant derivative	Raju et al. 2015
	Tyohoid *Salmonella typhi* (Njoya et al. 2021)	Enterocin LD3 (*Enterococcus hirae* LD3	Gupta et al. 2016
Fungi	Candidiasis *Candida albicans* (Romo et al. 2017)	Baicalein (*Scutellaria baicalensis*) Teasaponin (Tea) Resveratrol (polyphenolic compound) Gallic acid (polyphenol)	Da et al. 2019 Li et al. 2020 Houillé et al. 2014 Li et al. 2017
	Aspergillosis *Aspergillus fumigates* (van de Veerdonk et al. 2017)	Baicalein and wogonin (*Scutellaria baicalensis*)	Da et al. 2019
	Black fungus *Biatriospora mackinnonii* (Aoki et al. 2021)	Fluorometholone	Antibacterials/ antifungals/ fluorometholone 2022
Protozoa	Malaria (*Plasmodium malariae*) (Marteau et al. 2021)	ß-carboline alkaloids, indolactam alkaloids *Marinactinospora thermotolerans* SCSIO 00652 (Actinomycetes)	Huang et al. 2011
	Giardiasis (*Giardia duodenali*) (Ryan et al. 2019)	Metronidazole, Tinidazole, and Nitazoxanide	CDC 2022
Virus	Common cold Rhinovirus (Atkinson et al. 2016)/Covid (Fenollar et al. 2021)	Tetracycline, Penicillin, Erythromycin, Ampicillin, Co-trimoxazole, Amoxicillin, cephalosporin, and Amoxicillin/Clavulanic acid	Kenealy and Arroll 2013
	Dengue Dengue virus (Roy et al. 2021)	Stachybonoid (*Stachybotrys chartarum* 952)	Zhang et al. 2017
	Influenza virus A (H1N1) (Xiang et al. 2020, Xiang et al. 2021)	Anthranoside, Austalides (*Aspergillus aureolatus*) HDN14-107	Peng et al. 2016, Son et al. 2018

Communicable Diseases and Anti-Infective Compounds

Communicable diseases in humans can spread from one person to another through different modes. These were frequently reported in various forms, such as infectious diseases, outbreak diseases (foodborne disease outbreaks, waterborne disease Outbreaks) and other emergencies like aerial transfer, sexual transfer, etc. The California Department of Communicable Disease has reported some 80 most frequent communicable diseases that infect humans (CDPH 2022). The common bacterial communicable diseases are cholera, typhoid, tuberculosis, and other sexually transmitted diseases, such as gonorrhoea and syphilis, viral diseases such as influenza, HIV, Covid-19, fungal diseases such as ringworm, candidiasis and other dermatophytes, and protozoan diseases such as malaria and giardiasis. Carvacrol is an anti-infective compound isolated from *Origanum* spp. to treat bacterial infection caused by *V. cholera* which is involved in a number of processes, such as mucin penetration, adhesion, and reduced expression of virulence-associated genes; all of which resulted in a decreased fluid buildup (Das et al. 2021, Donoghue et al. 2017). Various anti-infective compounds were available for the treatment of communicable diseases. The most common group of microorganisms that creates bioactive chemicals is actinobacteria. They produce over two-thirds of all naturally produced antibiotics used in medicine, veterinary medicine, and agriculture today. The majority of these compounds are from the genus *Streptomyces* (Hughes et al. 2008, Jakubiec-Krzesniak et al. 2018). The herb mixture of *Brassica juncea, Forsythia suspensa, and Inula britannica* contained four types of phenylethyl glycosides and a lignan, which had a synergistic impact against avian pathogenic *Salmonella* produced by membrane damage and apoptosis. Cell growth was hindered and the cell membrane was damaged as a result of the herb mixture. Salmonella apoptosis was triggered by these cellular abnormalities (Bae et al. 2021). The endophytic fungus *Aspergillus terreus* produces metabolites that can be regarded as a potential source of natural bioactive compounds. Additionally, the metabolites aspergillide B1 and 3-Hydroxy-3, 5-dihydromonacolin L could be developed as phytopharmaceuticals for the treatment of COVID-19 (El-Hawary et al. 2021). In traditional medicine, Cnestis ferruginea is used to cure infectious disorders such as dysentery, bronchitis, eye problems, conjunctivitis, sinusitis, gonorrhoea, and syphilis. The plant leaves were found to be effective against *S. aureus* in both methanol and water extracts. Using a microdilution procedure on *S. aureus*, the anti-bacterial activity of hydroquinone was validated with hydroquinone outperforming caffeic acid methyl ester (Kouakou et al. 2019).

Tuberculosis (TB)

The bacteria, *Mycobacterium tuberculosis*, causes tuberculosis in people. It is highly resistant to most antibiotics, and it is recently resistant to powerful TB antibiotics, isoniazid, and rifampicin. Although many anti-TB medication candidates in clinical development are synthetic, nature-derived compounds (Table 2) make up the majority of approved therapeutic methods now employed in TB therapy (Cazzaniga et al. 2021).

Natural products (NPs) have had a significant impact on drug development and pharmacotherapy due to their structural and chemical variety. Many NPs have been discovered to have promising anti-tubercular properties over the years. There is a need for novel anti-infective compounds to treat TB. In *Mycobacterium tuberculosis,* dUTPase (Deoxyuridine 5-triphosphate nucleotidohydrolase) provides the sole source for thymidylate biosynthesis; hence, dUTPase has been regarded as a promising anti-TB drug target. A luminescence-based dUTPase assay to search for the inhibitors that target *M. tuberculosis* dUTPase (Mt-dUTPase) and identified compound F0414 as a potent Mt-dUTPase inhibitor F0414 has shown to have a stronger binding with Mt-dUTPase. F0414 exhibited anti-TB activity with low cytotoxicity and insensitivity Therefore, it suggests that F0414 is the first compound reported to have anti-TB activity by inhibiting Mt-dUTPase, which indicates the potential application in anti-TB therapy (Zhang et al. 2021). Moreover, the use of mycobacterial phages offers effective treatment without an increase in anti-microbial resistance (Diacon et al. 2021).

Table 2. Novel therapeutic methods for the treatment of TB.

Compound	Target	Biological Data	Function of Target	Reference
Epigallocatechin gallate (Phenolic compound), and plant derivative	DHFR	MIC = 4 mM	DNA, RNA, and protein synthesis and metabolism	Raju et al. 2015
Myxopyronin (bacterial derivative)	RNAP	MIC = 3.7 mM		Srivastava et al. 2011
Ecumicin (Actinomycetes derivative)	Clp protease	MIC = 160 nM		Gao et al. 2015

Acknowledgement

SNS thankfully acknowledges the Registrar of Yenepoya Deemed-to-be-University for awarding a Junior Research Fellowship (JRF) via letter no. Y/REG/ACA/JRF/Appointment/2021. BNR and PDR gratefully acknowledge Yenepoya Deemed-to-be University's funding support (Grant No. YU/SeedGrant/104-2021).

References

Alam, M. A., Parra-Saldivar, R., Bilal, M., Afroze, C. A., Ahmed, M. N., Iqbal, H., and Xu, J. (2021). Algae-derived bioactive molecules for the potential treatment of SARS-CoV-2. Molecules, 26: 2134.

Allemailem, K. S. (2021). Antimicrobial potential of naturally occurring bioactive secondary metabolites. Journal of Pharmacy & Bioallied Sciences, 13: 155–162.

Amaning Danquah, C., Minkah, P. A. B., Osei Duah Junior, I., Amankwah, K. B., and Somuah, S.O. (2022). Antimicrobial Compounds from Microorganisms. Antibiotics, 11: 285.

Antibacterials/antifungals/fluorometholone. (2022). Reactions Weekly 1889: 37. https://doi.org/10.1007/s40278-022-08716-4.

Anzai, Y., Sakai, A., Li, W., Iizaka, Y., Koike, K., and Kato, F. (2010). Isolation and characterization of 23-O-mycinosyl-20-dihydro-rosamicin: a newrosamicin analogue derived from engineered *Micromonospora rosaria*. Journal of Antibiotics, 63: 325–328.

Aoki, T., Fukuoka, H., Inatomi, T., Horiuchi, N. K., Amei, K., and Sotozono, C. A. (2021). Case of black fungal keratitis caused by *Biatriospora mackinnonii*. Cornea. 40: 1344–1347.

Asif, M., Saleem, M., Yaseen, H. S., Yehya, A. H., Saadullah, M., Zubair, H. M., Oon, C. E., Khaniabadi, P. M., Khalid, S. H., Khan, I. U., and Mahrukh. (2021). Potential role of marine species-derived bioactive agents in the management of SARS-CoV-2 infection. Future Microbiology, 16: 1289–1301.

Asokan, G. V., Ramadhan, T., Ahmed, E., and Sanad, H. (2019). WHO global priority pathogens list: a bibliometric analysis of medline-pubmed for knowledge mobilization to infection prevention and control practices in Bahrain. Oman Medical Journal, 34: 184–193.

Atkinson, S. K., Sadofsky, L. R., and Morice, A. H. (2016). How does rhinovirus cause the common cold cough? BMJ Open Respiratory Research, 3: e000118.

Bae, W. Y., Kim, H. Y., Yu, H. S., Chang, K. H., Hong, Y. H., Lee, N. K., and Paik, H. D. (2021). Antimicrobial effects of three herbs (*Brassica juncea, Forsythia suspensa, and Inula britannica*) on membrane permeability and apoptosis in *Salmonella*. Journal of Applied Microbiology, 130: 394–404.

Cazzaniga, G., Mori, M., Chiarelli, L. R., Gelatin, A., Meneghetti, F., and Villa, S. (2021). Natural products against key *Mycobacterium tuberculosis* enzymatic targets: Emerging opportunities for drug discovery. European Journal of Medicinal Chemistry, 224: 113732.

CDC. (2022). Centre for Disease Control. https://www.cdc.gov/ assessed on 6th May 2022 at 14:52:20 h.

CDPH. (2022). California Department of Public Heath, https://www.cdph.ca.gov/ assessed on 07-05-2022 at 10:33:51 h.

Chapeton-Montes, D., Plourde, L., Deneve, C., Garnier, D., Barbirato, F., Colombié, V., Demay, S., Haustant, G., Gorgette, O., Schmitt, C., Thouvenot, C., Brüggemann, H., and Popoff, M. R. (2020). Tetanus toxin synthesis is under the control of a complex network of regulatory genes in *Clostridium tetani*. Toxins, 12: 328.

Chapeton-Montes, D., Plourde, L., Deneve, C., Garnier, D., Barbirato, F., Colombié, V., Demay, S., Haustant, G., Gorgette, O., Schmitt, C., Thouvenot, C., Brüggemann, H., and Popoff, M. R. (2020). Tetanus toxin synthesis is under the control of a complex network of regulatory genes in *Clostridium tetani*. Toxins, 12: 328.

Chen, R., Wong, H. L., Kindler, G. S., MacLeod, F. I., Benaud, N., Ferrari, B. C., and Burns, B. P. (2020). Discovery of an abundance of biosynthetic gene clusters in shark bay microbial mats. Frontiers in Microbiology, 11: 1950. doi: 10.3389/fmicb.2020.01950.

Crowe, C. C., and Sanders, W. E. Jr. (1974). Rosamicin: evaluation *in vitro* and comparison with erythromycin and lincomycin. Antimicrobial Agents and Chemotherapy, 5: 272–275.

Da, X., Nishiyama, Y., Tie, D., Hein, K. Z., Yamamoto, O., and Morita, E. (2019). Antifungal activity and mechanism of action of Ou-Gon (*Scutellaria* root extract) components against pathogenic fungi. Scientific Reports, 9: 1683.

Das, S., Chourashi, R., Mukherjee, P., Kundu, S., Koley, H., Dutta, M., Mukhopadhyay, A. K., Okamoto, K., and Chatterjee, N. S. (2021). Inhibition of growth and virulence of *Vibrio cholerae* by carvacrol, an essential oil component of *Origanum* spp. Journal of Applied Microbiology, 131: 1147–1161.

Davies, J., and Davies, D. (2010). Origins and evolution of antibiotic resistance. Microbiology and molecular biology reviews: MMBR, 74: 417–433.

Diacon, A. H., Guerrero-Bustamante, C. A., Rosenkranz, B., Rubio Pomar, F. J., Vanker, N., and Hatfull, G. F. (2021). Mycobacteriophages to treat tuberculosis: dream or delusion? Thematic Review Series 101: 1–15.

Ding, T., Yang, L. J., Zhang, W. D., and Shen, Y. H. (2019). The secondary metabolites of rare actinomycetes: chemistry and bioactivity. RSC Advances, 9: 21964–21988.

Dixon, B. R. (2021). *Giardia duodenalis* in humans and animals—transmission and disease. Research in Veterinary Science, 135: 283–289.

Donoghue, H. D. (2017). Insights gained from ancient biomolecules into past and present tuberculosis—a personal perspective. International Journal of Infectious Diseases, 56: 176–180.

El-Hawary, S. S., Mohammed, R., Bahr, H. S., Attia, E. Z., El-Katatny, M. H., Abelyan, N., Al-Sanea, M. M., Moawad, A. S., and Abdelmohsen, U. R. (2021). Soybean-associated endophytic fungi as potential source for anti-COVID-19 metabolites supported by docking analysis. Journal of Applied Microbiology, 131: 1193–1211.

Fenollar, F., Bouam, A., Ballouche, M., Fuster, L., Prudent, E., Colson, P., Tissot-Dupont, H., Million, M., Drancourt, M., Raoult, D., and Fournier, P. E. (2021). Evaluation of the panbio COVID-19 rapid antigen detection test device for the screening of patients with COVID-19. Journal of Clinical Microbiology, 59: e02589-20.

Formiga, F. R., Leblanc, R., De, Souza, Rebouças, J., Farias, L. P., de, Oliveira, R. N. and Pena, L. (2021). Ivermectin: an award-winning drug with expected antiviral activity against COVID-19. Journal of Controlled Release, 329: 758–761.

Gao, W., Kim, J. Y., Anderson, J. R., Akopian, T., Hong, S., Jin, Y. Y., Kandror, O., Kim, J. W., Lee, I. A., Lee, S. Y., McAlpine, J. B., Mulugeta, S., Sunoqrot, S., Wang, Y., Yang, S. H., Yoon, T. M., Goldberg, A. L., Pauli, G. F., Suh, J. W., Franzblau, S. G., and Cho, S. (2015). The cyclic peptide ecumicin targeting CLpC1 is active against *Mycobacterium tuberculosis in vivo*. Antimicrobial Agents and Chemotherapy, 59: 880e889.

Gull, I, Saeed, M., Shaukat, H., Aslam, S. M., Samra, Z. Q., and Athar, A. M. (2012). Inhibitory effect of *Allium sativum* and *Zingiber officinale* extracts on clinically important drug resistant pathogenic bacteria. Annals of Clinical Microbiology and Antimicrobials, 11: 8.

Gupta, A., Tiwari, S. K., Netrebov, V., Chikindas and M. L. (2016). Biochemical properties and mechanism of action of enterocin LD3 purified from *Enterococcus hirae* LD3. Probiotics Antimicrob Proteins, 8: 161–169.

Hifnawy, M. S., Fouda, M. M., Sayed, A. M., Mohammed, R., Hassan, H. M., Abouzid, S. F., Rateb, M. E., Keller, A., Adamek, M., Ziemert, N., and Abdelmohsen, U. R. (2020). The genus *Micromonospora* as a model microorganism for bioactive natural product discovery. RSC Advances, 10: 20939–20959.

Houillé, B., Papon, N., Boudesocque, L., Bourdeaud, E., Besseau, S., Courdavault, V., Enguehard-Gueiffier, C., Delanoue, G., Guérin, L., Bouchara, J. P., Clastre, M., Giglioli-Guivarc'h, N., Guillard, J., and Lanoue, A. (2014). Antifungal activity of resveratrol derivatives against *Candida* species. Journal of Natural Products, 77: 1658–1662.

Huang, H., Yao, Y., He, Z., Yang, T., Ma, J., Tian, X., Li, Y., Huang., C., Chen, X., Li, W., Zhang, S., Zhang, C., and Ju, J. (2011). Antimalarial-carboline and indolactam alkaloids from *Marinactinospora thermotolerans*, a deep sea isolate. Journal of Natural Products, 74: 2122–2127.

Hughes, C. C. Prieto-Davo, A. Jensen, P. R., and Fenical, W. (2008). The marinopyrroles, antibiotics of an unprecedented structure class from a marine *Streptomyces* sp. Organic Letters, 10: 629–631.

Hui, M. L. Y., Tan, L. T. H., Letchumanan, V., He, Y. W., Fang, C. M., Chan, K. G., Law, J. W. F., and Lee, L. H. (2021). The extremophilic actinobacteria: from microbes to medicine. Antibiotics, 10: 682.

Huong, N. L., Hoang, N. H., Shrestha, A., Sohng, J. K., Yoon, Y. J. et al. (2014). Biotransformation of rosamicin antibiotic into 10, 11-dihydrorosamic in with enhanced *in vitro* antibacterial activity against MRSA. Journal of Microbiology and Biotechnology, 24: 44–47.

Jakubiec-Krzesniak, K., Rajnisz-Mateusiak, A., Guspiel, A., Ziemska, J., and Solecka, J. (2018). Secondary metabolites of actinomycetes and their antibacterial, antifungal and antiviral properties. Polish Journal of Microbiology, 67: 259–272. doi: 10.21307/pjm-2018-048.

Kenealy, T., and Arroll, B. (2013). Antibiotics for the common cold and acute purulent rhinitis. The Cochrane Database of Systematic Reviews, 2013: CD000247.

Kouakou, K., Panda, S. K., Yang, M.-R., Lu, J.-G., Jiang, Z.-H., Van, Puyvelde, L., and Luyten, W. (2019). Isolation of antimicrobial compounds from *Cnestis ferruginea* Vahl ex. DC (Connaraceae) leaves through bioassay-guided fractionation. Frontiers in Microbiology, 10: 705.

Li, Y., Shan, M., Li, S., Wang, Y., Yang, H., Chen, Y., Gu, B., and Zhu, Z. (2020). Teasaponin suppresses *Candida albicans* filamentation by reducing the level of intracellular cAMP. Annals of Translational Mmedicine, 8: 175.

Li, Z. J., Liu, M., Dawuti, G., Dou, Q., Ma, Y., Liu, H. G., and Aibai S. (2017). Antifungal activity of gallic acid *in vitro* and *in vivo*. Phytotherapy Research, 31: 1039–1045.

Liao, H., Ye, J., Gao, L., and Liu, Y. (2021). The main bioactive compounds of *Scutellaria baicalensis Georgi* for alleviation of inflammatory cytokines: A comprehensive review. Biomedicine & Pharmacotherapy, 133: 110917.

Ling, L., Schneider, T., Peoples, A. Spoering, A. L., Engels, I., Conlon, B. P., Mueller, A., Schäberle, T. F., Hughes, D. E., Epstein, S., Jones, M., Lazarides, L., Steadman, V. A., Cohen, D. R., Felix, C. R. Fetterman, K. A., Millett, W. P. Nitti, A. G., Zullo, A. M., Chen, C., and Lewis, K. (2015). A new antibiotic kills pathogens without detectable resistance. Nature. 517: 455–459.

Liu, M., El-Hossary, E. M., Oelschlaeger, T. A., Donia, M. S., Quinn, R. J., and Abdelmohsen U. R. (2019). Potential of marine natural products against drug-resistant bacterial infections. Lancet Infectious Diseases, 19: e237–e245.

Lockhart, S. R., and Guarner, J. (2019). Emerging and reemerging fungal infections. Seminars in Diagnostic Pathology, 36: 177–181.

Marteau, A., Ouedraogo, E., Van, der, Meersch, G., Akhoundi, M., Souhail, B., Cohen, Y., Bouchaud, O., and Izri, A. (2021). Severe long-delayed malaria caused by *Plasmodium malariae* in an elderly French patient. Malaria Journal, 20: 337.

Marteau, A., Ouedraogo, E., Van, der, Meersch, G., Akhoundi, M., Souhail, B., Cohen, Y., Bouchaud, O., and Izri, A. (2021). Severe long-delayed malaria caused by *Plasmodium malariae* in an elderly French patient. Malaria Journal, 20: 337.

Masuet-Aumatell, C., and Atouguia, J. (2021). Typhoid fever infection—antibiotic resistance and vaccination strategies: A narrative review. Travel Medicine and Infectious Disease, 40: 101946.

Njoya, H. F., Awolu, M. M., Christopher, T. B., Duclerc, J. F., Ateudjieu, J., Wirsiy, F. S., Atuhaire, C., and Cumber, S. N. (2021). Prevalence and awareness of mode of transmission of typhoid fever in patients diagnosed with Salmonella typhi and paratyphi infections at the Saint Elisabeth General Hospital Shisong, Bui Division, Cameroon. The Pan African Medical Journal, 40: 83.

Pal, B. B., Mohanty, A., Biswal, B., Nayak, S. R., Das, B. K., and Lenka, P. P. (2021). Haitian variant *Vibrio cholerae* O1 Ogawa caused cholera outbreaks in Odisha. Indian Journal of Medical Microbiology, 39: 513–517.

Peng, J., Zhang, X., Wang, W., Zhu, T., Gu, Q., and Li, D. (2016). Austalides S-U, new meroterpenoids from the sponge-derived fungus *Aspergillus aureolatus* HDN14-107. Marine Drugs, 14: 131.

Raju, A., Degani, M. S., Khambete, M. P., Ray, M. K., and Rajan, M. G. (2015). Antifolate activity of plant polyphenols against *Mycobacterium tuberculosis*. Phytotherapy Research, 29: 1646–1651.

Romo, J. A., Pierce, C. G., Chaturvedi, A. K., Lazzell, A. L., McHardy, S. F., Saville, S. P., and Lopez-Ribot, J. L. (2017). Development of anti-virulence approaches for candidiasis via a novel series of small-molecule inhibitors of *Candida albicans* filamentation. mBio., 8: e01991-17.

Roy, S. K., and Bhattacharjee, S. (2021). Dengue virus: epidemiology, biology, and disease aetiology. Canadian Journal of Microbiology, 67: 687–702.

Ryan, U., Hijjawi, N., Feng, Y., and Xiao, L. (2019). Giardia: an under-reported foodborne parasite. International Journal for Parasitology, 49: 1–11.

Shokeen, P., Ray, K., Bala, M., and Tandon, V. (2005). Preliminary studies on activity of *Ocimum sanctum*, *Drynaria quercifolia*, and *Annona squamosa* against *Neisseria gonorrhoeae*. Sexually Transmitted Diseases, 32: 106–111.

Skariyachan, S., Prakash, N., and Bharadwaj, N. (2012). *In silico* exploration of novel phytoligands against probable drug target of *Clostridium tetani*. Interdisciplinary Sciences: Computational Life Sciences, 4: 273–281.

Son, S. Y., Lee, S., Singh, D., Lee, N. R., Lee, D. Y. and Lee, C. H. (2018). Comprehensive secondary metabolite profiling toward delineating the solid and submerged-state fermentation of *Aspergillus oryzae* KCCM 12698. Frontiers in Microbiology, 9: 1076.

Soo, K. M., Khalid, B., Ching, S. M., and Chee, H. Y. (2016). Meta-analysis of dengue severity during infection by different dengue virus serotypes in primary and secondary infections. PLoS One, 11: e0154760.

Srivastava, A., Talaue, M., Liu, S., Degen, D., Ebright, R. Y., Sineva, E., Chakraborty, A., Druzhinin, S. Y., Chatterjee, S., Mukhopadhyay, J., Ebright, Y. W., Zozula, A., Shen, J., Sengupta, S., Niedfeldt, R. R., Xin, C., Kaneko, T., Irschik, H., Jansen, R., Donadio, S., Connell, N., and Ebright, R. H. (2011). New target for inhibition of bacterial RNA polymerase: 'switch region'. Current Opinion in Microbiology, 14: 532–543.

Takagi, R., Kawano, M., Nakagome, K., Hashimoto, K., Higashi, T., Ohbuchi, K., Kaneko, A., and Matsushita, S. (2014). Wogonin attenuates ovalbumin antigen-induced neutrophilic airway inflammation by inhibiting Th17 differentiation. International Journal of Inflammation, 571508: 8.

Tassakka, A. C. M. A. R., Sumule, O., Massi, M. N., Sulfahri, Manggau, M., Iskandar, I. S., Alam, J. F., Permana, A. D., and Liao, L. M. (2021). Potential bioactive compounds as SARS-CoV-2 inhibitors from extracts of the marine red alga *Halymenia durvillei* (Rhodophyta) – A computational study. Arabian Journal of Chemistry, 14: 103393.

Tompa, D. R., Immanuel, A., Srikanth, S., and Kadhirvel, S. (2021). Trends and strategies to combat viral infections: A review on FDA approved antiviral drugs. International Journal of Biological Macromolecules, 172: 524–541.

Ulaganathan, M., Ahmad, T., Rehman, R., Leishangthem, G. D., Dinda, A. K., Agrawal, A., Ghosh, B., and Sharma, S. K. (2020). Baicalein reduces airway injury in allergen and IL-13 induced airway inflammation. PLoS One, 8: e62916.

Unemo, M., and Shafer, W. M. (2014). Antimicrobial resistance in Neisseria gonorrhoeae in the 21st century: past, evolution, and future. Clinical Microbiology Reviews, 27: 587–613.

van de Veerdonk, F. L., Gresnigt, M. S., Romani, L., Netea, M. G., and Latgé, J. P. (2017). Aspergillus fumigatus morphology and dynamic host interactions. Nature Reviews Microbiology, 15: 661–674.

Xiang, L., Hu, X., Zhang, J., She, J., Li, M., and Zhou, T. (2020). Immunodepression induced by influenza A virus (H1N1) in lymphoid organs functions as a pathogenic mechanism. Clinical and Experimental Pharmacology and Physiology, 47: 1664–1673.

Xiang, X., Wang, Z. H., Ye, L. L., He, X. L., Wei, X. S., Ma, Y. L., Li, H., Chen, L., Wang, X. R., and Zhou, Q. (2021). Co-infection of SARS-COV-2 and Influenza A Virus: A Case Series and Fast Review. Current Medical Science, 41: 51–57.

Zamora-Quintero, A. Y., Torres-Beltrán, M., Guillén Matus, D. G., Oroz-Parra, I., and Millán-Aguiñaga, N. (2022). Rare actinobacteria isolated from the hypersaline Ojo de Liebre Lagoon as a source of novel bioactive compounds with biotechnological potential. Microbiology, 168: 001144.

Zhang, P., Li, Y., Jia, C., Lang, J., Niaz, S. I., Li, J., Yuan, J., Chen, S., and Liu, L. (2017). Antiviral and anti-inflammatory meroterpenoids: stachybonoids A–F from the crinoid-derived fungus *Stachybotrys chartarum* 952. RSC Advances, 7: 49910–49916.

Zhang, Y., Zhang, H., Chen, Y., Qiao, L., Han, Y., Lin, Y., Si, S., and Jiang, J. D. (2021). Screening and identification of novel anti-tuberculosis compound that targets deoxyuridine 5′-triphosphate nucleotidohydrolase. Frontiers in Microbiology, 12: 757914.

Chapter 2

Potential Biomolecules of Microbial Origin Against Infectious Diseases

Homem N.C.,[1,2,*] *Paixão R.M.*,[3] *Miranda C.S.*,[1] *Antunes J.C.*,[1,4] *Amorim M.T.P.*[1] and *Felgueiras H.P.*[1]

Introduction

Infectious diseases are a perpetual challenge as they are one of the main human health issues (Irshad et al. 2020). Several pathogens such as viruses, bacteria, fungi and parasites are behind these diseases, which are accountable for 10% of the global mortality annually. Recently, the severe acute respiratory syndrome coronavirus-2 (SARS-CoV-2) gave rise to a novel coronavirus disease, COVID-19, and triggered a pandemic situation which already caused more than 5 million deaths worldwide (WHO 2021). Among the causing agents of infectious diseases, bacteria are considered one of the most concerning classes of pathogens. This can be illustrated by the incidence of tuberculosis, a disease caused by the *Mycobacterium tuberculosis* bacterium, which is considered by the World Health Organization (WHO) as an infectious disease that has caused more deaths worldwide. It is estimated that tuberculosis caused 1.5 million deaths only in 2018 and during human history, tuberculosis is estimated

[1] Centre for Textile Science and Technology (2C2T), University of Minho, Campus Azurém, 4800-058, Portugal.
[2] Digital Transformation CoLab (DTx), University of Minho, Campus Azurém, Building 1, 4800-058, Portugal.
[3] Chemical Engineering Department, State University of Maringá, 87020-900, Brazil.
[4] Fibrenamics, Institute of Innovation on Fiber-based Materials and Composites, University of Minho, Guimarães, 4800-058, Portugal.
* Corresponding author: natalia.homem@dtx-colab.pt

to have killed over 1 billion individuals (WHO 2017, 2020). Similarly, a wide variety of emerging and re-emerging infectious diseases caused by bacteria with high replicative and mutational characteristics could be cited, thus positioning infectious diseases as a public health problem.

To handle/treat these infections, humanity has been applying a panoply of medicines developed during the last century. Categorized as antibiotics, these drugs play a key role in the bacterial infectious diseases combat once they are used as the main tool in the treatment of diseases previously considered lethal for humans. However, the rise of antimicrobial resistance, which has been aggravated since the discovery of penicillin, has set a different challenge for researchers all over the world (Nigam et al. 2014). The investigation of antimicrobial agents that could be effectively applied as alternatives to antibiotics in the treatment of infectious diseases is not recent. Biomolecules have been applied as antibacterial agents for many years (Li et al. 2020) and although the path to clinical trials is very long, several promising alternatives such as bacteriophages and antimicrobial peptides have proven to be efficient against several bacteria, including multidrug-resistant bacteria. This chapter offers a current overview of biomolecules of microbial origin applied in the treatment of bacterial infectious diseases and infers their potential for replacing antibiotics or serving as co-adjuvant therapies in such fights.

Infectious Diseases

An infectious disease is a disease caused by the presence and growth of pathogens (such as viruses and bacteria) in a host organism. Infectious diseases are named 'infectious' because they can be transmitted to other individuals. The transmission of infectious diseases may occur by different means and predominantly depends on the pathogen type.

Pathogens can be defined as organisms that cause diseases or illnesses to their host and can be classified as obligate or opportunistic. Obligate pathogens can cause diseases upon entering into contact with human bodies. Opportunistic pathogens are similar to microbes which are frequently in existence in the human body and only cause problems in very specific situations, such as immunosuppressed organisms. The human body is a nutrient-rich, warm (uniform temperature), and moist environment that constantly renews itself, being a perfect host for the proliferation of pathogens. Although our point of view toward pathogens is generally unfriendly, it is worth mentioning that they are just in search of a suitable environment that is able to provide them with conditions for survival and reproduction (Kaestli et al. 2019).

All pathogens have different levels of transmission and virulence, meaning that they can be more or less contagious or more or less likely to cause disease. In fact, not all highly contagious pathogens will be significantly virulent (and vice-versa). One example is the Ebola Virus Disease (EVD), identified also as Ebola Hemorrhagic Fever. The EVD is caused by a member of the *Filoviridae* family, which can kill up to 90% of infected patients but does not present high transmission rates (Salata et al. 2019). On the other hand, the bacterium causer of tuberculosis is known to be highly transmissible but not all people who have been contacted with it will develop symptoms and consequently the related disease (Ma et al. 2018). This occurs because each one of the pathogens responsible for infectious diseases holds intrinsic characteristics, such as transmissibility rate, replicability and the associated mechanisms, pathogenesis (i.e., the manner by which the pathogen causes disease), and finally the response elicited by the organism when in contact with the pathogen (Table 1).

The path that a pathogen must pursue from the entrance of a host to the development of clinical disease is very arduous. The contagion is the first step and occurs generally through ingestion (contaminated water and food) or contact with hands (which will direct the pathogen to other body sites) and orifices, such as the nose, eyes, and mouth. Contamination via sexual contact and through all types of open wounds is also very common. Once a pathogen invades an organism, it has to circumvent the immune system and reach specific host cells in which they will be able to establish themselves and secrete toxins while obtaining nutrients as a resource to replicate and start spreading to a new host (McArthur 2019).

In this search for nutrients, pathogens behave in different ways to survive. Microorganisms are usually classified from an immunopathological perspective as extracellular pathogens (which seize from extracellular supplies and thus have higher mobility within the body tissues) and intracellular pathogens, which are pathogens that seize from intracellular supplies, being capable of growing and replicating within the host cells. In fact, the host's immune system has the ability to destroy pathogens through processes, such as phagocytosis and lysosomal disruption, but intracellular pathogens can "escape" by growing inside macrophages and other cells (Thakur et al. 2019).

When a pathogen survives in the host's body for time enough to reproduce and cause infection, clinical signs will be developed. In some cases, the pathological responses elicited by the aid of microorganisms spread the pathogen or transmit it to other hosts. For instance, diarrheal infections caused by *E. coli* bacteria are easily spread from patient to hospital workers during cleaning. Likewise, viruses such as herpes simplex, which cause lesions on the oral or genital regions are easily transmitted due to the presence of the virus on the lesions. In a healthy host, most encounters with pathogens would only result in an acute infection within a few days and a proper immune response leads to pathogen elimination from the host's

Table 1. Examples and mechanism of action of some pathogens related to infectious diseases. Information is adapted and complemented by widely available publications (Mukesh et al. 2019, Recht et al. 2020, Tellier et al. 2019).

Pathogen Type	Pathogen Example	Type of Pathogen	Disease Caused	Reservoir (Host)	Mode of Transmission	Replication Mechanism	Pathogenesis
Viruses	Herpes simplex (DNA)	Obligate	Herpes	Humans	Via close contact with a contaminated person which holds a wound in a peripheral site, such as mouth, nose or genital and oral secretions.	Herpes simplex virus replicates by translocation of viral genome through the nuclear pore in which the virus has fixed itself, to the nucleoplasm of the host. Then, the virus is transcribed and replicates itself in order to propagate infection.	Mucocutaneous infections (oral or genital); herpetic simplex keratitis.
	SARS-CoV-2 (RNA)	Obligate	Covid-19	Humans; Animals	Direct mucous membrane contact with respiratory droplets from either infected people or fomites.	SARS-CoV-2 attaches itself to the host by binding to the angiotensin-converting enzyme 2 (ACE2), present in abundance in type II alveolar cells. After this step, the virion starts to release RNA into the cell, thus replicating itself.	Severe acute respiratory syndrome with pneumonia and acute respiratory distress syndrome.

Table 1 contd. ...

...Table 1 contd.

Pathogen Type	Pathogen Example	Type of Pathogen	Disease Caused	Reservoir (Host)	Mode of Transmission	Replication Mechanism	Pathogenesis
Bacteria	*Mycobacterium tuberculosis*	Obligate	Tuberculosis	Humans; Animals.	Contamination occurs via inhalation of droplets released by contaminated people (for both pulmonary and laryngeal TB). Droplets can be produced during cough or sneezing and then invasion occurs via mucous membranes or damaged skin.	When engulfed by alveolar macrophages, the bacilli are stored and can lay dormant (Latent Tuberculosis Infection) or reproduce inside macrophages, triggering the infection (Tuberculosis disease) and spreading to other organs.	Pulmonary tuberculosis causes chest pain, breathlessness, and severe coughing. Tuberculosis can also occur in extrapulmonary sites, such as brain, kidneys, larynx, and lymph nodes among others.
	Legionella pneumophila	Opportunistic	Legionary's disease (LD) or Pontiac fever (Pf)	Ubiquitous in the environment	Inhalation of contaminated aerosols of water or dust. No human-to-human transmission has been recorded.	In the environment, Legionella replicates within eukaryotic phagocytic cells as well as in human monocytes and alveolar macrophages.	LD: Pneumonia with cough, breathlessness, muscle and headaches; Pf: Fever and muscle aches.
	Escherichia coli	Opportunistic	*E. coli* Infections	Humans; Animals.	Consumption of contaminated foods and contaminated raw vegetables and sprouts.	This bacteria initiates replication bidirectionally from a single replication origin, usually incorporated in a chromosome.	The infection usually manifests as diarrhoea, stomach cramping or nausea and vomiting.

	Staphylococcus aureus	Opportunistic	*S. aureus* Infections	Humans	Direct contact with an infected person, by using a contaminated object, or by inhaling infected droplets.	Elevated salt concentrations in the surface fluid of respiratory epithelium and imbalances in the fluid flow lead to thickened mucus and impaired mucociliary, promoting persistent colonization.	Infections involving the skin, soft tissue, vasculature, bone and respiratory tract.
	Pseudomonas aeruginosa	Opportunistic	*P. aeruginosa* Infections	Ubiquitous in the environment. Commonly found in the soil, water, etc. *P. aeruginosa* is primarily a nosocomial pathogen	Can spread and infect people in healthcare situations through contact with hospital personnel, contaminated surfaces or equipment, or contaminated foods and water. In hospitals, is normally detected in respiratory equipment, sinks, taps, mops, and urine receptacles.	*P. aeruginosa* presents motility due to the presence of a polar component (flagella) and thus can move and reach proper environments for survival and DNA replication.	Contamination with *P. aeruginosa* can trigger infections in the respiratory and urinary tract in the blood and gastrointestinal system as well as in bone and joints. Also, bacteremia can occur.
	Staphylococcus epidermidis	Opportunistic	*S. epidermidis* Infections	Humans	Hands, blood, cough secretion, wound secretion, and skin contact or through contact with contaminated objects and surfaces.	This bacterium produces several molecules that resist the host defenses and many proteins contribute to Biofilm formations inhibiting phagocytosis and the activity of antimicrobial peptides.	Wound infections, boils, sinus infections, endocarditis, and other inflammations.

Table 1 contd. ...

...Table 1 contd.

Pathogen Type	Pathogen Example	Type of Pathogen	Disease Caused	Reservoir (Host)	Mode of Transmission	Replication Mechanism	Pathogenesis
Fungi	*Candida albicans*	Opportunistic	Candidiasis	Humans (normal human flora)	Contamination can occur via contact with excretions secreted by mouth, skin, or feces from contaminated people via the endogenous spread, from mother to child in childbirth as well as by the use of infected syringes and catheters.	When there is an imbalance between the external and internal macrophage environment, Candida albicans yeast starts to replicate itself via processes, such as hyphal germination and induction of pyroptosis and thus occasioning infection.	Mucocutaneous lesions, fungemia, and sometimes focal infection of multiple sites.
Parasites	*Toxoplasma gondii*	Obligate	Toxoplasmosis	Animals (generally domestic cats but also sheep, goats, rodents, cattle, swine, chicken and birds)	Ingestion of infected raw or undercooked meat, unpasteurized milk and unfiltered water in developing countries. In children, the ingestion of oocysts in dirt or sandpit sand can result in contamination, if there is previous fecal contamination. Also, blood transfusions, organ transplantation and transplacental transmissions are reported.	Right after ingestion, oocysts transform into tachyzoites and concentrate in neural and muscle tissue, developing tissue cyst bradyzoites.	Affects the central nervous system; tissue cysts can be formed, most commonly in skeletal muscle, myocardium, brain, and eyes, and remain throughout the life of the host.

body. Instead, in an immunocompromised host or when the pathogen is able to evade elimination by some mechanism, the response elicited by the pathogen can lead to chronic persistent infection, taking months to years to eliminate the pathogen and causing pathological responses that could lead to an increased morbidity and mortality rates (Thaljeh et al. 2019).

Bacterial Infections: Prevalence and Conventional Treatments

The relationship between humans and bacteria is controversial since many bacteria in existence in our bodies have important functions in helping, for instance, digesting food, modulating our immune system, and even producing proteins and vitamins. However, bacteria can cause serious damage to the human body if there is an imbalance in the microbiota or if a contagion with any pathogenic bacteria occurs.

Bacteria can be defined as unicellular prokaryotic organisms, meaning organisms with no organized internal membranous structures. They represent the oldest, simplest life form already identified and one of the smallest categories of microorganisms. According to their cell wall structures, bacteria are generally classified into two classes: Gram-positive bacteria (GPB)—microbial cells get purple following staining methodology—or Gram-negative bacteria (GNB)—microbial cells get pink following staining methodology. In a very simple way, their main difference consists in the fact that GP organisms possess a thicker peptidoglycan cell wall in comparison with GNB (Sizar and Unakal 2020). Regardless of their Gram-stain reaction, under ideal growth conditions, all bacteria present remarkable reproductive rates which enable them to easily dominate exposed surfaces and spread in the presence of infection. Besides, they possess very high resistance to extreme environmental conditions and thus, they can live and grow in the soil, water, plants, animals (including humans), and even in the earth's deepest areas (Zabel and Morrell 2020).

Bacterial infections have an enormous impact on public health as all human organs are susceptible to infections of these types. The degree of severeness varies with the type of bacteria involved, ranging from minor illnesses to more life-threatening infections. For instance, strep throat caused by GP *Streptococcus pyogenes* usually only causes soft symptoms of sore throats, fever, and difficulty swallowing. Differently, problems associated with infections caused by GNB *Pseudomonas aeruginosa* can be very severe, presenting an extremely high mortality rate (Montero et al. 2020).

In the pursuit of aiding the mechanics of our immune system to overcome these pathogen-related infections, mankind has been using medications containing antimicrobial agents from various sources. In bacterial infections, antibiotics play a very important role since Alexander Fleming discovered penicillin in 1928 (Breedlove 2019). Reports estimate that more than 100 antibiotics have been discovered since then, thus being a great weapon in the fight against infectious diseases. However, due to the growing human mobility throughout the world, the natural evolution of species and the overuse and misuse of antibiotics have risen a new challenge and spread antimicrobial agent resistance (Nigam et al. 2014).

Microbial resistance occurs when a bacterium becomes resistant to a determined antibiotic due to overexposure and starts multiplying among non-resistant bacteria. Contact between resistant and non-resistant bacteria can transform the former into becoming more resistant as well adaptable via DNA sequences exchange (Figure 1).

It is estimated that 2.4 million people in Europe, North America, and Australia may die from infections caused by resistant microorganisms by 2050, resulting in an expense of at least US$3.5 billion annually (Bianco et al. 2020). Twenty years ago, the WHO alerted that due to the increasing number of multiple drug resistance pathogens, some infectious diseases could become incurable. At that point, the WHO also warned that a very small percentage of all global health research and development funding was being devoted to finding new drugs or vaccines to stop HIV, acute respiratory infections, diarrheal diseases, malaria, and tuberculosis (WHO 2000). The levels of antimicrobial resistance continue rising and though novel antibiotics have been produced since then, in the majority of cases they are just improved versions of previously produced antibiotics which have already been proven to be ineffective for multidrug resistant microorganisms (Hutchings et al. 2019). Thus, both academic and industry sectors were forced to join in a search for alternatives to the use of antibiotics.

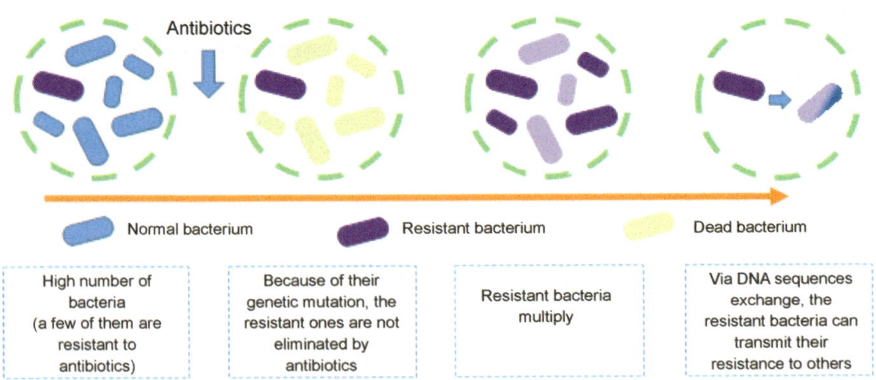

Figure 1. Schematic representation of microbial resistance development mechanisms against antibiotics.

Biomolecules: An Alternative to Conventional Antibiotics

Although bacteria can resist and survive in very extreme environmental conditions, bacterial growth can be inhibited when bacteria are left in contact with antimicrobial agents. Antibacterial agents are defined as materials which possess the capacity to interfere in bacteria growth and or survival, thus leading to their destruction. Usually, these agents act by disrupting the bacteria cell wall, which leads to imbalance and disruption of the membrane matrix and consequently to cell death (Tavares et al. 2020).

A broad array of materials possessing antibacterial activity has been discovered during the last century, amplifying the spectra of alternatives to antibiotics. Inorganic materials such as carbon nanotubes and graphene have been established as effective against a broad range of bacteria (Mohammed et al. 2020). However, their cytotoxicity against human cells is still not completely elucidated, hampering their generalized use in biomedicine. Biological materials such as biomolecules have been highlighted in this aim, once they are chemical compounds produced by living organisms, including plants and humans, and thus are less cytotoxic and less susceptive to induce side effects in comparison with inorganic materials (Li et al. 2020).

Biomolecules from numerous natural sources (animals, microorganisms, and plants) such as carbohydrates, protein, lipids, and nucleic acids are fundamental building blocks of living organisms, being vital for the structure and proper function of living cells. Antibacterial agents composed of biomolecules, combined with the proper chemical and biotechnological tools, have been broadly researched in the last century. For example, biomolecules of microbial origin such as antimicrobial peptides (AMPs), bacteriocins (bacteria displaying the ability to antagonize other bacteria), some organic nanoparticles, and even bacteriophages and enzymes have been established as promising antimicrobial agents once they can inhibit bacterial growth and are considered safe for biomedical purposes and are environment friendly. In the combat against infectious diseases, these biomolecule-based treatments can be administered orally or immobilized/incorporated within delivery systems for local or systemic distribution (Felgueiras 2021).

Antimicrobial Peptides

AMPs can be chemically synthesized or secreted by procaryotic and eukaryotic cells (e.g., the ones in human skin). They are considered natural antibiotics and work as defense mechanisms in most organisms, including humans, plants, insects, bacteria, and vertebrates as part of their innate immune system. Their impact and role in human immunology have initially been strongly neglected and underappreciated (Lai and Gallo 2009). However, recently they have been recognized as an essential route to our immune responses, providing defense against several invaders such as GPB and GNB, viruses, and fungi. Most AMPs display a fast action against microbial cells. More specifically, they bind to the cytoplasmic membrane and disrupt it and create pores or penetrate the membrane, interacting with many intracellular proteins. As a result, the capacity of those microorganisms to develop resistance is strongly reduced in comparison with antibiotics. Thus, AMPs have been increasingly considered in the search for new antimicrobial agents to fight infectious diseases (Felgueiras et al. 2020).

AMPs are molecules with relatively low molecular weights, composed of 5 to 100 amino acid residues that can be classified into three major classes according to their structure: Class I comprises the smallest peptides (< 5 kDa), which possess heat resistance; Class II entails the unmodified peptides, which possess stability to heat (< 10 kDa), and Class III, which encompasses the heat-labile, larger-size peptides (> 15 kDa) (Lai and Gallo 2009). Due to the strong presence of lysine, arginine and histidine amino acids in their structure, AMPs usually exhibit a cationic character. Moreover, they present an amphipathic structure which is considered an advantage since it empowers the AMPs with the ability to bind to lipid components and phospholipid groups—the hydrophobic and hydrophilic regions, respectively—making them effective for interactions with bio membranes (Lei et al. 2019).

One of the most interesting biological properties of AMPs is the fact that they are positively charged, allowing interactions with the negative parts of the cell wall and the plasma membrane of microorganisms. Once the interaction takes place, amphipathic AMPs are inserted into the membrane. Afterwards, membrane disruption takes place, damaging its integrity and resulting in cell death. Furthermore, AMPs display intracellular inhibitory activities by interfering with relevant processes, such as cell division, cell wall biosynthesis, nucleic acid metabolism, and protein biosynthesis. In addition, these peptides can modulate immune responses once they stimulate cytokine production, responding similarly to chemokines and promoting wound healing (Pizzolato-Cezar et al. 2019).

As biomolecules for fighting pathogens responsible for infectious diseases, the most significant advantage of the use of AMPs relies on the fact that they possess a very reduced propensity to induce resistance due to their multiplicity of targets and lack of interaction with specific receptors. Most AMPs act on the region of the cell membrane in which resistance development is less likely to happen. For these reasons, AMPs can be highlighted as versatile and multipurpose weapons that can be used to increase susceptibility and re-sensitizing of antibiotic-resistant microorganisms. These molecules can also display synergistic effects with antibiotics, increasing antibiotic activity above their effect (Yao et al. 2015). For instance, Reffuveille et al. reported that the peptide 1,018 acts in synergy with ceftazidime, ciprofloxacin, imipenem, and tobramycin, leading to a decrease in the concentration of antibiotics required to combat biofilms formed by bacteria, such as *Pseudomonas aeruginosa, Escherichia coli,* and *Staphylococcus aureus,* representing several multidrug-resistant pathogens (Der Torossian Torres and De La Fuente-Nunez 2019, Reffuveille et al. 2014).

Some preliminary *in vitro* studies pointed out that AMPs can be applied as tools in the battle against infectious diseases. For instance, according to AlMatar et al. AMPs were established to disturb the normal mycobacterial cell wall function and to interact with intracellular targets, embracing enzymes, nucleic acids, and organelles. Consequently, the authors stated that AMPs represent potential antibacterial agents to be used in the treatment of tuberculosis (AlMatar et al. 2018). Furthermore, in a study by Wang et al., JH-3, an AMP with broad bactericidal activity, has been proven effective for the treatment of *Salmonella*'s disease (Wang et al. 2019). Also, A24 and Ci-PAP-A22 antimicrobial peptides were reported as potential antimicrobial agents against the pathogen *Legionella pneumophila,* the bacteria which causes Legionnaires' disease (Schlusselhuber et al. 2015). In addition, the antimicrobial peptide L12 has been proven effective against the multidrug-resistant *S. aureus,* the pathogen which causes *S. aureus* infections and attacks skin, soft tissue, vasculature, bone, and respiratory tract (Xiong et al. 2019).

Besides their potential as antimicrobial agents in the combat of infectious diseases, the effectiveness of AMPs as re-sensitizers of multidrug-resistant bacteria has also been analyzed. Lazar et al. evaluated the effect of 24 AMPs against 60 resistant *E. coli* strains, and it was evident that the antibiotic-resistant bacteria presented a high frequency of collateral sensitivity to AMPs. Clinically relevant multidrug-resistance mutations were also identified, which caused an increase in the bacterial sensitivity to AMPs. As a result, the AMP-antibiotic combinations were seen to improve the responses against several multidrug-resistant bacteria. Consequently, those combinations may also reduce resistance evolution (Der Torossian Torres and De La Fuente-Nunez 2019).

Bacteriocins

Bacteriocins are a subset of AMPs produced by bacteria. These biomolecules were first discovered in 1925, and since then they had led to the development of numerous works identifying new bacteriocins. In fact, bacteriocins gained considerable attention as antimicrobial agents against bacterial, fungi and viral species, as well as bacterial biofilms (Akerey et al. 2009, Yasir et al. 2018). Bacteriocins are produced by GPB or by GNB. Nevertheless, most bacteriocins are produced by GPB and lactic acid bacteria (LAB) (Klaenhammer 1988). Since a wide range of bacteria can produce bacteriocins, many different bacteriocins have been discovered (Cintas et al. 2001). As a result, these molecules display a large variety of biotechnological and pharmaceutical applications (Balciunas et al. 2013) that fight against antibiotic-resistant bacteria and their main target (Singh and Abraham 2014).

Among the bacteriocins produced by GPB, they can be divided into four classes. The first class, the lantibiotics, is formed of small-sized and post-transcriptionally modified bacteriocins. That group of bacteriocins is often related to the inhibition of GPB and food-borne pathogens (Savadogo et al. 2006). In addition, the class contains unusual amino acids-dehydrated amino acids, lanthionine and 3-methyllanthionine with several ring structures being stable to heat, pH, and proteolysis (Simons et al. 2020). Class II, named non-lantibiotics, does not include unusual amino acids and post-translational modifications are reduced to the formation of disulfide bridges within a few members (Jack et al. 1995). These bacteriocins are heat-stable and small-sized (< 10 kDa). On the other hand, Class III is characterized by large peptides (> 30 kDa) and is often heat-labile lytic/non-lytic. The antibacterial functions of these peptides are associated with enzymatic activity, disrupting the cell wall of the bacteria. Examples of Class III bacteriocins include zoocin A, lysostaphin and helveticin J and V (Joerger and Klaenhammer 1986). Finally, the Class IV bacteriocins contain lipid or carbohydrate parts, like plantaricin S or leucocin S, also known to disrupt the bacterial cell membrane (Savadogo et al. 2006).

The mechanism of action of GPB bacteriocins is based on the disruption of the bacterial membrane integrity, ultimately leading to cell death. This effect results from direct interactions with lipid II components of bacterial membranes—Mannose PhosphoTransferase System (Man-PTS)—or without a specific receptor. For instance, nisins act by pore formation with the use of lipid II, enhancing the membrane permeability of the cell (Chatterjee et al. 2005).

Some studies have been conducted toward analyzing the effects of GPB bacteriocins against multidrug-resistant isolates. Bacteriocin BJ13 was isolated from *Pediococcus pentosaceus*, showing high antibacterial activity against *Listeria monocytogenes, S. aureus, Clostridium sp.,* and *Klebsiella pneumoniae*. Moreover, bacteriocin PJ4, produced by *Lactobacillus helveticus* showed activity against a wide range of GPB and some GNB, including *Enterococcus faecalis, E. coli,* and *P. aeruginosa* (Vidhyasagar and Jeevaratnam 2013).

Lantibiotics, including nisin, planisporicin, Pep5, epidermin, gallidermin, mutacin B-Ny266, lacticin 3147, and actagardine also display activity against GP pathogens, such as *S. pneumoniae, Staphylococci,* VRE, *Propionibacterium acnes,* and *Clostridium difficile.* Among those, nisin (produced by the non-pathogenic bacteria *Lactococcus lactis*) is currently the only bacteriocin classified as "Generally Recognized as Safe" (GRAS) by the Food and Drug Administration (FDA) (Niaz et al. 2018). Nisin is an oligopeptide with 34 amino acids which has two main variants, namely nisin A and nisin Z, which differ in the position of the amino acid at the 27th position (asparagine for nisin A and histidine for nisin Z). Although the antimicrobial activity of both variants is similar, the solubility at the neutral pH of nisin Z is eased due to the polarity of the asparagine amino acid. As a consequence, nisin Z presents a larger spectrum of applications in the combat of bacterial infectious diseases, which is also improved due to the electrostatic interactions that may occur between nisin Z and negatively charged natural-origin polymers, such as sodium alginate (SA) (Felgueiras and Amorim 2017). The incorporation of nisin Z at bio-based polymeric microfibers composed of SA and gelatin, neat and cross-linked, produced via wet-spinning technique was explored for applications in the fight against *S. aureus*-induced infections. The presence of nisin Z prevented the earlier disintegration of the fibers, while a controlled release of the peptide in physiological conditions was identified for up to 48 hours, thus maintaining the inhibitory effect against *S. aureus* for longer periods in comparison with the effect of the peptide alone (Homem et al. 2021). In addition, the use of bacteriocins jointly with other drugs to improve or restore antibacterial activity has also been gaining considerable attention. For instance, the use of nisin with polymyxin E and clarithromycin results in a synergic effect against *P. aeruginosa* (Giacometti et al. 2000).

Bacteriocins have also shown significant activity in fighting bacteria growth in biofilms. Biofilms constitute another mechanism of resistance in which bacteria protect themselves from host defenses and antimicrobial agents (Benech et al. 2002). In fact, infections related to biofilm formation are considered very serious problems, specifically for cystic fibrosis patients and patients in intensive care units that are at high risk for developing device-related infections (Høiby et al. 2011). Gallidermin, a lantibiotic, has been reported to effectively reduce biofilm formation by *S. aureus* and *S. epidermidis.* At sub-lethal concentrations, the lantibiotic inhibited the transcription of two genes—atl and ica—that are crucial for biofilm formation (Saising et al. 2012). Similar results were also found for nisin A and lacticin Q (Okuda et al. 2013).

Just like GPB, bacteriocins produced by GNB are also divided into different groups. The first group, named colicins, results from bacteriocins with high molecular weights (> 10 kDa) that are produced by *E. coli*. The mechanism of action of colicins is based on the formation of pores in the bacterial cell wall or on the degradation of nucleic acid-based structures. The latter mechanism is usually linked to colicin-like bacteriocins that are produced by different bacteria, such as *Klebsiella* spp. and *P. aeruginosa* (Chavan and Riley 2007). The third group, named microcins, includes small-sized peptides (< 10 kDa). This group of peptides comprises a subclass of microcins that are post-translationally modified with molecular weights lower than 5 kDa (e.g., B17, C7, J25, and D93) and a second subclass with unmodified/minimally modified peptides with molecular weights between 5 and 10 kDa (e.g., E492, V, LH47, and LH24) (Duquesne et al. 2007). Microcins present two distinct mechanisms of action, pore formation in the inner membrane (e.g., microcins E492, M and H47), or targeting intracellular enzymes (e.g., microcins J25, B17, and C) (Duquesne et al. 2007). Microcins enter the cell and use specific receptors at the outer membrane, such as receptors related to iron uptake and outer membrane porins (Mathavan and Beis 2012). Pore formation occurs as a result of interactions with specific components in the inner membrane, including the receptors involved in the absorption of mannose for microcins E492 and adenosine triphosphate (ATP) synthase complex for microcin H47 (Biéler et al. 2010). Microcins C, L, and B17 present the largest antibacterial spectrum, including *E. coli, Shigella* spp., *Salmonella* spp., *Klebsiella* spp., and *Pseudomonas* spp. (Pons et al. 2004). Nevertheless, microcin E492 presents a spectrum reduced to *E. coli, K. pneumoniae*, and *Salmonella enteritidis* strains (Thomas et al. 2004). Microcin J25 showed higher activity against *E. coli* strains in comparison with other microcins; however, its spectrum of activity only includes *Salmonella serovars* and *E. coli* strains (Sable et al. 2000). Furthermore, microcin S, when isolated from an *E. coli* strain successfully inhibited the adherence of enteropathogenic *E. coli* in human intestinal epithelial cellular models (Zschüttig et al. 2012). Aside from acting by pore formation in sensitive bacterial membranes or by enzymatic degradation of specific targets, colicins also make use of specific receptors in the outer membranes to enter targeted cells. Some colicin receptors include TonB-dependent vitamin B12 transporter, outer membrane proteins A and F, nucleoside transporter Tsx, and receptors involved in iron uptake, like FepA, GhuA, and Cir (Cascales et al. 2007).

Phages

Bacteriophages (or phages) are defined as viruses that infect and induce lyse in bacteria and do not cause harm to humans. The introduction of the investigation of these biomolecules as therapeutic tools against infectious diseases (also known as phage therapy) dates back to the 1920s when phages were identified and their clinical effectiveness was demonstrated by Félix d'Herelle. However, back in 1896, Ernest Hankin, a British bacteriologist, had already described the antibacterial potential of an unknown substance which was able to pass through bacterial filters. Regardless of the precise discovery date, it is undeniable that phages were applied as antibacterial agents way before the discovery of Penicillin, perhaps being one of the oldest biomolecules applied to this aim (Brives and Pourraz 2020).

The phages were produced on large scale and submitted to clinical trials during the 1940s. Yet, their efficacy and safety were questioned, mainly because they were organisms not fully understood at that point. Also, the discovery of antibiotics harmed the advances in this field. But since 1980 and after the researchers took conscience of the advent of antimicrobial resistance, more and more reports indicate that the administration of proper amounts of living phages can contribute to the treatment of infectious diseases. In this scenario, phages have been known for years to display an outstanding antimicrobial activity against both GNB and GPB and even against multidrug-resistant strains (Aslam et al. 2020).

The phages' action mechanism differs from antibiotics once they are parasites of bacteria, which implies that organisms not only infect but also replicate within bacteria. They can develop in two distinct cycles: lytic or lysogenic. For therapeutical purposes, lysogenic phages are considered unsuitable candidates since they have a longer infection cycle (A. Nigam et al. 2014). From the moment that they infect a bacterium, phages will replicate exponentially until host destruction is achieved. Consequently, in theory, even a small concentration of bacteriophage can reach a great level of host destruction (Chen et al. 2019).

The classification of phages in 13 families takes into consideration their morphology, nucleic acid type, and the presence or absence of lipid or envelope. Among the 5,100 phages discovered until the end of the 20th century, 96% are considered 'tailed phages' which are phages composed of an icosahedral head and tail (*Caudovirales* order). The remaining 4% can be divided into cubic, filamentous, or pleomorphic. The main difference between them is the fact that the tailed phages only have double-stranded DNA as the genome, while the cubic, filamentous, and pleomorphic can contain double-stranded or single-stranded DNA or RNA as genome. This difference affects its antibacterial efficacy and thus most therapeutic phages belong to the tailed family (Matsuzaki et al. 2005).

Isolating phages to attain targeted bacterial strains is a challenge as it demands time and deep knowledge of the phage morphology.

Thus, much research has been devoted to investigating phages that exhibit a broad host range. For example, recently researchers affirmed that the phage JD419, which has a similar morphology to other known phages (IME-EF1, P70, and 3A), was tested against 138 clinical strains of *S. aureus* from two different hospitals in Shangai, China. The JD419 phage was able to lyse 44.2 % of the tested strains, though with different lyse extents being incomplete for some strains. Besides, the JD419 phage was stable under different pH and temperature conditions, thus indicating that phages with morphology similar to the JD419 could be applied as prophylaxis agents against staphylococcal infections (Feng et al. 2021). On the other hand, it is stated that to avoid the rise of phage-resistant bacteria, phage cocktails should be considered a viable option. Phage cocktails inhibit the development of phage-resistant bacteria in the short term since the hosts are unlikely to become resistant to several phages simultaneously. However, controlling the stability of phages to withstand the conditions demanded until administration in clinical trials remains a challenge that yet needs to be overcome.

It is undoubtedly stated that living phages can hinder the growth of several bacteria such as *E. coli* and *P. aeruginosa* as well as *S. aureus* and *E. faecium*. According to Abedon et al., phage therapy has been investigated in several countries to act in the treatment of respiratory, urinary and gastrointestinal tract infections, eye infections, and even in the treatment of burns if infected with resistant bacteria (Abedon et al. 2011). In synergy with antibiotics, its effects are even more exceptional. In a recent study, this statement was proven. The effect of isolating a single phage from the Eliava PYO phage cocktail, which is commercially available, together with a total of nine antibiotics against *S. aureus* was reported. The results showed that at a low antibiotic concentration (2xMIC), the overall antibacterial activity of antibiotics was increased when phage was added (Dickey and Perrot 2019). Nevertheless, phage therapy is still in an experimental stage and many questions regarding ethical and safety reasons have been raised (Górski et al. 2018, Harada et al. 2018).

As mentioned before, the human body presents in its microbiota a panoply of microbial cells, including commensal bacteriophages. However, in the case of the application of therapeutic bacteriophages, it is critical to understand deeply their mechanisms of interaction to prevent disturbance in the dysregulation of the microbiota equilibrium. Currently, it is not well clarified how phages can affect in a long term on the human immune system. Consequently, further investigations for better targeting of bacteria by phages are required. Nevertheless, it is predictable that in the next years, research on the efficiency of phage therapy showing the safety of such therapies will be developed and also change the paradigm of the use of phages (Divya Ganeshan and Hosseinidoust 2019, Paule et al. 2018).

Microbial Enzymes

Enzymes are catalysts with crucial roles in several stages of metabolism and biochemical reactions. Microbial enzymes are considered superior enzymes obtained from a broad range of microorganisms. The number of studies on isolation, properties characterization, and applications of these enzymes has been constantly increasing. Many classes of enzymes are used to conduct catalytic reactions and possess applications in several bioprocesses. In addition, some molecular techniques have been developed to enhance the quality and performance of microbial enzymes for wider applications in industries (Nigam 2013).

Most microbial-origin enzymes are synthesized by bacteria strains, such as *Pseudomonas, Bacillus* and *Clostridium*, as well as some fungi. Microbial enzymes are usually able to present activity under abnormal conditions of temperature and pH and are often named thermophilic, acidophilic, or alkalophilic. In fact, microorganisms possessing thermostable enzymes, working at higher reaction temperatures, strongly decrease microbial contaminations (Wang et al. 2012).

Among hydrolytic enzymes, microbial proteases have been extensively studied in the past recent years (Vijayalakshmi et al. 2013). This type of enzyme, produced from microbial systems, can be acidic, neutral, or alkaline. Alkaline proteases are effective under basic pH conditions and contain a serine residue at their active site (Gupta et al. 2002). Their high activity and stability in abnormal conditions, for instance under high pH and temperatures, and in the presence of inhibitory compounds make them suitable for many different applications. Microbial keratinases, a group of proteases, are characterized by degrading keratins into amino acids. Keratin can be found in human and animal tissues and its structure usually resists degradation by common proteases. Keratinase enhances the cleavage of peptides, which increases the potential of the enzyme to be used in pharmaceutical industries (Gupta et al. 2008).

Enzymes derived from bacteriophages, also known as 'enzybiotics', have also been explored as therapeutic agents against human pathogenic bacteria. Although some authors suggest that the term 'enzybiotics' should be employed for all enzymes which exhibit antimicrobial activity, it is indeed applied largely to designate bacteriophage enzymes endowed with bacterial cell wall-degrading capacity (Borysowski and Górski 2009). Enzybiotics are derived from bacteriophage-encoded enzymes that have the ability to degrade the bacterial cell wall of infected cells at the end of the lytic replication cycle. Enzybiotics present a fast and unique mechanism, high selectivity, and are endowed with a low probability for bacterial resistance development as well as a proteinaceous nature (Dams and Briers 2019). These phage-derived enzymes have been applied as antibacterial agents for both GPB and GNB and can be divided into two classes: peptidoglycan hydrolases (PGHs) and polysaccharide depolymerases (PSDs). The PGHs act by degrading the bacterial peptidoglycan layer, while the latter targets the extracellular or surface polysaccharides and both classes to date have been only developed from phages belonging to the *Caudovirales* order. Currently, PGHs are eligible for FDA fast-track status, and clinical trials with these enzymes have been already implemented all of which target the *S. aureus* strains (Danis-Wlodarczyk et al. 2021). For instance, the PGHs LysMR-5 (the endolysin of phage MR-5) and the virion-associated HydH5 have proved to possess lytic activity against *S. aureus* at conditions similar to the physiological environment, showing great potential to be applied to biomedical applications (Kaur et al. 2020, Rodríguez et al. 2011).

Other Biomolecules

Aside from the former categories of biomolecules, there have been others with potential for therapeutical applications, namely nanoparticles (NPs) and biopolymers. NPs can be defined as solid colloidal particles with dimensions in the nanoscale, ranging from 1 to 100 nm, which are currently being evaluated to be applied as tools for targeted delivery of drugs and diagnostic and therapeutical purposes (Tavares et al. 2020). The main advantage of the use of NPs in biomedical applications is that in comparison with antibiotics, they present high efficiency with a low tendency to trigger side effects due to their specific targeting mechanism (Vallet-Regí et al. 2019). By endocytosis, NPs are absorbed within the cell, unlike the drug molecules. For this reason, NPs have the ability to deliver the drugs directly into the cell, decreasing the possibility of inactivation through penetration (Aparna et al. 2019).

In biomedical applications, NPs are seen as a strategy for overcoming pathogenic microorganisms' actions. For instance, polymeric NP surfaces have been functionalized with lectin for carbohydrates-binding in the bacterial cell walls, releasing biomolecules directly inside the bacteria and consequently killing the pathogen. The antimicrobial-capped NPs have synergically increased the efficacy of antibiotics by raising the concentration of drugs at the targeted site (Dakal et al. 2016). However, the synthesis of NPs using nontoxic and eco-friendly methods is a key role in biomedical applications, and this can be achieved using microorganisms.

The recent literature states that biological systems can act as a sort of 'bio-laboratory' to produce metal and metal oxide nanoparticles by biomimetic approach. Several microorganisms, such as bacteria, yeast, and fungi can be applied as eco-friendly precursors for the synthesis of NPs. In fact, NPs produced by biological processes do not require the use of chemicals and thus the process is less energy-intensive in comparison with the production of NPs by chemical methods. Biosynthesized NPs present several outstanding features, such as higher catalytic reactivity and greater specific surface area, which enables them to be applied as drug carriers for targeted delivery and as antibacterial therapeutic agents (Li et al. 2011).

Microorganisms usually grab target ions from their environment, turning the metal ions into element metal by enzymes that are being generated by cellular activities (Mann 2006). One example of such NPs relies on gold and silver NPs produced by *Verticillium* sp. This process can occur intracellularly by the transport of ions into the microbial cell with enzymes forming NPs or occurs extracellularly by trapping the metal ions at the surface of cells and reducing ions also in the presence of enzymes (Zhang et al. 2011). Heavy metallic NPs can be formed based on the development of metallophilic microorganisms and their genetic and proteomic responses under abnormal conditions. For instance, some metal ions, including Hg^{2+}, Cd^{2+}, Ag^{+}, Co^{2+}, Cu^{2+}, Ni^{2+}, Pb^{2+} and Zn^{2+}, can present toxic effects on some microorganisms. Therefore, the genetic and proteomic responses of those microorganisms can regulate metal homeostasis and count those effects (Reith et al. 2007).

Over 60 elements have been identified by several microorganisms and converted into complex nanoscale architectures. For instance, Sastry et al. reported the extracellular synthesis of gold nanoparticles by fungi *Fusarium oxysporum* and actinomycete *Thermomonospora* sp. and intracellular synthesis by fungus *Verticillium* sp. (Ahmad et al. 2003, Priyabrata Mukherjee et al. 2002). Also, according to Konishi et al. platinum nanoparticles can be produced by an ion-reducing bacteria *Shewanella* algae (Konishi et al. 2007). Silver nanoparticles (AgNPs) have also been reported to be synthesized as films or accumulated on the surface of cells by the presence of fungi *Verticillium, Fusarium oxysporum* or *Aspergillus flavus* (Jain et al. 2011, Patra et al. 2015). Moreover, Jha et al. conducted research in which the presence of Saccharomyces cerevisiae led to the synthesis of Sb_2O_3 nanoparticles (Jha et al. 2009).

The biocompatibility of NPs can be controlled through the fabrication process. The dimensions of the NPs and biomolecules are very close, which allows them to display similar functions and properties to biomolecules present in cellular systems, e.g., proteins, nucleic acids, and small peptides. NPs can be also readily coupled with biomolecules, such as vaccines, drugs, peptides, proteins, and nucleotides (Masri et al. 2019). A wide range of biomolecules, such as lipids, nucleic acids, proteins, and polysaccharides, can be used in the development of NPs (Torres et al. 2019). Among these, the recent literature states that nucleic acids and proteins possess very good compatibility when interacting with NP's surface. Due to their physicochemical stability, mechanical rigidity, easy accessibility and high specificity of base pairing, and nucleic acids are considered appropriate receptors to be applied in NPs production. As for the proteins, they have many different binding sites, which can be accessed via adsorption (specific or non-specific) (Auría-Soro et al. 2019). Additionally, liposome-based (LIP) NPs drug delivery systems are currently being commercialized as therapeutic treatments for various diseases, such as tuberculosis (Kerry et al. 2018). Inorganic NPs tuned biologically have presented remarkable discoveries, and in spite of inorganic NPs not being considered biomolecules due to their great potential to be applied in the biomedical field allied with the advantages of their combination with bio composites, most of the literature seems to have opened an exception and included them in this section (Auría-Soro et al. 2019, White et al. 2019).

Research on bacterial origin polymers has also arisen in most recent years, showing their potential for a wide range of applications. Polysaccharides produced by bacteria can be divided into exopolysaccharides (e.g., alginate, cellulose, hyaluronic acid, and dextrans), which are secreted or synthesized extracellularly by cell wall-anchored enzymes. This group of polysaccharides is produced by several bacteria and archaea (Rehm 2010). Cellulose can be produced by microorganisms, displaying different properties and applications, and can be compared with plant cellulose. Many bacteria have been reported to have the potential to produce cellulose, such as the genera Gluconacetobacter, Rhizobium, Agrobacterium, Rhodobacter, and Sarcina. Bacterial cellulose presents unique features, like a high degree of crystallinity, water retention, tensile strength, and moldability (and does not require extra processing methods to exclude unwanted impurities, unlike plant cellulose). This polymer can also be molded to a specific form, size, and thickness to be suitable for a wide range of applications (Becker et al. 1998, Rehm, 2010).

Capsular polysaccharides are secreted and stay attached to the cell, working as surface antigens and virulence factors. Many capsular polysaccharides have been produced by GNB, including *K. pneumoniae* and *S. pneumoniae* (Rahn et al. 1999). Such polysaccharides have gained attention for their potential applications as virulence factors. A further understanding of their biosynthesis could lead to their use in vaccines (Rehm 2010). Non-ribosomally synthesized polyamides contain two extracellular polyamides—poly-y-glutamate (PgA) and e-poly-l-lysine (Pl)—and an intracellular cyanophycin granule peptide, that is CgP 52. Pl presents antibacterial properties, whereas PgA function as a virulence factor in Bacillus anthracis (Rehm 2010).

Among polyanhidrides, inorganic polyphosphate is the only polymer present in all types of living cells. This polymer is also present in bacteria, forming intracellular storage particles or forming a membrane-anchored complex, enhancing the uptake of many ions (Reusch and Sadoff 1988).

Polyphosphate has been reported to influence many features related to bacterial physiology, including survival in the stationary growth phase, an enhanced response to stress, biofilm formation, and pathogenicity (Rehm 2010). Although there are still no reports on the use of these bacterial origin biomolecules in the treatment of infectious diseases, their unique properties, and antibacterial features enhance their potential for future innovative therapies for treating infectious diseases.

Conclusions

Despite the undoubtedly advances introduced by antibiotics in the treatment of infectious diseases, their widespread use has been worsening problems associated with antimicrobial resistance in the last decades. The search for alternatives to antibiotics has led researchers to investigate the antimicrobial potential of microbial origin biomolecules. AMPs, bacteriophages, and enzybiotics among others are biomolecules known to possess great antimicrobial potential with a reduced tendency to induce bacterial resistance, and their characteristics, mode of action, and effectivity against bacterial pathogens were briefly described in this chapter. All the alternatives presented have disclosed promising results that support their use in the treatment of infectious diseases. In fact, it is thought that synergistic methods coupling some of the biomolecules presented with conventional antibiotics may accelerate clinical developments. However, additional tests need to be conducted; for instance, the development of models that evaluate the pharmacological applicability of these antimicrobial agents. Furthermore, there is still a lack of information related to the metabolism, clearance, and toxicity of the NPs as well as the right dose for therapeutic activity at the site of the pathogen. Therefore, before passing to a more advanced phase of the combination of biomolecules with antibiotics, it is mandatory to study the pharmacokinetics and pharmacodynamics of each component individually. Additionally, researchers must find a way to make use of all the information enclosed in the literature as an instrument in order to develop original alternatives which could match epidemiological and clinical data as solutions for the treatment of not only infectious but all emergent diseases.

Acknowledgement

This work was supported by the Portuguese Foundation for Science and Technology (FCT), project PEPTEX n° PTDC/CTM-TEX/28074/2017 (POCI-01-0145-FEDER-028074). The authors would also like to thank the project UID/CTM/00264/2020 of the Centre for Textile Science and

Technology (2C2T), funded by FCT/MCTES. C.S.M. acknowledges FCT for PhD funding via scholarship 2020.08547.BD.

References

Abedon, S. T., Kuhl, S. J., Blasdel, B. G., and Kutter, E. M. (2011). Phage treatment of human infections. Bacteriophage, 1(2): 66–85. https://doi.org/10.4161/bact.1.2.15845.

Ahmad, A., Senapati, S., Khan, M. I., Kumar, R., and Sastry, M. (2003). Extracellular Biosynthesis of Monodisperse Gold Nanoparticles by a Novel Extremophilic Actinomycete, *Thermomonospora* sp., 15: 3550–3553.

Akerey, B., Le-Lay, C., Fliss, I., Subirade, M., and Rouabhia, M. (2009). *In vitro* efficacy of nisin Z against *Candida albicans* adhesion and transition following contact with normal human gingival cells. Journal of Applied Microbiology, 107(4): 1298–1307. https://doi.org/10.1111/j.1365-2672.2009.04312.x.

AlMatar, M., Makky, E. A., Yakıcı, G., Var, I., Kayar, B., and Köksal, F. (2018). Antimicrobial peptides as an alternative to anti-tuberculosis drugs. In Pharmacological Research. https://doi.org/10.1016/j.phrs.2017.10.011.

Aparna, V. A., Biswas, R., and Jayakumar, R. (2019). Chapter 12 - Targeted nanoparticles for treating infectious diseases. pp. 169–185. *In*: Unnithan, A. R., Sasikala, A. R. K., Park, C. H., and Kim, C. S. (Eds.), Biomimetic Nanoengineered Materials for Advanced Drug Delivery. Elsevier. https://doi.org/https://doi.org/10.1016/B978-0-12-814944-7.00012-6.

Aslam, S., Lampley, E., Wooten, D., Karris, M., Benson, C., Strathdee, S., and Schooley, R. T. (2020). Lessons learned from the first 10 consecutive cases of intravenous bacteriophage therapy to treat multidrug-resistant bacterial infections at a single center in the United States. Open Forum Infectious Diseases. https://doi.org/10.1093/ofid/ofaa389.

Auría-Soro, C., Nesma, T., Juans-Velasco, P., Landeira-Viñuela, A., Fidlgo-Gomez, H., Acebes-Fernandez, V., Gongora, R., Parra, M. J. A., Manzano-Roman, R., and Fuentes, M. (2019). Interactions of nanoparticles and biosystems: microenvironmentl of nanoparticles and biomolecules in nanomedicine. Nanomaterials, 9(10). https://doi.org/10.3390/nano9101365.

Balciunas, E. M., Castillo Martinez, F. A., Todorov, S. D., Franco, B. D. G. de M., Converti, A., and Oliveira, R. P. de S. (2013). Novel biotechnological applications of bacteriocins: A review. Food Control, 32(1): 134–142. https://doi.org/10.1016/j.foodcont.2012.11.025.

Becker, A., Katzen, F., Pühler, A., and Ielpi, L. (1998). Xanthan gum biosynthesis and application: A biochemical/genetic perspective. Applied Microbiology and Biotechnology, 50(2): 145–152. https://doi.org/10.1007/s002530051269.

Benech, R. O., Kheadr, E. E., Laridi, R., Lacroix, C., and Fliss, I. (2002). Inhibition of *Listeria innocua* in cheddar cheese by addition of nisin Z in liposomes or by *in situ* production in mixed culture. Applied and Environmental Microbiology, 68(8): 3683–3690. https://doi.org/10.1128/AEM.68.8.3683-3690.2002.

Bianco, A., Licata, F., Zucco, R., Papadopoli, R., and Pavia, M. (2020). Knowledge and practices regarding antibiotics use: Findings from a cross-sectional survey among Italian adults. Evolution, Medicine, and Public Health, 2020(1): 129–138. https://doi.org/10.1093/emph/eoaa028.

Biéler, S., Silva, F., and Belin, D. (2010). The polypeptide core of Microcin E492 stably associates with the mannose permease and interferes with mannose metabolism. Research in Microbiology, 161(8): 706–710. https://doi.org/10.1016/j.resmic.2010.07.003.

Borysowski, J., and Górski, A. (2009). Enzybiotics and their potential applications in medicine. pp. 1–26. *In*: Enzybiotics. John Wiley and Sons, Ltd. https://doi.org/https://doi.org/10.1002/9780470570548.ch1.

Breedlove, B. (2019). Repurpose and reuse: artistic perspectives on antimicrobial resistance. Emerging Infectious Disease Journal, 25(1): 198. https://doi.org/10.3201/eid2501.ac2501.

Brives, C., and Pourraz, J. (2020). Phage therapy as a potential solution in the fight against AMR: obstacles and possible futures. Palgrave Communications, 6(1): 100. https://doi.org/10.1057/s41599-020-0478-4.

Cascales, E., Buchanan, S. K., Duché, D., Kleanthous, C., Lloubès, R., Postle, K., Riley, M., Slatin, S., and Cavard, D. (2007). Colicin biology. Microbiology and Molecular Biology Reviews, 71(1): 158–229. https://doi.org/10.1128/mmbr.00036-06.

Chatterjee, C., Paul, M., Xie, L., and van der Donk, W. A. (2005). Biosynthesis and mode of action of lantibiotics. Chemical Reviews, 105(2): 633–683. https://doi.org/10.1021/cr030105v.

Chavan, M. A., and Riley, M. A. (2007). Molecular evolution of bacteriocins in gram-negative bacteria. Bacteriocins, 19–43. https://doi.org/10.1007/978-3-540-36604-1_3.

Chen, Y., Batra, H., Dong, J., Chen, C., Rao, V. B., and Tao, P. (2019). Genetic engineering of bacteriophages against infectious diseases. Frontiers in Microbiology, 10: 954. https://doi.org/10.3389/fmicb.2019.00954.

Cintas, L. M., Casaus, M. P., Herranz, C., Nes, I. F., and Hernández, P. E. (2001). Review: bacteriocins of lactic acid bacteria. Food Science and Technology International, 7(4): 281–305. https://doi.org/10.1106/R8DE-P6HU-CLXP-5RYT.

Dakal, T. C., Kumar, A., Majumdar, R. S., and Yadav, V. (2016). Mechanistic basis of antimicrobial actions of silver nanoparticles. Frontiers in Microbiology, 7: 1831. https://doi.org/10.3389/fmicb.2016.01831.

Dams, D., and Briers, Y. (2019). Enzybiotics: enzyme-based antibacterials as therapeutics. pp. 233–253 *In*: Labrou, N. (Ed.). Therapeutic Enzymes: Function and Clinical Implications. Springer Singapore. https://doi.org/10.1007/978-981-13-7709-9_11.

Danis-Wlodarczyk, K. M., Wozniak, D. J., and Abedon, S. T. (2021). Treating bacterial infections with bacteriophage-based enzybiotics: *in vitro, in vivo* and clinical application. Antibiotics, 10(12). https://doi.org/10.3390/antibiotics10121497.

Der Torossian Torres, M., and De La Fuente-Nunez, C. (2019). Reprogramming biological peptides to combat infectious diseases. Chemical Communications. https://doi.org/10.1039/c9cc07898c.

Dickey, J., and Perrot, V. (2019). Adjunct phage treatment enhances the effectiveness of low antibiotic concentration against Staphylococcus aureus biofilms *in vitro*. PloS One, 14(1): e0209390–e0209390. https://doi.org/10.1371/journal.pone.0209390.

Divya Ganeshan, S., and Hosseinidoust, Z. (2019). Phage therapy with a focus on the human microbiota. Antibiotics (Basel, Switzerland), 8(3): 131. https://doi.org/10.3390/antibiotics8030131.

Duquesne, S., Destoumieux-Garzón, D., Peduzzi, J., and Rebuffat, S. (2007). Microcins, gene-encoded antibacterial peptides from enterobacteria. Natural Product Reports, 24(4), 708–734. https://doi.org/10.1039/b516237h.

Felgueiras, H. P., and Amorim, M. T. P. (2017). Functionalization of electrospun polymeric wound dressings with antimicrobial peptides. Colloids and Surfaces B: Biointerfaces, 156, 133–148. https://doi.org/10.1016/j.colsurfb.2017.05.001.

Felgueiras, H. P., Teixeira, M. A., Tavares, T. D., Homem, N. C., Zille, A., and Amorim, M. T. P. (2020). Antimicrobial action and clotting time of thin, hydrated poly(vinyl alcohol)/cellulose acetate films functionalized with LL37 for prospective wound-healing applications. Journal of Applied Polymer Science, 137(18): 48626. https://doi.org/10.1002/app.48626.

Felgueiras, H. P. (2021). An insight into biomolecules for the treatment of skin infectious diseases. Pharmaceutics, 13(7). https://doi.org/10.3390/pharmaceutics13071012.

Feng, T., Leptihn, S., Dong, K., Loh, B., Zhang, Y., Stefan, M. I., Li, M., Guo, X., and Cui, Z. (2021). JD419, a staphylococcus aureus phage with a unique morphology and broad host range. Frontiers in Microbiology, 12: 840. https://doi.org/10.3389/fmicb.2021.602902.

Giacometti, A., Cirioni, O., Barchiesi, F., and Scalise, G. (2000). *In-vitro* activity and killing effect of polycationic peptides on methicillin-resistant *Staphylococcus aureus* and interactions with clinically used antibiotics. Diagnostic Microbiology and Infectious Disease, 38(2): 115–118. https://doi.org/10.1016/S0732-8893(00)00175-9.

Górski, A., Międzybrodzki, R., Łobocka, M., Głowacka-Rutkowska, A., Bednarek, A., Borysowski, J., Jończyk-Matysiak, E., Łusiak-Szelachowska, M., Weber-Dabrowska, B., Bagińska, N., Letkiewicz, S., Dabrowska, K., and Scheres, J. (2018). Phage therapy: What have we learned? Viruses, 10(6): 1–28. https://doi.org/10.3390/v10060288.

Gupta, R., Beg, Q., and Lorenz, P. (2002). Bacterial alkaline proteases: Molecular approaches and industrial applications. Applied Microbiology and Biotechnology, 59(1): 15–32. https://doi.org/10.1007/s00253-002-0975-y.

Gupta, A., Joseph, B., Mani, A., and Thomas, G. (2008). Biosynthesis and properties of an extracellular thermostable serine alkaline protease from *Virgibacillus pantothenticus*. World Journal of Microbiology and Biotechnology, 24(2): 237–243. https://doi.org/10.1007/s11274-007-9462-z.

Harada, L. K., Silva, E. C., Campos, W. F., Del Fiol, F. S., Vila, M., Dąbrowska, K., Krylov, V. N., and Balcão, V. M. (2018). Biotechnological applications of bacteriophages: State of the art. Microbiological Research, 212–213(February), 38–58. https://doi.org/10.1016/j.micres.2018.04.007.

Høiby, N., Ciofu, O., Johansen, H. K., Song, Z. J., Moser, C., Jensen, P. Ø., Molin, S., Givskov, M., Tolker-Nielsen, T., and Bjarnsholt, T. (2011). The clinical impact of bacterial biofilms. International Journal of Oral Science, 3(2): 55–65. https://doi.org/10.4248/IJOS11026.

Homem, N. C., Tavares, T. D., Miranda, C. S., Antunes, J. C., Amorim, M. T. P., and Felgueiras, H. P. (2021). Functionalization of Crosslinked Sodium Alginate/Gelatin Wet-Spun Porous Fibers with Nisin Z for the Inhibition of Staphylococcus aureus-Induced Infections. International Journal of Molecular Sciences, 22(4). https://doi.org/10.3390/ijms22041930.

Hutchings, M. I., Truman, A. W., and Wilkinson, B. (2019). Antibiotics: past, present and future. Current Opinion in Microbiology, 51: 72–80. https://doi.org/https://doi.org/10.1016/j.mib.2019.10.008.

Irshad, A., Sarwar, N., Sadia, H., Malik, K., Javed, I., Irshad, A., Afzal, M., Abbas, M., and Rizvi, H. (2020). Comprehensive facts on dynamic antimicrobial properties of polysaccharides and biomolecules-silver nanoparticle conjugate. In International Journal of Biological Macromolecules. https://doi.org/10.1016/j.ijbiomac.2019.12.089.

Jack, R. W., Tagg, J. R., and Ray, B. (1995). Bacteriocins of gram-positive bacteria. Microbiological Reviews, 59(2): 171–200. https://doi.org/10.1128/mmbr.59.2.171-200.1995.

Jain, N., Bhargava, A., Majumdar, S., Tarafdar, J. C., and Panwar, J. (2011). Extracellular biosynthesis and characterization of silver nanoparticles using *Aspergillus flavus* NJP08: A mechanism perspective. Nanoscale, 3(2): 635–641. https://doi.org/10.1039/c0nr00656d.

Jha, A. K., Prasad, K., and Prasad, K. (2009). A green low-cost biosynthesis of Sb2O3 nanoparticles. Biochemical Engineering Journal, 43(3): 303–306. https://doi.org/10.1016/j.bej.2008.10.016.

Joerger, M. C., and Klaenhammer, T. R. (1986). Characterization and purification of helveticin J and evidence for a chromosomally determined bacteriocin produced by *Lactobacillus helveticus* 481. Journal of Bacteriology, 167(2): 439–446. https://doi.org/10.1128/jb.167.2.439-446.1986.

Kaestli, M., O'Donnell, M., Rose, A., Webb, J. R., Mayo, M., Currie, B. J., and Gibb, K. (2019). Opportunistic pathogens and large microbial diversity detected in source-to-distribution drinking water of three remote communities in Northern Australia. PLoS Neglected Tropical Diseases. https://doi.org/10.1371/journal.pntd.0007672.

Kaur, J., Singh, P., Sharma, D., Harjai, K., and Chhibber, S. (2020). A potent enzybiotic against methicillin-resistant *Staphylococcus aureus*. Virus Genes, 56(4): 480–497. https://doi.org/10.1007/s11262-020-01762-4.

Kerry, R. G., Gouda, S., Sil, B., Das, G., Shin, H.-S., Ghodake, G., and Patra, J. K. (2018). Cure of tuberculosis using nanotechnology: an overview. Journal of Microbiology, 56(5): 287–299. https://doi.org/DOI 10.1007/s12275-018-7414-y.

Klaenhammer, T. R. (1988). Bacteriocins of lactic acid bacteria. Biochimie, 70(3): 337–349. https://doi.org/10.1016/0300-9084(88)90206-4.

Konishi, Y., Ohno, K., Saitoh, N., Nomura, T., Nagamine, S., Hishida, H., Takahashi, Y., and Uruga, T. (2007). Bioreductive deposition of platinum nanoparticles on the bacterium *Shewanella algae*. Journal of Biotechnology, 128(3): 648–653. https://doi.org/10.1016/j.jbiotec.2006.11.014.

Lai, Y., and Gallo, R. L. (2009). AMPed up immunity: how antimicrobial peptides have multiple roles in immune defense. In Trends in Immunology. https://doi.org/10.1016/j.it.2008.12.003.

Lei, J., Sun, L. C., Huang, S., Zhu, C., Li, P., He, J., Mackey, V., Coy, D. H., and He, Q. Y. (2019). The antimicrobial peptides and their potential clinical applications. In American Journal of Translational Research.

Li, D., Zhou, B., and Lv, B. (2020). Antibacterial therapeutic agents composed of functional biological molecules. In Journal of Chemistry. https://doi.org/10.1155/2020/6578579.

Li, X., Xu, H., Chen, Z. S., and Chen, G. (2011). Biosynthesis of nanoparticles by microorganisms and their applications. Journal of Nanomaterials, 2011. https://doi.org/10.1155/2011/270974.

Ma, Y., Horsburgh, C. R., White, L. F., and Jenkins, H. E. (2018). Quantifying TB transmission: A systematic review of reproduction number and serial interval estimates for tuberculosis. Epidemiology and Infection. https://doi.org/10.1017/S0950268818001760.

Mann, R. (2006). Book Reviews. Black Theology, 4(1): 116–124. https://doi.org/10.1558/blth.2006.4.1.116.

Masri, A., Anwar, A., Khan, N. A., and Siddiqui, R. (2019). The use of nanomedicine for targeted therapy against bacterial infections. Antibiotics, 8: 260. https://doi.org/10.3390/antibiotics8040260.

Mathavan, I., and Beis, K. (2012). The role of bacterial membrane proteins in the internalization of microcin MccJ25 and MccB17. Biochemical Society Transactions, 40(6): 1539–1543. https://doi.org/10.1042/BST20120176.

Matsuzaki, S., Rashel, M., Uchiyama, J., Sakurai, S., Ujihara, T., Kuroda, M., Ikeuchi, M., Tani, T., Fujieda, M., Wakiguchi, H., and Imai, S. (2005). Bacteriophage therapy: A revitalized therapy against bacterial infectious diseases. In Journal of Infection and Chemotherapy. https://doi.org/10.1007/s10156-005-0408-9.

McArthur, D. B. (2019). Emerging infectious diseases. In Nursing Clinics of North America. https://doi.org/10.1016/j.cnur.2019.02.006.

Mohammed, H., Kumar, A., Bekyarova, E., Al-Hadeethi, Y., Zhang, X., Chen, M., Ansari, M. S., Cochis, A., and Rimondini, L. (2020). Antimicrobial mechanisms and effectiveness of graphene and graphene-functionalized biomaterials. A Scope Review. In Frontiers in Bioengineering and Biotechnology. https://doi.org/10.3389/fbioe.2020.00465.

Montero, M. M., López Montesinos, I., Knobel, H., Molas, E., Sorlí, L., Siverio-Parés, A., Prim, N., Segura, C., Duran-Jordà, X., Grau, S., and Horcajada, J. P. (2020). Risk factors for mortality among patients with pseudomonas aeruginosa bloodstream infections: What is the influence of XDR phenotype on outcomes? Journal of Clinical Medicine, 9(2): 514. https://doi.org/10.3390/jcm9020514.

Mukesh, M., Swapnil, P., Barupal, T., and Sharma, K. (2019). A review on infectious pathogens and mode of transmission. Journal of Plant Pathology and Microbiology, 10(1). https://doi.org/10.4172/2157-7471.1000472.

Niaz, T., Shabbir, S., Noor, T., Rahman, A., Bokhari, H., and Imran, M. (2018). Potential of polymer stabilized nano-liposomes to enhance antimicrobial activity of nisin Z against foodborne pathogens. LWT. https://doi.org/10.1016/j.lwt.2018.05.029.

Nigam, A., Gupta, D., and Sharma, A. (2014). Treatment of infectious disease: Beyond antibiotics. Microbiological Research, 169(9): 643–651. https://doi.org/https://doi.org/10.1016/j.micres.2014.02.009.

Nigam, P. S. (2013). Microbial enzymes with special characteristics for biotechnological applications. Biomolecules, 3(3): 597–611. https://doi.org/10.3390/biom3030597.

Okuda, K. I., Zendo, T., Sugimoto, S., Iwase, T., Tajima, A., Yamada, S., Sonomoto, K., and Mizunoe, Y. (2013). Effects of bacteriocins on methicillin-resistant Staphylococcus aureus biofilm. Antimicrobial Agents and Chemotherapy, 57(11): 5572–5579. https://doi.org/10.1128/AAC.00888-13.

Patra, S., Mukherjee, S., Barui, A. K., Ganguly, A., Sreedhar, B., and Patra, C. R. (2015). Green synthesis, characterization of gold and silver nanoparticles and their potential application for cancer therapeutics. Materials Science and Engineering C, 53: 298–309. https://doi.org/10.1016/j.msec.2015.04.048.

Paule, A., Frezza, D., and Edeas, M. (2018). Microbiota and Phage therapy: future challenges in medicine. Medical Sciences, 6(4). https://doi.org/10.3390/medsci6040086.

Pizzolato-Cezar, L. R., Okuda-Shinagawa, N. M., and Teresa Machini, M. (2019). Combinatory therapy antimicrobial peptide-antibiotic to minimize the ongoing rise of resistance. In Frontiers in Microbiology. https://doi.org/10.3389/fmicb.2019.01703.

Pons, A. M., Delalande, F., Duarte, M., Benoit, S., Lanneluc, I., Sablé, S., Van Dorsselaer, A., and Cottenceau, G. (2004). Genetic analysis and complete primary structure of *Microcin L*. Antimicrobial Agents and Chemotherapy, 48(2): 505–513. https://doi.org/10.1128/AAC.48.2.505-513.2004.

Priyabrata Mukherjee, Satyajyoti Senapati, Deendayal Mandal, Absar Ahmad, M. Islam Khan, RajivKumar, and M. S. (2002). Extracellular Synthesis of Gold.pdf. Chem. Bio. Chem., 5(5): 461–463.

Rahn, A., Drummelsmith, J., and Whitfield, C. (1999). Conserved organization in the cps gene clusters for expression of *Escherichia coli* group 1 K antigens: Relationship to the colanic acid biosynthesis locus and the cps genes from Klebsiella pneumoniae. Journal of Bacteriology, 181(7): 2307–2313. https://doi.org/10.1128/jb.181.7.2307-2313.1999.

Recht, J., Schuenemann, V. J., and Sánchez-Villagra, M. R. (2020). Host diversity and origin of zoonoses: The ancient and the new. Animals, 10(9): 1–14. https://doi.org/10.3390/ani10091672.

Reffuveille, F., De La Fuente-Nűnez, C., Mansour, S., and Hancock, R. E. W. (2014). A broad-spectrum antibiofilm peptide enhances antibiotic action against bacterial biofilms. Antimicrobial Agents and Chemotherapy. https://doi.org/10.1128/AAC.03163-14.

Rehm, B. H. A. (2010). Bacterial polymers: Biosynthesis, modifications and applications. Nature Reviews Microbiology, 8(8): 578–592. https://doi.org/10.1038/nrmicro2354.

Reith, F., Lengke, M. F., Falconer, D., Craw, D., and Southam, G. (2007). The geomicrobiology of gold. ISME Journal, 1(7): 567–584. https://doi.org/10.1038/ismej.2007.75.

Reusch, R. N., and Sadoff, H. L. (1988). Putative structure and functions of a poly-beta-hydroxybutyrate/calcium polyphosphate channel in bacterial plasma membranes. Proceedings of the National Academy of Sciences, 85(12): 4176–4180. https://doi.org/10.1073/pnas.85.12.4176.

Rodríguez, L., Martínez, B., Zhou, Y., Rodríguez, A., Donovan, D. M., and García, P. (2011). Lytic activity of the virion-associated peptidoglycan hydrolase HydH5 of *Staphylococcus aureus* bacteriophage vB_SauS-phiIPLA88. BMC Microbiology, 11: 138. https://doi.org/10.1186/1471-2180-11-138.

Sable, S., Pons, A. M., Gendron-Gaillard, S., and Cottenceau, G. (2000). Antibacterial activity evaluation of microcin J25 against diarrheagenic *Escherichia coli*. Applied and Environmental Microbiology, 66(10): 4595–4597. https://doi.org/10.1128/AEM.66.10.4595-4597.2000.

Saising, J., Dube, L., Ziebandt, A. K., Voravuthikunchai, S. P., Nega, M., and Götz, F. (2012). Activity of gallidermin on *Staphylococcus aureus* and *Staphylococcus epidermidis* biofilms. Antimicrobial Agents and Chemotherapy, 56(11): 5804–5810. https://doi.org/10.1128/AAC.01296-12.

Salata, C., Calistri, A., Alvisi, G., Celestino, M., Parolin, C., and Palù, G. (2019). Ebola virus entry: From molecular characterization to drug discovery. Viruses. https://doi.org/10.3390/v11030274.

Savadogo, A., Ouattara, C. A. T., Bassole, I. H. N., and Traore, S. A. (2006). Bacteriocins and lactic acid bacteria—A minireview. African Journal of Biotechnology, 5(9): 678–684. https://doi.org/10.4314/ajb.v5i9.42771.

Schlusselhuber, M., Humblot, V., Casale, S., Méthivier, C., Verdon, J., Leippe, M., and Berjeaud, J. M. (2015). Potent antimicrobial peptides against *Legionella pneumophila* and its environmental host, *Acanthamoeba castellanii*. Applied Microbiology and Biotechnology. https://doi.org/10.1007/s00253-015-6381-z.

Simons, A., Alhanout, K., and Duval, R. E. (2020). Bacteriocins, antimicrobial peptides from bacterial origin: Overview of their biology and their impact against multidrug-resistant bacteria. Microorganisms, 8(5). https://doi.org/10.3390/microorganisms8050639.

Singh, N., and Abraham, J. (2014). Ribosomally synthesized peptides from natural sources. Journal of Antibiotics, 67(4): 277–289. https://doi.org/10.1038/ja.2013.138.

Sizar, O., and Unakal, C. G. (2020). Gram Positive Bacteria. StatPearls. https://www.ncbi.nlm.nih.gov/books/NBK470553/.

Tavares, T. D., Antunes, J. C., Ferreira, F., and Felgueiras, H. P. (2020). Biofunctionalization of natural fiber-reinforced biocomposites for biomedical applications. Biomolecules, 10(1). https://doi.org/10.3390/biom10010148.

Tavares, T. D., Antunes, J. C., Padrão, J., Ribeiro, A. I., Zille, A., Amorim, M. T. P., Ferreira, F., and Felgueiras, H. P. (2020). Activity of specialized biomolecules against gram-positive and gram-negative bacteria. Antibiotics, 9(6): 1–16. https://doi.org/10.3390/antibiotics9060314.

Tellier, R., Li, Y., Cowling, B. J., and Tang, J. W. (2019). Recognition of aerosol transmission of infectious agents: A commentary. In BMC Infectious Diseases. https://doi.org/10.1186/s12879-019-3707-y.

Thakur, A., Mikkelsen, H., and Jungersen, G. (2019). Intracellular pathogens: host immunity and microbial persistence strategies. Journal of Immunology Research, 2019, 1356540. https://doi.org/10.1155/2019/1356540.

Thaljeh, L. F., Rothschild, J. A., Naderi, M., Coghill, L. M., Brown, J. M., and Brylinski, M. (2019). Hinge region in DNA packaging terminase pUL15 of herpes simplex virus: A potential allosteric target for antiviral drugs. Biomolecules. https://doi.org/10.3390/biom9100603.

Thomas, X., Destoumieux-Garzón, D., Peduzzi, J., Afonso, C., Blond, A., Birlirakis, N., Goulard, C., Dubost, L., Thai, R., Tabet, J. C., and Rebuffat, S. (2004). Siderophore peptide, a new type of post-translationally modified antibacterial peptide with potent activity. Journal of Biological Chemistry, 279(27): 28233–28242. https://doi.org/10.1074/jbc.M400228200.

Torres, F. G., Troncoso, O. P., Pisani, A., Gatto, F., and Bardi, G. (2019). Natural polysaccharide nanomaterials: An overview of their immunological properties. In International Journal of Molecular Sciences. https://doi.org/10.3390/ijms20205092.

Vallet-Regí, M., González, B., and Izquierdo-Barba, I. (2019). Nanomaterials as promising alternative in the infection treatment. Int. J. Mol. Sci., 20(15): 386. https://doi.org/10.3390/ijms20153806.

Vidhyasagar, V., and Jeevaratnam, K. (2013). Bacteriocin activity against various pathogens produced by *Pediococcus pentosaceus* VJ13 isolated from Idly batter. Biomedical Chromatography, 27(11): 1497–1502. https://doi.org/10.1002/bmc.2948.

Vijayalakshmi, S., Venkat, K. S., and Thankamani, V. (2013). Optimization and cultural characterization of *Bacillus* RV.B2.90 producing alkalophilic thermophilic protease. Research Journal of Biotechnology, 8(5): 37–43.

Wang, L., Zhao, X., Xia, X., Zhu, C., Qin, W., Xu, Y., Hang, B., Sun, Y., Chen, S., Zhang, H., Jiang, J., Hu, J., Fotina, H., and Zhang, G. (2019). Antimicrobial peptide JH-3 effectively kills salmonella enterica serovar *Typhimurium* strain CVCC541 and reduces its pathogenicity in mice. Probiotics and Antimicrobial Proteins. https://doi.org/10.1007/s12602-019-09533-w.

Wang, X., Li, D., Watanabe, T., Shigemori, Y., Mikawa, T., Okajima, T., Mao, L., and Ohsaka, T. (2012). A glucose/O2 biofuel cell using recombinant thermophilic enzymes. International Journal of Electrochemical Science, 7(2): 1071–1078.

White, B. D., Duan, C., and Townley, H. E. (2019). Nanoparticle activation methods in cances treatment. Biomolecules, 9(5). https://doi.org/10.3390/biom9050202.

WHO. (2017). WHO | Global Health Estimates 2015: Disease Burden by Cause, Age, Sex, by Country and by Region. http://www.who.int/%0Dhealthinfo/global_burden_disease/estimates/en/index2.html.

WHO. (2000). WHO Report on Infectious diseases: Overcoming antimicrobial resistance (p. 69). World Health Organization. https://apps.who.int/iris/handle/10665/66672.

WHO. (2020). WHO | Global tuberculosis report 2019. In World Health Organization. https://doi.org/.1037//0033-2909.I26.1.78.

WHO. (2021). WHO Coronavirus Disease (COVID-19) Dashboard. https://covid19.who.int.

Xiong, F., Dai, X., Li, Y., Wei, R., An, L., Wang, Y., and Chen, Z. (2019). Effects of the antimicrobial peptide L12 against multidrug-resistant *Staphylococcus aureus*. Mol. Med. Rep., 19(4): 3337–3344. https://doi.org/10.3892/mmr.2019.9988.

Yao, L., Zhang, L., Zhang, Y., Wang, R., Wongchitphimon, S., and Dong, Z. (2015). Self-assembly of rare-earth *Anderson polyoxometalates* on the surface of imide polymeric hollow fiber membranes potentially for organic pollutant degradation. Separation and Purification Technology, 151: 155–164. https://doi.org/http://dx.doi.org/10.1016/j.seppur.2015.05.045.

Yasir, M., Willcox, M. D. P., and Dutta, D. (2018). Action of antimicrobial peptides against bacterial biofilms. Materials, 11(12). https://doi.org/10.3390/ma11122468.

Zabel, R. A., and Morrell, J. J. (2020). The characteristics and classification of fungi and bacteria. In Wood Microbiology. https://doi.org/10.1016/b978-0-12-819465-2.00003-6.

Zhang, X., Yan, S., Tyagi, R. D., and Surampalli, R. Y. (2011). Synthesis of nanoparticles by microorganisms and their application in enhancing microbiological reaction rates. Chemosphere, 82(4): 489–494. https://doi.org/10.1016/j.chemosphere.2010.10.023.

Zschüttig, A., Zimmermann, K., Blom, J., Goesmann, A., Pöhlmann, C., and Gunzer, F. (2012). Identification and characterization of microcin S, a new antibacterial peptide produced by probiotic *Escherichia coli* G3/10. PLoS ONE, 7(3): 1–9. https://doi.org/10.1371/journal.pone.0033351.

Chapter 3

Biomolecules and Natural Macromolecules Against Tuberculosis

Awalagaway Dhulappa,[4] *Shuang Wang*[3] and *Manik Prabhu Narsing Rao*[1,2*]

Introduction

Tuberculosis is one of the most ancient diseases of mankind and has co-evolved with humans for many thousands of years or perhaps for several million years (Hirsh et al. 2004, Sandhu 2011). The oldest known molecular evidence of tuberculosis in humans was found in 9,000 years old human remains, which were recovered from a neolithic settlement in the Eastern Mediterranean (Hershkovitz et al. 2008). Robert Koch isolated *Mycobacterium tuberculosis*, the causative agent of human tuberculosis (Koch 1882). Despite 90 years of vaccination and 60 years of chemotherapy, it is still considered one of the biggest killers of infectious diseases (Smith 2003).

[1] Key Laboratory of Environmental Pollution Monitoring and Disease Control, Ministry of Education, Guizhou Talent Base of Microbiology and Human Health of Guizhou Province, School of Basic Medical Sciences, Guizhou Medical University, Guiyang, 550025, PR China.
[2] State Key Laboratory of Biocontrol, Guangdong Provincial Key Laboratory of Plant Resources and Southern Marine Science and Engineering Guangdong Laboratory (Zhuhai), School of Life Sciences, Sun Yat-Sen University, Guangzhou, 510275, PR China.
[3] Heilongjiang Academy of Black Soil Conservation & Utilization, Harbin 150086, People's Republic of China.
[4] Department of Microbiology, Maharani's Science College for Women, Bangalore 560001, India.
* Corresponding author: deene.manik@gmail.com

Tuberculosis is spread by the inhalation of infectious droplet nuclei containing live bacilli. When a patient with active pulmonary tuberculosis coughs, mycobacteria-laden droplet nuclei develop and can remain suspended in the air for several hours. Bacilli can be expelled by sneezing. The bacillary burden of the source case as well as the closeness and length of exposure influence the likelihood of transmission (Heemskerk et al. 2015). Tuberculosis usually affects the lungs, but it can also damage the brain, intestines, kidneys, or spine. The symptoms of tuberculosis vary depending on where the bacteria are developing in the body. In the case of pulmonary tuberculosis, symptoms may include a persistent cough, chest discomfort, hemoptysis, weakness or weariness, weight loss, fever, and night sweats (Zaman 2010).

Several challenges need to be addressed for effective control of tuberculosis. Tuberculosis is usually treated with a 6-month course of rifampicin, isoniazid, pyrazinamide, and ethambutol (Maiolini et al. 2020). Many countries use the BCG vaccine as part of their tuberculosis-control program (Bannon 1999). The WHO recommended a directly observed therapy short-course (DOTS) strategy, which was built on five pillars, viz., political commitment and continued funding for tuberculosis control programs, diagnosis by sputum smear examinations, uninterrupted supply of high-quality anti-tuberculosis drugs, drug intake under direct observation, and accurate reporting and recording of all registered cases (Davies 2003). Between 1900 and 1980, there was a steady decline in tuberculosis infections and this can be linked to socioeconomic improvements, the introduction of anti-TB medications (including ethambutol, isoniazid, pyrazinamide, rifampicin, and streptomycin), and the BCG immunization program (Newton et al. 2002). However, the number of cases worldwide is now increasing rapidly due to multi-drug resistant strains of *M. tuberculosis* (Newton et al. 2002). Multidrug-resistant tuberculosis is a major concern at present. To overcome this issue, many biomolecules and natural macromolecules from plants and microbial sources against tuberculosis were reported. In this chapter, we will discuss tuberculosis pathogenesis and anti-tuberculosis mechanisms of some natural macromolecules derived from plants and microbial sources.

Tuberculosis Pathogenesis and Anti-Tuberculosis Mechanisms

The risk of tuberculosis infection is determined by various factors, including the infectiousness of the source case, the proximity of contact, the bacillary load inhaled, and the prospective host's immunological condition (Moule and Cirillo 2020, Smith 2003). Tuberculosis infection occurs when a few air-dispersed tubercle bacilli from a patient with active pulmonary tuberculosis reach the host's alveoli (Mishra et al. 2017). It is then phagocytized by alveolar macrophages, but it is also possible that bacteria can be initially ingested by alveolar epithelial type II pneumocytes because this cell type is found in greater numbers (Bermudez and Goodman 1996). It then starts to replicate and disperse to epithelial and endothelial cells, and soon it spreads to other organs through lymph and blood which affects other cells as well (Mishra et al. 2017). Human toll-like receptors mediated the production of cytokines and other chemical mediators that act as a signal for mycobacterial infection. This causes monocytes associated with macrophages and dendritic cells to migrate to the infection site in the lungs (Means et al. 1999). The dendritic cells with engulfed bacilli mature and migrate to the regional lymph node and prime T cells (both CD4+ and CD8+) against mycobacterial antigens (Gonzalez-Juarrero et al. 2001). The developed T-cells proliferate and move back to the site of infection in response to the mediators produced by infected cells. These movements culminated in the creation of granuloma, a characteristic of tuberculosis (Mishra et al. 2017). This granuloma development maintains bacilli dormancy for an extended period and prevents its disruption within macrophages. Dormant bacilli contained within granuloma can be liberated under favourable conditions, resulting in infection relapse (Mishra et al. 2017).

To control tuberculosis, various drugs have been reported and these drugs have a different modes of action. *Mycobacterium tuberculosis* has a unique cell envelope structure and composition, containing peptidoglycan, mycolic acid, and arabinogalactan (Maitra et al. 2019). The anti-tuberculosis drugs have been reported to disrupt bacterial cell membrane (targets mycolic acids, arabinogalactan, and peptidoglycan synthesis), permeabilization and leakage of cellular contents leading to cell death (Ignacimuthu and Shanmugam 2010, Shetty and Dick 2018). Biofilm formation is a common pathogenic factor of *Mycobacterium* (Esteban and García-Coca 2018). The anti-tuberculosis drugs have been reported to inhibit biofilm formation (Arai et al. 2013). Drug efflux is an important survival strategy in *M. tuberculosis* (Gupta et al. 2014); however, some drugs have been reported to inhibit drug efflux pumps (Sharma et al. 2010). The anti-tuberculosis drugs have been reported to inhibit protein synthesis by binding to 23S RNA in the 50S ribosomal subunit of bacteria (Gordeev and Yuan 2014). The anti-tuberculosis drugs have been reported to inhibit DNA synthesis by binding to DNA gyrase (Locher et al. 2015). Andries et al. (2005) suggested anti-tuberculosis activity by inhibiting ATP synthase. The anti-tuberculosis drugs are also known to block *M. tuberculosis* growth by targeting the respiratory cytochrome bc_1 complex (Pethe et al. 2013). Detailed anti-tuberculosis activity is mentioned in Figure 1.

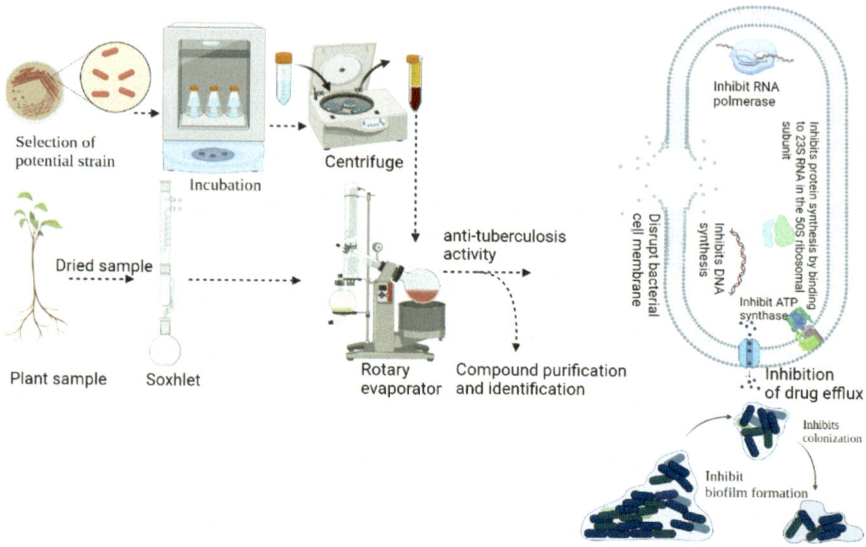

Figure 1. Procedure to obtain anti-tuberculosis molecule (plant and microbial source) and its mode of action. The figure is created using BioRender.com.

Plant-Based Molecules Against Tuberculosis

Plant-based drugs have been used worldwide and approximately 60% of the world's population still relies on medicinal plants for their healthcare (Gautam et al. 2007). Directly or indirectly, more than 25% of modern drugs and 60% of total anti-cancer drugs, are derived from plant secondary metabolites (Kumar et al. 2021). Plants contain a variety of secondary metabolites and they needed to be extracted. The plant sample was washed, dried, and powdered. It was then extracted using Soxhlet. The liquid extracts obtained will be filtered and concentrated using a rotary evaporator (Fauziyah et al. 2017). A detailed extraction procedure is mentioned in Figure 1.

Plants compounds, such as tannins, curcumin, genistein, quercetin, etc., have been reported to have anti-TB activity (Dey et al. 2015, Bai et al. 2016, Maiolini et al. 2020).

Historically, curcumin has traditionally been used in medicine (Slika and Patra 2020). It was reported that curcumin enhances the clearance of *M. tuberculosis* in differentiated THP-1 human monocytes and primary human alveolar macrophages. It was also noticed that curcumin acts as an inducer of caspase-3-dependent apoptosis and autophagy. Curcumin inhibited nuclear factor-kappa B (NFB) activity, which facilitated these anti-tuberculosis cellular actions (Bai et al. 2016).

Tea, next to the water, is the cheapest beverage humans consume. Drinking tea has been considered a health-promoting habit since ancient times (Khan and Mukhtar 2013). Epigallocatechin gallate from green tea impact the integrity of the mycobacterial cell wall and is likely to be a better prophylactic agent against tuberculosis (Sun et al. 2015).

Costus speciosus, *Cymbopogon citratus*, and *Tabernaemontana coronaria* are herbal plants traditionally used as remedies for symptoms of tuberculosis, including cough. The *C. speciosus*, *C. citratus*, and *T. coronaria* exhibited anti-tuberculosis activity. The majority of the identified compounds in the plant materials belonged to lipophilic fatty acid groups (Mohamad et al. 2018). Leaves of these medicinal legumes *Kingiodendron pinnatum* Rox. Hams., *Humboldtia brunonis* Wall., *Indigofera cassioides* Rottl.ex DC., *Derris scandens* Benth and *Ceasalpinia mimosoides* Lamk revealed the presence of saponins, steroids, anthro-quinones, terpinods, flavonoids, and phlabotanins exhibited potential anti-tubercular activity against *M. tuberculosis* (Kumar et al. 2014). Ellagitannin and punicalagin obtained from *Combretum molle* stem bark showed activity against *M. tuberculosis* (Asres et al. 2001). *Acalypha indica*, *Adhatoda vasica*, *Allium cepa*, *Allium sativum*, and *Aloe vera* plant extract were used against multidrug-resistant *M. tuberculosis*. All these plants exhibited anti-tuberculosis activity, but *A. vera*, *A. vasica* and *A. sativum* confirm earlier results (Gupta et al. 2010). These results suggest that different plant extracts have different activities.

The combination of antituberculosis drugs and medicinal plant extracts also showed good results. The antituberculosis drugs were combined with *Hibiscus sabdariffa*, *Kaempferia galanga*, and *Piper crocatum* plant extract to evaluate their activity against MDR *M. tuberculosis* isolates. *Kaempferia galanga*, *Piper crocatum*, and *Hibiscus sabdariffa* showed good combination effects with rifampicin to combat the rifampicin/streptomycin-resistant *M. tuberculosis* strain (Fauziyah et al. 2017).

The anti-tuberculosis activity also depends on the solvent used for the extraction. Khlifi et al. (2011) extracted tannins from a perennial flowering plant (*Globularia alypum*) using two extraction methods; the first utilized methanol, petroleum, and dichloromethane mixture and the second used methanol, acetone, and water mixture. The study showed that the methanol, petroleum, and dichloromethane mixture demonstrated improved activity against tuberculosis (Khlifi et al. 2011). Similarly, Oloya et al. (2022) extracted alkaloids, flavonoids, tannins, saponins, steroids, terpenoids, resins, cardiac glycosides, phenolic compounds, and coumarins using aqueous and methanol/DCM (1:1) solvents. It was noticed that the methanol/DCM (1:1) extracts showed higher antimycobacterial activity than aqueous extracts. This finding could be attributed to the ability of the medium polar (methanol/DCM, 1:1) solvent system to extract less-polar and lipophilic bioactive molecules with higher permeability across the lipid cell membranes of *M. tuberculosis*, inhibiting its growth and resulting in higher antimycobacterial activity; in contrast, aqueous extracts contain polar molecules with reduced permeability across the membranes, resulting in low antimycobacterial activity (Kevin et al. 2018, Oloya et al. 2022). Furthermore, the methanol/DCM solvent combination with a broad solvent polarity range may have allowed for a significant chemical diversity of the extracted chemicals, resulting in synergism in the crude extract and hence enhanced antimycobacterial activity of the extracts (Oloya et al. 2022). Table 1 list some plants and their active compounds used against *M. tuberculosis*.

Table 1. List some plants and their active compounds used against *M. tuberculosis*.

Plant Extract	Against	Active Compounds	References
Tetrapleura tetraptera	MDR *M. tuberculosis*	Tannins, alkaloids, saponins, flavonoids, phenols, and resins	Izebe et al. (2020)
Artemisia capillaris	*M. tuberculosis*	Ursolic acid and hydroquinone	Jyoti et al. (2016)
Cinnamomum verum and *Solanun surattense*	*M. tuberculosis*	Flavanoids, tannins, steroids and essential-oils	Vaidya et al. (2016)
Zanthoxylum leprieurii	*M. tuberculosis*	Acridone alkaloids	Bunalema et al. (2017)
Erythrina abyssinica	Rifampicin-resistant *M. tuberculosis*	Alkaloids, tannins, and flavones	Bunalema et al. (2011)
Euclea natalensis	Drug-resistant *M. tuberculosis*	Binaphthoquinone and diospyrin	Lall and Meyer (2001)
Haplopappus sonorensis	*M. tuberculosis*	5-hydroxy-3,7,4'-trimethoxyflavone, 5,7-dihydroxy-3,4'-dimethoxyflavone, and 5,4'-dihydroxy-3,7-dimethoxyflavone	Murillo et al. (2003)
Tithonia diversifolia	*M. tuberculosis*	Zingiberene and Bis 2- (ethyl hexyl) phthalate	Priyadarshini et al. (2022)

Microbial Molecules against Tuberculosis

Microbes are a good source of anti-tuberculosis molecules. However, it is important to isolate only potential strains that can produce anti-tuberculosis molecules. The detailed procedure for the extraction of anti-tuberculosis molecules from microbial sources is mentioned in Figure 1.

Although the first natural antibiotic, penicillin was discovered in a mold, *Penicillium notatum*, the majority of other clinically useful antibiotics, such as streptomycin, gentamycin, chloramphenicol, tetracycline, erythromycin, neomycin, and others, were discovered in the members of the phylum *Actinomycetota*, particularly the genus *Streptomyces* (Arefa et al. 2021). Different types of anti-tuberculous antibiotics are produced by *Streptomyces* sp. Streptomycin was the first antibiotic used against *tuberculosis* (Schatz et al. 1944) and later many anti-tuberculous antibiotics have been reported by *Streptomyces.* A polyketide macrolide identified as treponemycin was obtained from *Streptomyces* strain MS-6-6 showed promising anti-tuberculous activity (Yassien et al. 2015). Strain LS462 showed the highest 16S rRNA gene sequence similarity to *Streptomyces fuscichromogenes* (99.3%). *Streptomyces fuscichromogenes* under nonoptimized culture conditions produce high echinomycin (172 mg/l) yield. This compound has been reported to exhibit activity against *M. tuberculosis* H37Rv (Chen et al. 2021). The antibiotic success story prompted researchers to examine a wide range of ecological niches for actinomycetotal strains capable of producing novel antibiotics. A marine-derived *Streptomyces* sp. MS449, which showed the high 16S rRNA gene sequence similarity to *Streptomyces avermitilis* (99.7%), produced actinomycin X2 and actinomycin D in substantial quantities that showed strong anti-tuberculosis activity (Chen et al. 2012). *Streptomyce*s sp. G248. associated with marine sponges showed activity against *M. tuberculosis* (Cao et al. 2019). Endophytic actinomycetotal strain BCC72023 was isolated from rice (*Oryza sativa* L.). The strain BCC72023 was a member of the genus *Streptomyces* and produced macrolides, efomycins, and oxohygrolidin along with polyethers, abierixin, and 29-*O*-methylabierixin. The compounds showed activity against *M. tuberculosis* (Supong et al. 2016). Apart from *Streptomyces,* other microbes were also reported to have anti-tuberculosis activity.

Penicillium fellutanum, a marine fungus, produced peptide aldehydes such as lipopeptide aldehyde (fellutamide B), which has shown to be a potent proteasome inhibitor of *M. tuberculosis.* Fellutamide B was shown to be almost 1,000-fold more effective against *M. tuberculosis* than other peptide aldehydes (Lin et al. 2010). Polyketides from an endophyte, *Aspergillus fumigatus,* were reported to inhibit the growth of *M. tuberculosis* (Flewelling et al. 2015). Soft coral-associated fungus *Simplicillium* sp. SCSIO 41209 produced simplicilliumtide B and cyclo (L-Val-L-Pro) compound which showed anti-tuberculosis activity (Dai et al. 2018).

Staphylococcus hominis produced 30 kDa protein having anti-tuberculosis activity. The protein was stable over a wide range of temperatures and pH. This property will be beneficial in providing a longer shelf life at room temperature as well as a vital characteristic for any medicine that is delivered orally since it will be protected from the acidic and alkaline pH of the gastrointestinal system (Hussain et al. 2022). Further, amycobactin obtained from uncultured bacteria was active against *M. tuberculosis.* For the first time, amycobactin was reported to target the

Table 2. List some microorganisms and their active compounds used against *M. tuberculosis*.

Microorganisms	Against	Active Compounds	References
Streptomyces avermitilis	*M. tuberculosis* H37Rv	Actinomycin X2 and actinomycin D	Chen et al. (2012)
Streptomyces sp. BCC26924	*M. tuberculosis*	Cyclomarin C	Intaraudom et al. (2011)
Streptomyces sp. YIM65484	*M. tuberculosis*	(2*E*,4*E*)-5-(3-hydroxyphenyl)- penta-2,4-dienamide, ergosterol, ergosterol peroxide, and halolitoralin B	Hao et al. (2015)
Aspergillus sp. LS57	*M. tuberculosis*	Aspergilluone A	Liu et al. (2021)
Pseudoplectania nigrella	*M. tuberculosis*	Plectasin	Tenland et al. (2018)
Aspergillus sp.	*M. tuberculosis* protein tyrosine phosphatase B	Ergosterdiacids A and B	Liu et al. (2018)
Aspergillus terreus SCSIO 41008	*M. tuberculosis* protein tyrosine phosphatase B	Butyrolactone I	Luo et al. (2019)

Sec protein secretion machinery in *M. tuberculosis* (Quigley et al. 2020). A list of some microorganisms and their active compounds used against *M. tuberculosis* was mentioned in Table 2.

Conclusion

M. tuberculosis, a highly infectious bacterial pathogen and the causative agent of tuberculosis, affects approximately one-third of the world's population. Taking preventative measures and utilizing anti-tuberculosis drugs helped to manage the disease, but continued use of the drugs resulted in a drug resistance problem. Several medicinal plants and microbial products have been alternatively used for the development of new anti-tuberculosis drugs. The plants and microbial products were reported to have anti-tuberculosis activity and have different strategies to inhibit it. The plants and microbial compounds also show a synergistic effect with the antibiotics used to treat tuberculosis. Looking at the volume of plants and microbial present, only a tip of an iceberg was used against *M. tuberculosis* and hence further studies are needed to be carried out to combat tuberculosis.

Acknowledgment

The authors thank Dr. Bhagwan Rekadwad (Yenepoya, Deemed to be University, India) for his valuable suggestions.

References

Andries, K., Verhasselt, P., Guillemont, J., Göhlmann, H. W., Neefs, J. M., Winkler, H., Van Gestel, J., Timmerman, P., Zhu, M., Lee, E., Williams, P., de Chaffoy, D., Huitric, E., Hoffner, S., Cambau, E., Truffot-Pernot, C., Lounis, N., and Jarlier, V. (2005). A diarylquinoline drug active on the ATP synthase of *Mycobacterium tuberculosis*. Science, 307: 223–227.

Arai, M., Niikawa, H., and Kobayashi, M. (2013). Marine-derived fungal sesterterpenes, ophiobolins, inhibit biofilm formation of *Mycobacterium* species. J. Nat. Med., 67: 271–275.

Arefa, N., Sarker, A. K., and Rahman, M. A. (2021). Resistance-guided isolation and characterization of antibiotic-producing bacteria from river sediments. BMC Microbiology, 21: 116.

Asres, K., Bucar, F., Edelsbrunner, S., Kartnig, T., Höger, G., and Thiel, W. (2001). Investigations on antimycobacterial activity of some Ethiopian medicinal plants. Phytother Res., 15: 323–326.

Bai, X., Oberley-Deegan, R. E., Bai, A., Ovrutsky, A. R., Kinney, W. H., Weaver, M., Zhang, G., Honda, J. R., and Chan, E. D. (2016). Curcumin enhances human macrophage control of *Mycobacterium tuberculosis* infection. Respirology, 21: 951–957.

Bannon, M. J. (1999). BCG and tuberculosis. Arch. Dis. Child, 80: 80–83.

Bermudez, L. E., and Goodman, J. (1996) *Mycobacterium tuberculosis* invades and replicates within type II alveolar cells. Infect. Immun., 64: 1400–1406.

Bunalema, L., Fotso, G. W., Waako, P., Tabuti, J., and Yeboah, S. O. (2017). Potential of *Zanthoxylum leprieurii* as a source of active compounds against drug resistant *Mycobacterium tuberculosis*. BMC Complementary and Alternative Medicine, 17: 89.

Bunalema, L., Kirimuhuzya, C., Tabuti, J., Waako, P., Magadula, J., Otieno, N., Orodho, J., and Okemo, P. (2011). The efficacy of the crude root bark extracts of *Erythrina abyssinica* on rifampicin resistant *Mycobacterium tuberculosis*. African Health Sciences, 11: 587–593.

Cao, D. D., Trinh, T. T. V., Mai, H. D. T., Vu, V. N., Le, H. M., Thi, Q. V., Nguyen, M. A., Duong, T. T., Tran, D. T., Chau, V. M., Ma, R., Shetye, G., Cho, S., Murphy, B. T., and Pham, V. C. (2019). Antimicrobial Lavandulylated Flavonoids from a Sponge-Derived *Streptomyces* sp. G248 in East Vietnam Sea. Mar Drugs 17.

Chen, C., Chen, X., Ren, B., Guo, H., Abdel-Mageed, W. M., Liu, X., Song, F., and Zhang, L. (2021). Characterization of *Streptomyces* sp. LS462 with high productivity of echinomycin, a potent antituberculosis and synergistic antifungal antibiotic. Journal of Industrial Microbiology and Biotechnology, 48.

Chen, C., Song, F., Wang, Q., Abdel-Mageed, W. M., Guo, H., Fu, C., Hou, W., Dai, H., Liu, X., Yang, N., Xie, F., Yu, K., Chen, R., and Zhang, L. (2012). A marine-derived *Streptomyces* sp. MS449 produces high yield of actinomycin X2 and actinomycin D with potent anti-tuberculosis activity. Appl. Microbiol. Biotechnol. 95: 919–927.

Dai, Y., Lin, Y., Pang, X., Luo, X., Salendra, L., Wang, J., Zhou, X., Lu, Y., Yang, B., and Liu, Y. (2018). Peptides from the Soft Coral-associated Fungus *Simplicillium* sp. SCSIO41209. Phytochemistry, 154: 56–62.

Davies, P. D. (2003). The role of DOTS in tuberculosis treatment and control. Am. J. Respir. Med., 2: 203–209.

Dey, D., Ray, R., and Hazra, B. (2015). Antimicrobial activity of pomegranate fruit constituents against drug-resistant *Mycobacterium tuberculosis* and β-lactamase producing *Klebsiella pneumoniae*. Pharmaceutical Biology, 53: 1474–1480.

Esteban, J., and García-Coca, M. (2018). *Mycobacterium* Biofilms. Front Microbiol., 8: 2651.

Fauziyah, P. N., Sukandar, E. Y., and Ayuningtyas, D. K. (2017). Combination effect of antituberculosis drugs and ethanolic extract of selected medicinal plants against multi-drug resistant *Mycobacterium tuberculosis* Isolates. Sci. Pharm., 85 (1): 14.

Flewelling, A. J., Bishop, A. I., Johnson, J. A., and Gray, C.A. (2015). Polyketides from an Endophytic *Aspergillus fumigatus* Isolate Inhibit the Growth of *Mycobacterium tuberculosis* and MRSA. Nat.. Prod. Commun. 10: 1661–1662.

Gautam, R., Saklani, A., and Jachak, S. M. (2007). Indian medicinal plants as a source of antimycobacterial agents. Journal of Ethnopharmacology, 110: 200–234.

Gonzalez-Juarrero, M., Turner, O. C., Turner, J., Marietta, P., Brooks, J. V., and Orme, I. M. (2001). Temporal and spatial arrangement of lymphocytes within lung granulomas induced by aerosol infection with Mycobacterium tuberculosis. Infect. Immun., 69: 1722–1728.

Gordeev, M. F., and Yuan, Z. Y. (2014). New potent antibacterial oxazolidinone (MRX-I) with an improved class safety profile. J. Med. Chem., 57: 4487–4497.

Gupta, R., Thakur, B., Singh, P., Singh, H. B., Sharma, V. D., Katoch, V. M., and Chauhan, S. V. (2010). Anti-tuberculosis activity of selected medicinal plants against multi-drug resistant *Mycobacterium tuberculosis* isolates. Indian J. Med. Res., 131: 809–813.

Gupta, S., Cohen, K. A., Winglee, K., Maiga, M., Diarra, B., and Bishai, W. R. (2014). Efflux inhibition with verapamil potentiates bedaquiline in *Mycobacterium tuberculosis*. Antimicrob. Agents Chemother., 58: 574–576.

Hao Z., Lixing, Z., Wei, L., Yabin, Y., Lihua, X., and Zhongtao, D. (2015). Anti-Mycobacterium tuberculosis active metabolites from an endophytic *Streptomyces* sp. YIM65484. Natural Product Reports, 9: 196–200.

Heemskerk, D., Caws, M., Marais, B., and Farrar, J. (2015). "Pathogenesis" in Tuberculosis in Adults and Children. (Cham: Springer International Publishing), 9–16.

Hershkovitz, I., Donoghue, H. D., Minnikin, D. E., Besra, G. S., Lee, O. Y., Gernaey, A. M., Galili, E., Eshed, V., Greenblatt, C. L., Lemma, E., Bar-Gal, G. K., and Spigelman, M. (2008). Detection and molecular characterization of 9,000-year-old *Mycobacterium tuberculosis* from a Neolithic settlement in the Eastern Mediterranean. PLoS One, 3: e3426.

Hirsh, A. E., Tsolaki, A. G., Deriemer, K., Feldman, M. W., and Small, P. M. (2004). Stable association between strains of *Mycobacterium tuberculosis* and their human host populations. Proc. Natl. Acad. Sci. U S A., 101: 4871–4876.

Hussain, M. S., Vashist, A., Kumar, M., Taneja, N. K., Gautam, U. S., Dwivedi, S., Tyagi, J. S., and Gupta, R. K. (2022). Anti-mycobacterial activity of heat and pH stable high molecular weight protein(s) secreted by a bacterial laboratory contaminant. Microbial Cell Factories, 21: 15.

Intaraudom, C., Rachtawee, P., Suvannakad, R., and Pittayakhajonwut, P. (2011). Antimalarial and antituberculosis substances from *Streptomyces* sp. BCC26924. Tetrahedron 67: 7593–7597.

Ignacimuthu, S., and Shanmugam, N. (2010). Antimycobacterial activity of two natural alkaloids, vasicine acetate and 2-acetyl benzylamine, isolated from Indian shrub Adhatoda vasica Ness. leaves. J. Biosci., 565–570.

Izebe, K., Ibrahim, K., Onaolapo, J., Oladosu, P., Ya'aba, Y., Njoku, M., Shehu, M., Ezeunala, M., and Ibrahim, Y. (2020) Evaluation of *In-Vitro* Anti-Tuberculosis Activity of *Tetrapleura tetraptera* Crude and Fractions on Multidrug Resistant *Mycobacterium tuberculosis*. Journal of Tuberculosis Research, 8: 165–176.

Jyoti, M. A., Nam, K. W., Jang, W. S., Kim, Y. H., Kim, S. K., Lee, B. E., and Song, H. Y. (2016). Antimycobacterial activity of methanolic plant extract of *Artemisia capillaris* containing ursolic acid and hydroquinone against *Mycobacterium tuberculosis*. J. Infect. Chemother., 22: 200–208.

Kevin, K., John, K., Carolyn, N., Derrick, S., and Lubega, A. (2018). *In vitro* anti-tuberculosis activity of total crude extract of *Echinops amplexicaulis* against multi-drug resistant *Mycobacterium tuberculosis*. J. Health Sci., 6: 296–303.

Khan, N., and Mukhtar, H. (2013). Tea and health: studies in humans. Curr. Pharm. Des., 19: 6141–6147.

Khlifi, D., Hamdi, M., Hayouni, A. E., Cazaux, S., Souchard, J. P., Couderc, F., and Bouajila, J. (2011). Global chemical composition and antioxidant and anti-tuberculosis activities of various extracts of *Globularia alypum* L. (*Globulariaceae*) Leaves. Molecules, 16: 10592–10603.

Koch, R. (1882). Die aetiologie der tuberculose. Berl. Klin. Wochenschr., 15: 221–236.

Kumar, J. K., Devi Prasad, A. G., and Chaturvedi, V. (2014). Phytochemical screening of five medicinal legumes and their evaluation for in vitro anti-tubercular activity. Ayu, 35: 98–102.

Kumar, M., Singh, S. K., Singh, P. P., Singh, V. K., Rai, A. C., Srivastava, A. K., Shukla, L., Kesawat, M. S., Kumar Jaiswal, A., Chung, S. M., and Kumar, A. (2021). Potential anti-mycobacterium tuberculosis activity of plant secondary metabolites: insight with molecular docking interactions. Antioxidants (Basel) 10.

Lall, N., and Meyer, J. J. M. (2001). Inhibition of drug-sensitive and drug-resistant strains of *Mycobacterium tuberculosis* by diospyrin, isolated from *Euclea natalensis*. Journal of Ethnopharmacology, 78: 213–216.

Locher, C. P., Jones, S. M., Hanzelka, B. L., Perola, E., Shoen, C. M., Cynamon, M. H., Ngwane, A. H., Wiid, I. J., van Helden, P. D., Betoudji, F., Nuermberger, E. L., and Thomson, J. A. (2015). A novel inhibitor of gyrase B is a potent drug candidate for treatment of tuberculosis and nontuberculosis mycobacterial infections. Antimicrob. Agents Chemother., 59: 1455–1465.

Lin, G., Li, D., Chidawanyika, T., Nathan, C., and Li, H. (2010). Fellutamide B is a potent inhibitor of the *Mycobacterium tuberculosis* proteasome. Arch. Biochem. Biophys., 501: 214–220.

Liu, Y., Ding, L., He, J., Zhang, Z., Deng, Y., He, S., and Yan, X. (2021). A new antibacterial chromone from a marine sponge-associated fungus *Aspergillus* sp. LS57. Fitoterapia, 154: 105004.

Liu, Z., Dong, Z., Qiu, P., Wang, Q., Yan, J., Lu, Y., Wasu, P.-A., Hong, K., and She, Z. (2018). Two new bioactive steroids from a mangrove-derived fungus *Aspergillus* sp. Steroids, 140: 32–38.

Luo, X.-W., Lin, Y., Lu, Y.-J., Zhou, X.-F., and Liu, Y.-H. (2019). Peptides and polyketides isolated from the marine sponge-derived fungus *Aspergillus terreus* SCSIO 41008. Chinese Journal of Natural Medicines, 17: 149–154.

Maiolini, M., Gause, S., Taylor, J., Steakin, T., Shipp, G., Lamichhane, P., Deshmukh, B., Shinde, V., Bishayee, A., and Deshmukh, R. R. (2020). The War against Tuberculosis: A Review of Natural Compounds and Their Derivatives. Molecules, 25.

Maitra, A., Munshi, T., Healy, J., Martin, L. T., Vollmer, W., Keep, N. H., and Bhakta, S. (2019). Cell wall peptidoglycan in *Mycobacterium tuberculosis*: An Achilles' heel for the TB-causing pathogen. FEMS Microbiol. Rev., 43: 548–575.

Means, T. K., Wang, S., Lien, E., Yoshimura, A., Golenbock, D. T., and Fenton, M. J. (1999). Human toll-like receptors mediate cellular activation by *Mycobacterium tuberculosis*. J Immunol., 163: 3920–3927.

Mishra, S. K., Tripathi, G., Kishore, N., Singh, R. K., Singh, A., and Tiwari, V. K. (2017). Drug development against tuberculosis: Impact of alkaloids. Eur. J. Med. Chem., 137: 504–544.

Moule, M. G., and Cirillo, J. D. (2020). *Mycobacterium tuberculosis* Dissemination Plays a Critical Role in Pathogenesis. Front Cell Infect. Microbiol., 10: 65.

Mohamad, S., Ismail, N. N., Parumasivam, T., Ibrahim, P., Osman, H., and Wahab, H. A. (2018). Antituberculosis activity, phytochemical identification of *Costus speciosus* (J. Koenig) Sm., *Cymbopogon citratus* (DC. Ex Nees) Stapf., and *Tabernaemontana coronaria* (L.) Willd. and their effects on the growth kinetics and cellular integrity of *Mycobacterium tuberculosis* H37Rv. BMC Complementary and Alternative Medicine, 18: 5.

Murillo, J. I., Encarnación-Dimayuga, R., Malmstrøm, J., Christophersen, C., and Franzblau, S. G. (2003). Antimycobacterial flavones from *Haplopappus sonorensis*. Fitoterapia, 74: 226–230.

Newton, S. M., Lau, C., Gurcha, S. S., Besra, G. S., and Wright, C. W. (2002). The evaluation of forty-three plant species for *in vitro* antimycobacterial activities; isolation of active constituents from *Psoralea corylifolia* and *Sanguinaria canadensis*. Journal of Ethnopharmacology, 79: 57–67.

Oloya, B., Namukobe, J., Ssengooba, W., Afayoa, M., and Byamukama, R. (2022). Phytochemical screening, antimycobacterial activity and acute toxicity of crude extracts of selected medicinal plant species used locally in the treatment of tuberculosis in Uganda. Tropical Medicine and Health, 50: 16.

Pethe, K., Bifani, P., Jang, J., Kang, S., Park, S., Ahn, S., Jiricek, J., Jung, J., Jeon, H. K., Cechetto, J., Christophe, T,. Lee, H., Kempf, M., Jackson, M., Lenaerts, A. J., Pham, H., Jones, V., Seo, M. J., Kim, Y. M., Seo, M., Seo, J. J., Park, D., Ko, Y., Choi, I., Kim, R., Kim, S. Y., Lim, S., Yim, S. A., Nam, J., Kang, H., Kwon, H., Oh, C. T., Cho, Y., Jang, Y., Kim, J., Chua, A., Tan, B. H., Nanjundappa, M. B., Rao, S. P., Barnes, W. S., Wintjens, R., Walker, J. R., Alonso, S., Lee, S., Kim, J., Oh, S., Oh, T., Nehrbass, U., Han, S. J., No, Z., Lee, J., Brodin, P., Cho, S. N., Nam, K., and Kim, J. (2013). Discovery of Q203, a potent clinical candidate for the treatment of tuberculosis. Nat. Med., 19: 1157–1160.

Priyadarshini, N., Veeramani, A., Chinnathambi, P., Palanichamy, A., Al-Dosary, M.A., Ali, M.A., Lee, J., and Paulraj, B. (2022). Antimycobacterial effect of plant derived phthalate against *Mycobacterium tuberculosis* H37Ra. Physiological and Molecular Plant Pathology, 117: 101761.

Quigley, J., Peoples, A., Sarybaeva, A., Hughes, D., Ghiglieri, M., Achorn, C., Desrosiers, A., Felix, C., Liang, L., Malveira, S., Millett, W., Nitti, A., Tran, B., Zullo, A., Anklin, C., Spoering, A., Ling, L. L., and Lewis, K. (2020). Novel Antimicrobials from Uncultured Bacteria Acting against *Mycobacterium tuberculosis*. mBio, 11.

Sandhu, G. K. (2011). Tuberculosis: current situation, challenges and overview of its control programs in India. J. Glob. Infect. Dis., 3: 143–150.

Schatz, A., Bugle, E., and Waksman, S. A. (1944). Streptomycin, a substance exhibiting antibiotic activity against gram-positive and gram-negative bacteria. Proceedings of the Society for Experimental Biology and Medicine, 55: 66–69.

Sharma, S., Kumar, M., Sharma, S., Nargotra, A., Koul, S., and Khan, I. A. (2010). Piperine as an inhibitor of Rv1258c, a putative multidrug efflux pump of *Mycobacterium tuberculosis*. J. Antimicrob. Chemother., 651694–651701.

Shetty, A., and Dick, T. (2018). Mycobacterial Cell Wall Synthesis Inhibitors Cause Lethal ATP Burst. Front Microbiol., 9: 1898.

Slika, L., and Patra, D. (2020). Traditional uses, therapeutic effects and recent advances of curcumin: a mini-review. Mini, Rev. Med. Chem., 20: 1072–1082.

Smith, I. (2003). *Mycobacterium tuberculosis* pathogenesis and molecular determinants of virulence. Clin. Microbiol. Rev., 16: 463–496.

Sun, T., Qin, B., Gao, M., Yin, Y., Wang, C., Zang, S., Li, X., Zhang, C., Xin, Y., and Jiang, T. (2015). Effects of epigallocatechin gallate on the cell-wall structure of *Mycobacterial smegmatis* mc^2155. Nat. Prod. Res., 29: 2122–2124.

Supong, K., Thawai, C., Choowong, W., Kittiwongwattana, C., Thanaboripat, D., Laosinwattana, C., Koohakan, P., Parinthawong, N., and Pittayakhajonwut, P. (2016). Antimicrobial compounds from endophytic *Streptomyces* sp. BCC72023 isolated from rice (*Oryza sativa* L.). Research in Microbiology, 167: 290–298.

Tenland, E., Krishnan, N., Rönnholm, A., Kalsum, S., Puthia, M., Mörgelin, M., Davoudi, M., Otrocka, M., Alaridah, N., Glegola-Madejska, I., Sturegård, E., Schmidtchen, A., Lerm, M., Robertson, B. D., and Godaly, G. (2018). A novel derivative of the fungal antimicrobial peptide plectasin is active against *Mycobacterium tuberculosis*. Tuberculosis, 113: 231–238.

Vaidya, S., Sharma, J., Maniar, J., Prabhu, N., Mamawala, M., Joshi-Pundit, S., and Chowdhary, A. (2016). Assessment of anti-tuberculosis activity of extracts of cinnamomum verum and solanun surattense along with isoniazid. European Respiratory Journal, 48: PA2691.

Yassien, M. A., Abdallah, H. M., El-Halawany, A. M., and Jiman-Fatani, A. A. (2015). Anti-tuberculous activity of treponemycin produced by a streptomyces strain MS-6-6 isolated from Saudi Arabia. Molecules, 20: 2576–2590.

Zaman, K. (2010). Tuberculosis: a global health problem. J. Health Popul. Nutr., 28: 111–113.

Chapter 4

Biomolecules for Treatment of Communicable Diseases

Chamma Gupta, Abhishek Byahut, Karma G. Dolma and *Mingma L. Sherpa**

Introduction

When the twenty-first century began, the globe was split into nations burdened with morbidity and mortality related to communicable and/or infectious diseases and other nations that were battling non-communicable diseases, like Ischemic disease, cancer and stroke. However, the emergence of drug-resistant pathogens like multidrug-resistant TB (MDR TB) and extreme drug-resistant TB (XDR TB) along with emerging pathogens like the severe acute respiratory syndrome (SARS) coronavirus, MERS and the current SARS Covid-19 have transcended all boundaries and grew to a pandemic scale. Despite the tremendous economic progress, technical expertise and improvement in major health indices, India still faces challenges with communicable diseases such as human immunodeficiency virus (HIV), tuberculosis (TB), hepatitis, influenza (IAV) and other chronic diseases as well as the current COVID-19 pandemic. This trend is predicted to prevail as newly evolved pathogens (multi-resistant) emerge and infect with abilities to bypass or overwhelm existing host defense and cross transmitting across species like is seen for swine flu, coronavirus, etc. (Bloom et al. 2017). To help combat the communicable disease burden and infection, basic biology, molecular transcriptional and epidemiological

Sikkim Manipal Institute of Medical Sciences, Sikkim Manipal University (SMU), Gangtok, Sikkim, India, 737102.

* Corresponding author: mingmals@yahoo.com

research are required to gain a deeper understanding of the spectrum of human pathogens, biomolecules and their interactions with host species in order to develop novel and efficient diagnostics and therapeutics for communicable diseases (DBT 2022). A plethora of biomolecules of plant and animal origin have contributed significantly to current practice in clinical medicine and diagnostic technologies. It has widespread application in the prophylaxis, detection, therapy and management of human disease, and it is also proving to be useful in personalized medicine.

Following the discovery of penicillin, several antibiotics were synthesized and proved effective in the treatment of communicable diseases and infections (Bennett and Chung 2001). However, antimicrobial resistance has caused severe global public health problems and has confronted biologists with a major hurdle. Scientists are currently assessing alternative innovative approaches to address problems associated with communicable and infectious diseases. This has sparked interest in biomolecule-based therapies with low resistance, such as antimicrobial peptides (AMPs), essential oils (Eos), bacteriocins and aptamer (Nigam et al. 2014). There are several other specific groups of biomolecules detected to have significant importance in the prevention of the progression and treatment of such diseases. These include an array of biological molecules, including antibiotics, peptides, galectin, vaccines, bacteriocins, glycomes and natural extracts (Iwasaki-Hozumi et al. 2021).

Communicable Diseases and Infection: According to the World Health Organization (WHO) and Centres for Disease Control and Prevention (CDC), infectious disease is defined as a condition caused by the existence and proliferation of pathogenic microorganisms such as viruses, bacterium, parasites, and/or fungi or its lethal toxins (WHO 2022, CDC 2022). Communicable diseases are viral or bacterial causing illnesses that can be transmitted directly or indirectly from an individual to another or a susceptible host by an infected animal, vector, an infected inanimate object, body fluids or through the air (WHO 2022, CDC 2022). Some of the prevalent communicable diseases are malaria, tuberculosis (TB), HIV, Hepatitis A, B, C, Measles, Syphilis, Influenza and candidiasis as shown in Figure 1 caused by microbial agents, that is protozoa, bacteria, virus and fungi, respectively. Almost every year, new infectious diseases emerge or re-emerge and are largely rising due to socioeconomic, physiological and ecological factors. The type and severity of the injury caused by these microbes are closely linked to the synthesis as well as the release of toxoids that disrupt the normal functions of each organ/system and may result in the emergence of distinctive clinical manifestations (Van Seventer and Hochberg 2017, Jones et al. 2008, Rohr et al. 2019).

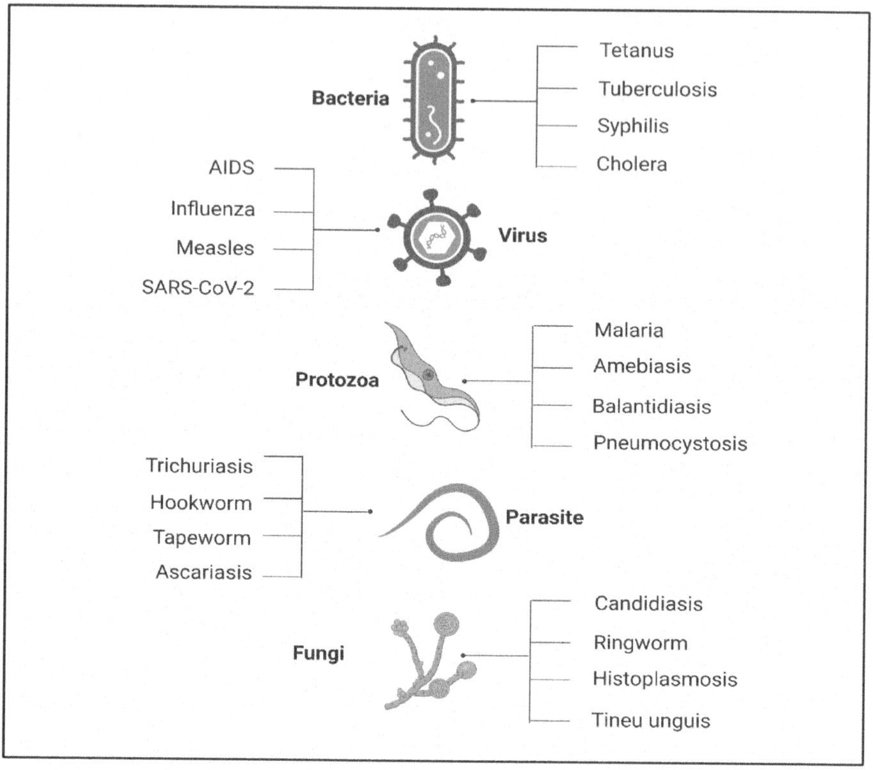

Figure 1. Classification of microorganisms involved in the spread of communicable diseases.

Biomolecules

Ancient biomolecules (DNA, proteins and lipids) have paved the way for researchers to better understand the evolutionary history and address problems associated with the spread of specific pathogenic microbes around the world. It has provided an insight into the mechanisms of host and/or pathogen alteration and investigated microbiomes that serve as indicators of health and disease.

Biomolecules are organic molecules that consist of more than 25 pre-existing elements (carbon, hydrogen, oxygen, phosphorus and sulfur) serving as the major component that forms the basis of life. Biomolecules have diverse biological functions due to their varied shape and size. It comprises several complex compounds including hydrocarbon derivatives which may covalently bind to different functional groups by replacing H atoms to form alcohols, amines, aldehydes, ketones or carboxylic acid (Verma and Verma 2022, Singh 2021). Primarily, carbohydrates, proteins, lipids and nucleic acid are the four classified biomolecules as shown in

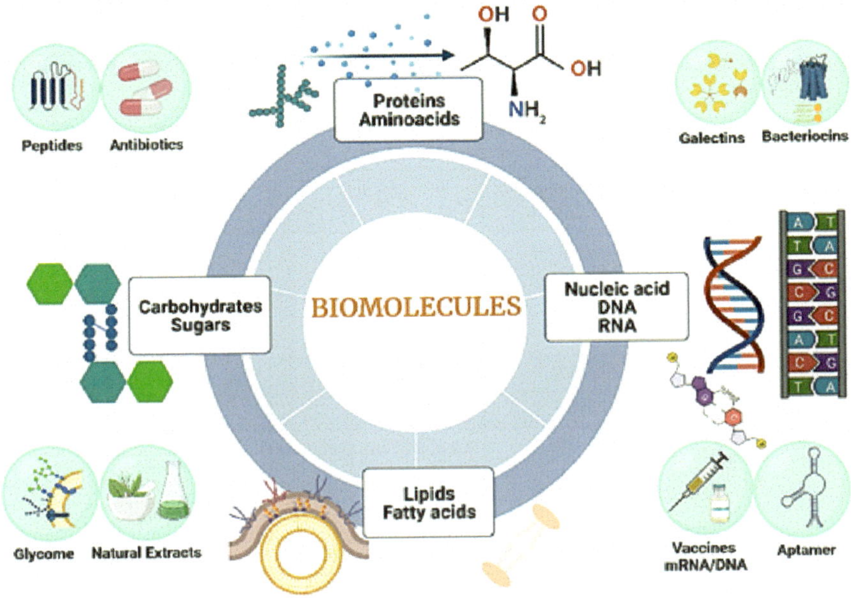

Figure 2. Classified biomolecules and associated therapeutics.

Figure 2. They are biological macromolecules that can be degraded into simple forms. Simple sugar, amino acids, fatty acid and glycerol and nucleotides are the monomeric forms (biomicromolecules) of respective biomacromolecules: carbohydrates, proteins, lipids and nucleic acid. These biological molecules are frequently given orally or incorporated into delivery agents (carriers) for topical and systemic perfusion in infectious disorders (Tavares et al. 2020). Therapeutic biomolecules as shown in Figure 1 with an antibacterial and antiviral activity that has been found to be effective in various infectious diseases and includes antibiotics, peptides, galectin, vaccines, aptamers, glycomes, bacteriocins and natural extracts as summarized in Table 1. In addition to these, the increased attributes of a therapeutic agent when lipids (liposomes and micelles) are incorporated into therapy have sparked a lot of interest in lipid-based drug delivery systems. Lipids offer various benefits as drug carriers, improving the bioavailability of lipid-soluble compounds, lower toxicity, high drug release potential, more regulated targeting, and other characteristics (Puri et al. 2009).

Protein and Peptide-Based Biomolecules

Proteins are needed for many biological activities in the human body and their absence may pose complications to the normal functioning

Table 1. Summary of common biomolecules used as a therapeutic-drugs.

Nature of Biomolecule	Drugs	Administrations	Clinical Implication
Antibiotics	Fusidic acid and Mupirocin	Topical	Atopic dermatitis
	Acyclovir, Famciclovir and Valacyclovir	Oral	Herpes simplex virus
	Levofloxacin and Rifampicin	Oral	Tuberculosis
AMP	Bacitracin	Topical	Localized skin and eye infections, wound infections
	Enfuvirtide	Subcutaneous	HIV-1 infection
	Telaprevir	Oral	Hepatitis C
	Vancomycin	Oral and Intravenous	Bacterial Infection
Vaccine	Havrix, Vaqta	Intramuscular	Hepatitis
	OPP/Ipol	Oral	Polio
	BCG	Intradermal	Bacillus
	FluMist and Quadrivalent	Intranasal	Influenza
Natural Extracts	Thymol	Topical	Helminthic infection

of a cell/tissue. Protein-based therapeutics are often big and have a peptide-like structure and a long half-life. The structure is composed of around 1,000 amino acids, linked by peptide bonds resulting in its large molecular weight, which can cause problems. Therapeutic peptides have significant advantages over protein-based molecules due to their smaller size and short half-life which reduces the threat of systemic toxicity by reducing peptide aggregation in tissues. They have the ability to permeate efficiently to the site-specific targeted tissues, are less immunogenic and have a reduced production cost due to their smaller size (Vlieghe et al. 2010, Holowacz et al. 2017).

Antibiotics

Antibiotics are defined as natural or synthetic substances endowed with antibacterial or antimicrobial properties, used to treat or prevent and control severe communicable and/or infectious diseases. Depending upon the mode of action, it may be bactericidal (aminoglycosides, beta-lactams, fluoroquinolones, glycopeptides, cyclic lipopeptides, nitroimidazoles); act by killing bacteria and/or bacteriostatic (glycylcyclines, tetracyclines, lincosamides, macrolides, erythromycin, oxazolidinones, linezolid, sulfonamides) and act by blocking their growth (FDA 2019, Grennan et al. 2020). There are hundreds or more different antibiotics that may be administered on the basis of pathogenic bacteria and associated infection. (Leekha et al. 2011). As per CDC and NIH, the following are the six classified groups of commonly used antibiotics (NHS 2021, NIH 2022, CDC 2020):

1. Penicillin, also known as beta-lactams (amoxicillin, co-amoxiclav, flucloxacillin and phenoxymethylpenicillin) are commonly prescribed to treat infectious diseases, such as infection related to skin, lung and urinary tract.
2. Aminoglycosides (gentamicin and tobramycin) are commonly used in hospitals to treat life-threatening infections, including septicaemia. They are usually given as injections, but some ear and eye infections can be treated with drops.
3. Cephalosporin (cephalexin) are antibiotics with expanded coverage of Gram-Negative bacteria used to treat septicaemia, meningitis infections, etc.
4. Tetracycline (minocycline and doxycycline) may be used to treat a variety of infections of the urinary, intestinal tract, gum or eye, but they are most typically designed to control and treat acute to chronic acne and other skin-related infections.
5. Macrolides (azithromycin, erythromycin and clarithromycin) are notably used for the treatment of respiratory infection (lung and chest) as well as penicillin-resistant bacteria in patients allergic to penicillin.
6. Fluoroquinolones (ciprofloxacin, moxifloxacin and levofloxacin) are broad-ranging synthetic antibiotics that may be used in adults (not in children) to treat several other forms of infections associated with the respiratory or urinary tract. These classes are no longer in use due to several adverse reactions posed by the antibiotics.

Lincosamides (clindamycin), sulfonamides, aminoglycosides (tobramycin, gentamicin and amikacin), chloramphenicol, fusidic acid, nitrofurantoin and trimethoprim are the list of other groups of antibiotics administered for the treatment of infection of ear, eye or urinary tract. Rifapentine-moxifloxacin is one of the treatment regimens directed by the CDC for TB. In addition to treating an infection, antibiotics may be used prophylactically to prevent infections among people at risk. For example, it might be administered to HIV/AIDS or immunosuppressed patients to avoid any post complications, including bacterial infection. This is effective for patients suffering from cancer with reduced immune mechanisms, those requiring major procedures and with dental issues with the possibility of bacterial endocarditis (Flower et al. 2013).

Limitation: Misuse and overuse of antibiotics as well as inadequate infection prevention and control have led to bacterial tolerance and has largely contributed to the spread of antibiotic resistance. The emergence of antibiotic-resistant bacteria has posed to be a serious threat globally, reducing the effectiveness of life-saving medications. As a result, alternative medicines based on antimicrobial peptides (AMPs) or natural extracts are becoming more popular in the treatment of patients. Antibiotics used to treat infectious diseases, like pneumonia, TB, gonorrhoea, and salmonellosis, are becoming less effective making treatment and control of these diseases more difficult (WHO 2022).

Antimicrobial Peptides (AMPs)

AMPs have gained a lot of momentum as a promising therapeutic option that may be used topically and systemically for the treatment of ocular infections in ophthalmology and skin infections, respectively. AMPs are low molecular weight oligopeptides made up of smaller sequences of amino acids (12–15) and may be found in almost all organisms (e.g., keratinocyte in the skin) with varied structural and functional properties. AMPs have immunomodulatory capabilities in addition to direct antibacterial action, making them potential molecules for the development of new therapies (Nestle et al. 2009). They interact with bacterial membranes and fold into an amphiphilic structure because of their short size, cationic, amphiphilic and hydrophobic nature. To date, all AMPs were classified into four groups based on their conformations, which are beta-sheet and alpha-helix (most prevalent, prolonged and in a loop). They specifically target lipopolysaccharide membranes of a microbe, which is distinct to them, and so do not affect low-anionic, cholesterol-rich biological systems. The development of antibiotic resistance by microbes is deterred or delayed due to its efficient microbicidal action. They can be classed as antimicrobial, antifungal, antiviral or antiparasitic based on the targeted microbial agent (Mirski et al. 2017, Lei et al. 2019).

The following are the advantages of AMPs (Magana et al. 2020):

i. **Higher potency:** Peptide-based therapeutics are highly active at their target receptor site, resulting in a strong effect at a comparatively lower dose compared to the antibiotics in use.

ii. **Highly specific:** AMPs have higher selectivity than their existing counterparts as they adhere extremely closely to the targeted receptor and are less likely to trigger adverse reactions, like immune response and toxicity.

iii. **Naturally occurring biologics provide better safety:** Circulatory proteases spontaneously break down peptides into their constituent amino acid residues. Due to their low doses and natural origin, these therapeutics are accompanied by minor accumulation in tissues with lowered adverse effects.

Limitations: There are a few drawbacks to using peptides, such as their short half-life, inability to be delivered intravenously and poor product consistency. According to recent research, microbes have been found to have resistance mechanisms as opposed to skin AMPs that exist naturally. Peptides that are synthesised artificially had been developed to bypass this limitation.

Galectin

Galectins are a non-glycosylated group of soluble and β-galactoside animal lectins (carbohydrate-binding proteins) with carbohydrate recognition domain (CRD) (130–140 amino acids). In mammals, 15 classes have been recognised so far and have been recorded in the order of their discovery and categorised according to their structural characteristics. Galectins are divided into three groups categorized as those with a single CRD (-1, -2, -5, -7, -10, -11, -13, -14 and -15), Gal-3 and a 31 ~ 35 kDa chimeric protein form has unusual tandem of proline and glycine-rich short segments with a non-carbohydrate domain at N-terminus and a carbohydrate domain at C-terminus and lastly, tandem-repeat type galectins made up of duplets of homologous CRDs linked together by a spacer peptide (-4, -6, -8, -9 and -12) (Iwasaki-Hozumi et al. 2021).

Galectins have recently emerged to have a significant role in the control of communicable diseases including HIV-1, enterovirus, HSV-1, adenovirus, influenza A virus (IAV) and dengue virus infections. Galectin-1, 3 and 9 have been shown to pose anti-influenza therapeutic potential having antiviral activity with the ability to reduce the severity of IAV infection (Chernyy et al. 2011, Lin et al. 2021). Galectin-9 (hGal-9) was initially recognized as a potential antigenic protein of 37kDa in Hodgkin's disease (Türeci et al. 1997). Gal-9 is found in many cells and organs of the immune system playing a pivotal role in the regulation of intracellular and extracellular molecular interaction (cellular or cell-matrix) and processes including cell growth, specialization and cell death. RNA splicing, signal transduction, cellular movement and cellular immunity (Hirashima et al. 2002). Gal-9, also designated as an eosinophil chemoattractant (Matsumoto et al. 1998) is a ligand for T Cell Ig and mucin moieties-3 (Tim-3), which is observed in infected and non-functional T cells with anti-allergic effect. Gal-9 concentration in the blood is elevated in persons infected with a variety of viruses, bacteria and parasites (Chagan-Yasutan et al. 2013, Dembele et al. 2016, Merani et al. 2015). Galectines interact with multiple receptors and biologics which are prime targets for the therapy of immune diseases as shown in Figure 3.

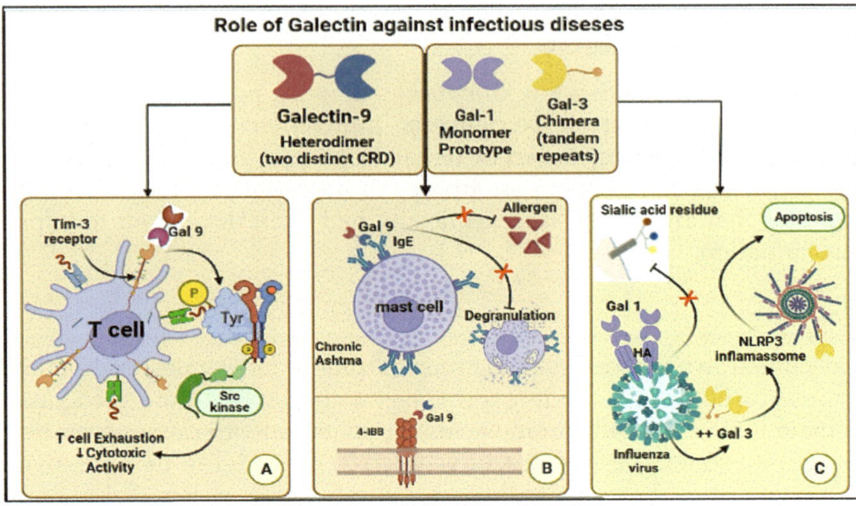

Figure 3. Role of Galectins against IAV Infection and their interactions with different receptors: (A) Gal-9 with Tim3 receptor (B) Gal-9 with IgE and 4-IBB receptor (C) Gal-1 and Gal- 3 with NLRP3 inflammasome and Sialic acid receptor, respectively (created with BioRender.com).

A brief explanation of the role of Galectins in communicable and/or infectious diseases:

1. Tim-3 is a T-cell inhibitory molecule and a biomarker for T-cell depletion (seen in chronic viral fever) and is the most explored receptor of Gal-9 that acts by a negative feedback mechanism. Gal-9 binds to Tim-3 in T cells and monocytes causing phosphorylation of tyrosine at Tim-3's cytoplasmic tail which in turn activates Src family kinases (Lee et al. 2011). Furthermore, Gal-9 modulates mast cell activity by attaching to the highly glycosylated immunoglobulin (Ig)E and preventing excessive degranulation (Niki et al. 2009).

   Gal-9 binds to an agonist 4-1BB (CD 137), cell surface protein containing cysteine that communicates with its tumor necrosis factor (TNF) group ligand to boost immune responses against tumor and viruses, while suppressing allergic and autoimmune disorders. The direct binding of Gal-9 and 4-1BB at a specific site different from the binding region of the CD137 monoclonal abs results in 4-1BB augmentation, accumulation, signalling, T cells activation, dendritic cells (DCs) and natural killer cells. Other Gal-9-dependent mechanism is the CD4+ forkhead box P3 (Foxp3) + regulatory T cell (Treg) proliferation and related downregulation of allergic reaction by death receptor 3 (DR3), another receptor of the TNF family.

2. The binding of Gal-1 to the hemagglutinin (HA) protein of IAV blocks its interaction with sialic acid host receptors, which act as a major phase to suppress the influenza infection. Gal-1 was also found to regulate the cytoplasmic mechanism generated by the H1N1pdm09 virus by interacting with cyclin-dependent kinases (CDKs) and cyclins to trigger a cell cycle arrest at G0/G1 phase resulting in virus-induced cell death (Chernyy et al. 2011, Fang et al. 2014).
3. Endogenous Gal-3 expressed in IAV infection binds to the NLRP3 inflammasome and aids in the assembly and oligomerization of the NLRP3 apoptosis-related speck-like protein that contains a CARD (ASC) inflammasome. The interaction activates the NLRP3 inflammasome and causes apoptosis to progress. Gal-3 expression induces antiviral gene and cytokines expression to combat IAV infection. The use of recombinant Gal-3 exhibited antiviral efficacy and reduced IAV replication (Li et al. 2014).

Limitation: Since galectins are biomolecules of animal origin, there might be some limitations associated with immune reaction and some galectins are prone to be inactive (within a day) in absence of any reducing buffers. This is due to the oxidised cysteine or the tryptophan residues cross-linked at the CRD.

Bacteriocins

Bacteriocins (333aa), produced by lactobacillus helveticus inhibit the synthesis of closely related lactobacillus species and were discovered by Gratia in 1925. Bacteriocins are bactericidal peptides generated by a variety of bacteria to eliminate pathogenic microbe's competitors in the microenvironment (Guder et al. 2000, Sahl et al. 1995). Bacteriocins have been discovered to be effective against a wide range of human infections. In contrast to other antimicrobials, they target a small range of bacteria, but their primary benefit is that they can operate without altering much of the body's natural microbiome (Gillor et al. 2005). Bacteriocins are categorized into four primary groups (Heng et al. 2007, Sablon et al. 2000):

i. Class I (lantibiotics): Small, heat-resistant peptides of 5 kDa size.
ii. Class II: Small (15 kDa), thermally stable, membrane-active, non-modified peptides.
iii. Class III: Heat-labile proteins with sizes greater than 15 kDa.
iv. Class IV: Complex bacteriocins conjugated with lipid or carbohydrate moieties.

The mode of action varies for bacteriocins, as it may act by binding to lipid-II, which is required for transferring cytoplasmic peptidoglycan subunits to the cell wall, halting cell wall formation and causing cell death. Alternatively, employing lipid-II as a docking molecule, bacteriocin can trigger pore development in the cell wall (Cotter et al. 2005). Cell death may be caused by several mechanisms including de-energizing of the cell membrane, dissipation of the proton motive force (PMF) or it may function by blocking amino acid uptake causing their release from the cell (Hugenholtz and de Veer 1991) or triggering potassium ion exclusion, cytoplasmic membrane depolarization, hydrolysis and partial outflow of cellular ATP (Hugenholtz and de Veer 1991). Bacteriocins have also been shown to possess bactericidal (endonuclease) activity in sensitive cells (Pugsley 1984). Epidermin, Lanthiopeptin, Nisin and S-35 are a few examples of bacteriocins that act as potential antibiotics in infectious diseases: dermal infections, herpes simplex virus, antimicrobial for inhibiting multi-drug resistant pathogens and treating pulmonary infections respectively (Gillor et al. 2005).

Nucleic Acid and Oligopeptide-Based Biomolecules

Aptamer, antisense oligonucleotides (ASOs) and small interfering RNA (siRNA) are among the several classes of therapeutic oligonucleotides consisting of 15-60, 16-20 and 20-25 nucleotides respectively.

Aptamers

An oligonucleotide aptamer is a genomic molecule consisting of a short sequence of single-stranded DNA (ssDNA) or RNA (ssRNA) which can fold autonomously to form a 3D-spatial configuration. Aptamers are often known as 'chemical antibodies' and they provide several benefits including group uniformity, endurance at room temperature, lower immunogenicity, low cost and chemical modification options to improve transport properties. With reference to the detection, prevention and diagnosis of communicable diseases, aptamers can be produced promptly against various pathogenic antigens and nonimmunogenic constituents (Dunn et al. 2017). The single labeled aptamer can be utilized over antibodies to detect targeted microbes in the form of diagnostic assays, including enzyme-linked oligonucleotide assay (ELONA), aptamer-based lateral flow assay (ALFA), aptasensors and fluorophotometry. Blocking of virus entry, pathogen replication control and toxin neutralisation are three aptamer-based therapeutic techniques for the prevention of pathogenic infection, where the targeted molecule by aptamers are as follows (Wan et al. 2021) (Figure 4):

i. Surface proteins on the host or viral cell that control the invasion of the host cell.
ii. Pathogenic polypeptides to reduce the propagation.
iii. Microbial toxins to relieve symptoms.
iv. Immune-related molecules are produced by the host in order to improve the operation of the host immune system.

Aptamers can also be used as carriers for microbicidal (against bacterial, viral and parasitic diseases) molecules to specifically targeted sites. It is to be noted that small interfering RNA (SiRNA) are shown to have the potential to act as a therapeutic for infectious diseases like hepatitis B (Esposito et al. 2018, Levin 2017).

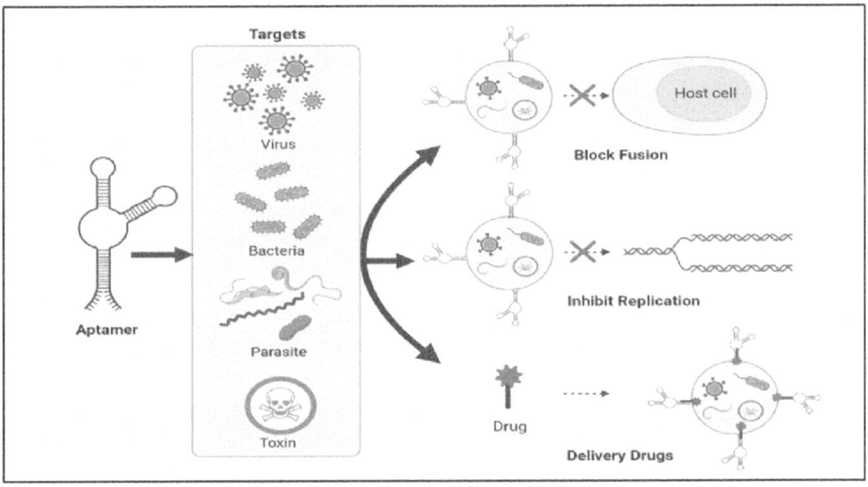

Figure 4. Therapeutic approach to Aptamer.

Carbohydrate and Lipid-Based Biomolecules

Glycomes

The glycomes of an organism are defined as an array of modified carbohydrate molecules (glycan) in conjugation with genomes (DNA/ RNA), lipids and polypeptides (Vigerust 2011). Proteins play an important role in biological processes, regulating the functional activities of a cell by a series of events known to be 'post-translational modification'. Glycosylation is one of the key events in the modification of cellular proteins. It is an important post-translational process that involves the attachment of glycans to amino acids of a polymer. Glycosylation is found to be essential for cell adhesion, receptor activation, cell signalling, molecular transportation and endocytosis. Assembly of a cellular glycoprotein is achieved by conjugation of 13 monosaccharides and 8 amino acids involving more than 40 glycosidic linkages that bond glycans to protein targets in N-linked, O-linked, C-linked, phosphoglycosyl and glypiation processes.

The cyanobacterial cyanovirin-N (CV-N), plant lectins Urticadioica agglutinin, Galanthus nivalia, concanavalin A and Pradimicin A have all been classed as carbohydrate-binding agents (CBAs) that carry the potential in inhibiting virus transmission (François et al. 2012). The N-linked glycosylated compound in which a large mannose core is coupled to the amide nitrogen of asparagine at the conserved motif Asn-X-Ser/Thr is the most prevalent variant in terms of infectious disease (Spiro 2002, Stepper et al. 2011). A polypeptide from the red algae Griffithsia was recently discovered to bind oligo mannose N-linked glycans on the membrane of HIV with exceptional specificity, blocking primary isolates and strains of T- and M-tropic HIV from invading tissues (Williams et al. 2005, O'Keefe et al. 2003, Bertaux et al. 2007).

Biomolecules Derived from other Sources

Vaccines

Vaccines are made up of a diverse range of organic and inorganic molecules, including peptides, glycoproteins, immunogenic antigens, adjuvants, preservatives and stabilizers. For decades, it has been used as a prophylactic and therapeutic agent aiding the immune system to recognise pathogenic microbes and acting as a means for combating infectious and/or communicable diseases. Vaccine development has been slowed by a number of issues, including formulation safety, immunogenicity, production capacity limitation and storage concerns. Genomic-based DNA/mRNA vaccine has effectively taken over the challenge by surpassing conventional vaccination constraints in preclinical and clinical studies. These biomolecule-based vaccines use chemically modified polymeric proteins and hydrogels as well as biomimetic virus-like nanoparticles and cell membranes (lipoproteins, micelle, dendrimers, etc.) to qualify the necessary levels of both safety and immunogenicity using polymer chemistry and bioscience principles (Ravi et al. 2021).

Natural Extracts

A significant majority of plant extracts are said to have anti-inflammatory properties. Antioxidants, free radical-scavenging, UV light absorbents and protectants against foreign microbes are all functions of secondary plant metabolites. Phenolics, terpenoids and alkaloids are the three major classes of natural extracts (Felgueiras 2021). For years, natural extracts have been thought to be beneficial in the treatment of wound infection. Aloe vera is a highly effective compound with antibacterial and fungicidal properties and relieves inflammation, irritation and scarring caused by a variety of skin infections. Salicylic acid, a major component, is said to have analgesic and anti-inflammatory properties by suppressing prostaglandin formation. The immunoregulatory properties of the gel polysaccharides present, specifically the acetylated mannans, may be responsible for this outcome. Resveratrol is a silent information regulator that is typically found in grape seeds. Thymol and tyrosol, which are derived from essential oils, have been recommended as suitable antimicrobial treatment additives since they have both bactericidal and anti-inflammatory properties (García-Salinas et al. 2020).

References

Abee, T., Rombouts, F. M., Hugenholtz, J., Guihard, G., and Letellier, L. (1994). Mode of action of nisin Z against Listeria monocytogenes Scott A grown at high and low temperatures. Applied and Environmental Microbiology, 60(6): 1962–1968.

Bennett, J. W., and Chung, K.-T. (2001). Alexander Fleming and the discovery of penicillin. pp. 163–184. In Advances in Applied Microbiology (Vol. 49). Academic Press. https://doi.org/10.1016/S0065-2164(01)49013-7.

Bernard, J. J., and Gallo, R. L. (2011). Protecting the boundary: the sentinel role of host defense peptides in the skin. Cellular and molecular life sciences: CMLS, 68(13): 2189–2199. https://doi.org/10.1007/s00018-011-0712-8.

Bertaux, C., Daelemans, D., Meertens, L., Cormier, E. G., Reinus, J. F., Peumans, W. J., Van Damme, E. J., Igarashi, Y., Oki, T., Schols, D., Dragic, T., and Balzarini, J. (2007). Entry of hepatitis C virus and human immunodeficiency virus is selectively inhibited by carbohydrate-binding agents but not by polyanions. Virology, 366(1): 40–50. https://doi.org/10.1016/j.virol.2007.04.008.

Bloom, D. E., Black, S., and Rappuoli, R. (2017). Emerging infectious diseases: a proactive approach. Proceedings of the National Academy of Sciences, 114(16): 4055–4059.

Carnes, E. C., Lopez, D. M., Donegan, N. P., Cheung, A., Gresham, H., Timmins, G. S., and Brinker, C. J. (2010). Confinement-induced quorum sensing of individual Staphylococcus aureus bacteria. Nature Chemical Biology, 6(1): 41–45.

CDC (2022, March 22) National Center for Emerging and Zoonotic Infectious Diseases (NCEZID). https://www.cdc.gov/ncezid/index.html.

CDC fact sheet (2020) All antibiotic classes. https://arpsp.cdc.gov/profile/antibiotic-use/all-classes.

Chagan-Yasutan, H., Ndhlovu, L. C., Lacuesta, T. L., Kubo, T., Leano, P. S., Niki, T., Oguma, S., Morita, K., Chew, G. M., Barbour, J. D., Telan, E. F., Hirashima, M., Hattori, T., and Dimaano, E. M. (2013). Galectin-9 plasma levels reflect adverse hematological and immunological features in acute dengue virus infection. Journal of clinical virology: the official publication of the Pan American Society for Clinical Virology, 58(4): 635–640. https://doi.org/10.1016/j.jcv.2013.10.022.

Chernyy, E. S., Rapoport, E. M., Andre, S. et al. (2011). Galectins promote the interaction of influenza virus with its target cell. Biochemistry Moscow 76, 958. https://doi.org/10.1134/S0006297911080128.

Commissioner, O. (2022, March 1). U.S. Food and Drug Administration. FDA; FDA. https://www.fda.gov/home.

Cotter, P. D., Hill, C., and Ross, R. P. (2005). Bacteriocins: developing innate immunity for food. Nature Reviews Microbiology, 3(10): 777–788.

DBT, Infectious Disease Biology | Department of Biotechnology. (n.d.). Retrieved March 6, 2022, from https:// dbtindia.gov.in/ schemes-programmes/ research- development/ medical- biotechnology/ infectious-disease- biology.

Dembele, B. P., Chagan-Yasutan, H., Niki, T., Ashino, Y., Tangpukdee, N., Shinichi, E., Krudsood, S., Kano, S., and Hattori, T. (2016). Plasma levels of Galectin-9 reflect disease severity in malaria infection. Malaria Journal, 15(1): 403. https://doi.org/10.1186/s12936-016-1471-7.

Dunn, M. R., Jimenez, R. M., and Chaput, J. C. (2017). Analysis of aptamer discovery and technology. Nature Reviews Chemistry, 1(10): 1–16.

Esposito, C. L., Catuogno, S., Condorelli, G., Ungaro, P., and de Franciscis, V. (2018). Aptamer chimeras for therapeutic delivery: the challenging perspectives. Genes, 9(11): 529. https://doi.org/10.3390/genes9110529.

Fang, S., Zhang, K., Wang, T., Wang, X., Lu, X., Peng, B., Wu, W., Zhang, R., Chen, S., Zhang, R., Xue, H., Yu, M., and Cheng, J. (2014). Primary study on the lesions and specific proteins in BEAS-2B cells induced with the 2009 A (H1N1) influenza virus. Applied Microbiology and Biotechnology, 98(23): 9691–9701. https://doi.org/10.1007/s00253-014-5852-y.

FDA (2019, October 28) U.S. Food and Drug Administration. Antibiotics and antibiotic resistance, Drugs.https://www.fda.gov/drugs/buying-using-medicine-safely/antibiotics-and-antibiotic-resistance.

Felgueiras, H. P. (2021). An Insight into Biomolecules for the Treatment of Skin Infectious Diseases. Pharmaceutics, 13(7), 1012. MDPI AG. Retrieved from http://dx.doi.org/10.3390/pharmaceutics13071012

Flowers, C. R., Seidenfeld, J., Bow, E. J., Karten, C., Gleason, C., Hawley, D. K., Kuderer, N. M., Langston, A. A., Marr, K. A., Rolston, K. V., and Ramsey, S. D. (2013). Antimicrobial prophylaxis and outpatient management of fever and neutropenia in adults treated for malignancy: American Society of Clinical Oncology clinical practice guideline. Journal of clinical oncology: official journal of the American Society of Clinical Oncology, 31(6): 794–810. https://doi.org/10.1200/JCO.2012.45.8661.

François, K. O., and Balzarini, J. (2012). Potential of carbohydrate-binding agents as therapeutics against enveloped viruses. Medicinal research reviews, 32(2): 349–387. https://doi.org/10.1002/med.20216.

García-Salinas, S., Evangelopoulos, M., Gámez-Herrera, E., Arruebo, M., Irusta, S., Taraballi, F., Mendoza, G., and Tasciotti, E. (2020). Electrospun anti-inflammatory patch loaded with essential oils for wound healing. International Journal of Pharmaceutics, 577: 119067. https://doi.org/10.1016/j.ijpharm.2020.119067.

Gillor, O., Nigro, L. M., and Riley, M. A. (2005). Genetically engineered bacteriocins and their potential as the next generation of antimicrobials. Current Pharmaceutical Design, 11(8): 1067–1075. https://doi.org/10.2174/1381612053381666.

Grennan, D., Varughese, C., and Moore, N. M. (2020). Medications for Treating Infection. JAMA, 323(1): 100–100. https://doi.org/10.1001/jama.2019.17387.

Guder, A., Wiedemann, I., and Sahl, H.-G. (2000). Posttranslationally modified bacteriocins—The lantibiotics. Peptide Science, 55(1): 62–73. https://doi.org/10.1002/1097-0282(2000)55:1<62::AID-BIP60>3.0.CO;2-Y.

Heng, N. C., Wescombe, P. A., Burton, J. P., Jack, R. W., and Tagg, J. R. (2007). The diversity of bacteriocins in Gram-positive bacteria. In bacteriocins (pp. 45–92). Springer, Berlin, Heidelberg.

Hirashima, M., Kashio, Y., Nishi, N., Yamauchi, A., Imaizumi, T. A., Kageshita, T., Saita, N., and Nakamura, T. (2002). Galectin-9 in physiological and pathological conditions. Glycoconjugate Journal, 19(7–9): 593–600. https://doi.org/10.1023/B:GLYC.0000014090.63206.2f.

Holowacz, M., Krans, A., Wallén, C., Martinez, A., and Mohammadi, N. (2017). A survey of commercial biomolecules, delimited to pharmaceuticals and medical devices (Dissertation). Retrieved from http://urn.kb.se/resolve?urn=urn:nbn:se:uu:diva-324833.

Hugenholtz, J., and de Veer, G. J. C. M. (1991). Application of nisin A and nisin Z in dairy technology. Nisin and Novel Lantibiotics, 440–447.

Iwasaki-Hozumi, H., Chagan-Yasutan, H., Ashino, Y., and Hattori, T. (2021). Blood Levels of Galectin-9, an Immuno-Regulating Molecule, Reflect the Severity for the Acute and Chronic Infectious Diseases. Biomolecules, 11(3): 430. https://doi.org/10.3390/biom11030430.

Jones, K. E., Patel, N. G., Levy, M. A., Storeygard, A., Balk, D., Gittleman, J. L., and Daszak, P. (2008). Global trends in emerging infectious diseases. Nature, 451(7181): 990–993. https://doi.org/10.1038/nature06536.

Lee, J., Su, E. W., Zhu, C., Hainline, S., Phuah, J., Moroco, J. A., Smithgall, T. E., Kuchroo, V. K., and Kane, L. P. (2011). Phosphotyrosine-dependent coupling of Tim-3 to T-cell receptor signaling pathways. Molecular and Cellular Biology, 31(19): 3963–3974. https://doi.org/10.1128/MCB.05297-11.

Leekha, S., Terrell, C. L., and Edson, R. S. (2011). General principles of antimicrobial therapy. Mayo Clinic proceedings, 86(2): 156–167. https://doi.org/10.4065/mcp.2010.0639.

Lei, J., Sun, L., Huang, S., Zhu, C., Li, P., He, J., Mackey, V., Coy, D. H., and He, Q. (2019). The antimicrobial peptides and their potential clinical applications. American Journal of Translational Research, 11(7): 3919–3931.

Levin, A. A. (2017). Targeting therapeutic oligonucleotides. The New England Journal of Medicine, 376(1): 86–88. https://doi.org/10.1056/NEJMcibr1613559.

Li, S. W., Yang, T. C., Lai, C. C., Huang, S. H., Liao, J. M., Wan, L., Lin, Y. J., and Lin, C. W. (2014). Antiviral activity of aloe-emodin against influenza A virus via galectin-3 up-regulation. European journal of pharmacology, 738: 125–132. https://doi.org/10.1016/j.ejphar.2014.05.028.

Lin, C. Y., Yang, Z. S., Wang, W. H., Urbina, A. N., Lin, Y. T., Huang, J. C., Liu, F. T., and Wang, S. F. (2021). The antiviral role of galectins toward influenza A Virus Infection-An Alternative Strategy for Influenza Therapy. Pharmaceuticals (Basel, Switzerland), 14(5), 490. https://doi.org/10.3390/ph14050490.

Magana, M., Pushpanathan, M., Santos, A. L., Leanse, L., Fernandez, M., Ioannidis, Giulianotti, A., Apidianakis, Y., Bradefute, S., Ferguson, A. L., Cherkasov, A., Seleem M. N., Pinilla, C., Fuente, C., Lazaridis, T., Dai, T., Houghten, R. A., Robert, E. W., Hancock, R. E., and Tegos, G. P. (2020). The value of antimicrobial peptides in the age of resistance. The Lancet Infectious Diseases, 20(9), e216-e230.

Matsumoto, R., Matsumoto, H., Seki, M., Hata, M., Asano, Y., Kanegasaki, S., Stevens, R. L., and Hirashima, M. (1998). Human ecalectin, a variant of human galectin-9, is a novel

eosinophil chemoattractant produced by T lymphocytes. The Journal of Biological Chemistry, 273(27): 16976–16984. https://doi.org/10.1074/jbc.273.27.16976.
Merani, S., Chen, W., and Elahi, S. (2015). The bitter side of sweet: the role of Galectin-9 in immunopathogenesis of viral infections. Reviews in Medical Virology, 25(3): 175–186. https://doi.org/10.1002/rmv.1832.
Mirski, T., Niemcewicz, M., Bartoszcze, M., Gryko, R., and Michalski, A. (2017). Utilisation of peptides against microbial infections - a review. Annals of Agricultural and Environmental Medicine: AAEM, 25(2): 205–210. https://doi.org/10.26444/aaem/74471.
Nestle, F. O., Di Meglio, P., Qin, J. Z., and Nickoloff, B. J. (2009). Skin immune sentinels in health and disease. Nature reviews. Immunology, 9(10): 679–691. https://doi.org/10.1038/nri2622.
NHS 24 (July 5, 2021) https://www.nhsinform.scot/tests-and-treatments/medicines-and-medical-aids/types-of-medicine/antibiotics.
Nigam, A., Gupta, D., and Sharma, A. (2014). Treatment of infectious disease: Beyond antibiotics. Microbiological Research, 169(9): 643–651. https://doi.org/10.1016/j.micres.2014.02.009.
NIH, Medline plus (January 14, 2022) https://medlineplus.gov/antibiotics.html.
Niki, T., Tsutsui, S., Hirose, S., Aradono, S., Sugimoto, Y., Takeshita, K., Nishi, N., and Hirashima, M. (2009). Galectin-9 is a high affinity IgE-binding lectin with anti-allergic effect by blocking IgE-antigen complex formation. The Journal of Biological Chemistry, 284(47): 32344–32352. https://doi.org/10.1074/jbc.M109.035196.
O'Keefe, B. R., Smee, D. F., Turpin, J. A., Saucedo, C. J., Gustafson, K. R., Mori, T., Blakeslee, D., Buckheit, R., and Boyd, M. R. (2003). Potent anti-influenza activity of cyanovirin-N and interactions with viral hemagglutinin. Antimicrobial Agents and Chemotherapy, 47(8): 2518–2525. https://doi.org/10.1128/AAC.47.8.2518-2525.2003.
Pugsley, A. P. (1984). The ins and outs of colicins. Part II. Lethal action, immunity and ecological implications. Microbiological Sciences, 1(8): 203–205.
Puri, A., Loomis, K., Smith, B., Lee, J. H., Yavlovich, A., Heldman, E., and Blumenthal, R. (2009). Lipid-based nanoparticles as pharmaceutical drug carriers: from concepts to clinic. Critical Reviews in Therapeutic Drug Carrier Systems, 26(6): 523–580. https://doi.org/10.1615/critrevtherdrugcarriersyst.v26.i6.10.
Ravi, S. P., Shamiya, Y., Chakraborty, A., Elias, C., and Paul, A. (2021). Biomaterials, biological molecules, and polymers in developing vaccines. Trends in Pharmacological Sciences, 42(10): 813–828.
Rohr, J. R., Barrett, C. B., Civitello, D. J., Craft, M. E., Delius, B., DeLeo, G. A., Hudson, P. J., Jouanard, N., Nguyen, K. H., Ostfeld, R. S., Remais, J. V., Riveau, G., Sokolow, S. H., and Tilman, D. (2019). Emerging human infectious diseases and the links to global food production. Nature Sustainability, 2(6): 445–456.
Sablon, E., Contreras, B., and Vandamme, E. (2000). Antimicrobial peptides of lactic acid bacteria: mode of action, genetics and biosynthesis. New Products and New Areas of Bioprocess Engineering, 21–60.
Sahl, H.-G., Jack, R. W., and Bierbaum, G. (1995). Biosynthesis and biological activities of lantibiotics with unique post-translational modifications. European Journal of Biochemistry, 230(3): 827–853. https://doi.org/10.1111/j.1432-1033.1995.0827g.x.
Singh, A, 2021. Biomolecules: Types and Functions – Conduct Science. (n.d.). Retrieved March 12, 2022, from https://conductscience.com/biomolecules-types-and-functions/.
Spiro, R. G. (2002). Protein glycosylation: nature, distribution, enzymatic formation, and disease implications of glycopeptide bonds. Glycobiology, 12(4): 43R–56R. https://doi.org/10.1093/glycob/12.4.43r.
Stepper, J., Shastri, S., Loo, T. S., Preston, J. C., Novak, P., Man, P., Moore, C. H., Havlíček, V., Patchett, M. L., and Norris, G. E. (2011). Cysteine S-glycosylation, a new post-

translational modification found in glycopeptide bacteriocins. FEBS Letters, 585(4): 645–650. https://doi.org/10.1016/j.febslet.2011.01.023.

Tavares, T. D., Antunes, J. C., Padrão, J., Ribeiro, A. I., Zille, A., Amorim, M. T. P., Ferreira, F. et al. (2020). Activity of specialized biomolecules against gram-positive and gram-negative bacteria. Antibiotics, 9(6): 314. MDPI AG. Retrieved from http://dx.doi.org/10.3390/antibiotics9060314.

Teixeira, M. O., Antunes, J. C., and Felgueiras, H. P. (2021). Recent advances in fiber–hydrogel composites for wound healing and drug delivery systems. Antibiotics, 10(3): 248.

Türeci, O., Schmitt, H., Fadle, N., Pfreundschuh, M., and Sahin, U. (1997). Molecular definition of a novel human galectin which is immunogenic in patients with Hodgkin's disease. The Journal of Biological Chemistry, 272(10): 6416–6422. https://doi.org/10.1074/jbc.272.10.6416.

Van Seventer, J. M., and Hochberg, N. S. (2017). Principles of infectious diseases: transmission, diagnosis, prevention, and control. pp. 22–39. *In:* Quah, S. R. (Ed.). International Encyclopedia of Public Health (Second Edition). Academic Press. https://doi.org/10.1016/B978-0-12-803678-5.00516-6.

Verma, C., and Verma, D. (June 1, 2022) Handbook of Biomolecules—1st Edition. Retrieved March 25, 2022, from https://www.elsevier.com/books/handbook-of-biomolecules/verma/978-0-323-91684-4.

Vigerust, D. J. (2011). Protein glycosylation in infectious disease pathobiology and treatment. Central European Journal of Biology, 6(5): 802. https://doi.org/10.2478/s11535-011-0050-8.

Vlieghe, P., Lisowski, V., Martinez, J., and Khrestchatisky, M. (2010). Synthetic therapeutic peptides: science and market. Drug Discovery Today, 15(1-2): 40–56. https://doi.org/10.1016/j.drudis.2009.10.009.

Wan, Q., Liu, X., and Zu, Y. (2021). Oligonucleotide aptamers for pathogen detection and infectious disease control. Theranostics, 11(18): 9133–9161. https://doi.org/10.7150/thno.61804.

WHO (July 31 2020) Antibiotic Resistance. https://www.who.int/news-room/fact-sheets/detail/antibiotic-resistance.

WHO EMRO (2022) Infectious diseases. Health topics. World Health Organization - Regional Office for the Eastern Mediterranean. Retrieved March 24, 2022, from http://www.emro.who.int/health-topics/infectious-diseases/index.html.

Williams, D. C., Jr., Lee, J. Y., Cai, M., Bewley, C. A., and Clore, G. M. (2005). Crystal structures of the HIV-1 inhibitory cyanobacterial protein MVL free and bound to Man3GlcNAc2: structural basis for specificity and high-affinity binding to the core pentasaccharide from N-linked oligomannoside. The Journal of Biological Chemistry, 280(32): 29269–29276. https://doi.org/10.1074/jbc.M504642200.

Chapter 5

Marine Organisms-Derived Bioactive Molecules for Infectious Disease Treatment

Challenges and Future Prospect

Abhay B. Fulke,[1,*] *Atul Kotian*,[1] *Santosh K. Gothwal*,[3] *Parth Sharma*[1] and *Manisha D. Giripunje*[2]

Introduction

Incidences of infectious diseases globally have seen rising numbers in recent years. According to the statistics of the World Health Organization (WHO) on HIV/AIDS infections from 2018, 1,700,000 new infections were reported worldwide of which globally 770,000 deaths occurred due to the infection, particularly in Africa that saw maximum death toll figure stood at 470,000. An estimated 2.86 million people contract cholera in the at-risk region around the world and as many as 95,000 deaths occur due to its infection (Ali et al. 2015). As of the 15th of November 2021, a total of 253,163,330 confirmed cases of SARS-COVID-19 were observed worldwide, including 5,098,174 deaths reported to WHO with a steady increase in numbers in subsequent weeks.

[1] Microbiology Division, CSIR-National Institute of Oceanography (CSIR-NIO), Regional Centre, Mumbai-400053, Maharashtra, India.
[2] Sevadal Mahila Mahavidyalaya, Nagpur 440009, India.
[3] Department of Respiratory Medicine, Graduate School of Medicine, Kyoto University, Japan.
* Corresponding author: afulke@nio.org

Several factors have been attributed to the steady spread of these infections. Close physical contact, sexual intercourse, exposure to body fluids and exposure to contaminated personal items and vectors (Chadburn 2013, Edemekong and Huang 2017, Shaw and Domachowske 2019, Suryadevara 2019). Other factors such as age, sex, location, socio-economic condition, availability to clean food and water and access to healthcare and medicines also contribute to the spread of disease (Mossong et al. 2008, Dimarco et al. 2020). All of these factors determine the course of infection in a particular region. Particularly susceptible are developing countries where millions of people die every year, causing such communicable diseases to be a matter of major concern. With increased human mobilization, immigration and global trade, these infections no longer restrict themselves to any particular region and transcend globally. Other factors such as drug resistance and newly emerging diseases are contributing factors to the rapid spread of these diseases (Saxena 2018).

Since ancient times, there have been several instances of epidemics that have often been mentioned in biblical as well as medieval literature. Often referred to as plagues, their causes were not known and improper methods to cure these led to massive deaths. With the invention of the microscope, the causative microorganisms were identified and studied to understand their life cycles. This resulted in a more systematic approach to their cure that led to drug discoveries. Diseases like smallpox were thus completely eradicated worldwide using intensive vaccination and antibiotic therapy (Pollock 2012).

Drug tolerance and resistance in bacteria, protozoa, fungi and viruses have been observed to occur more frequently in recent times. It has been seen that this pattern holds true for all living organisms, including humans, animals, fishes and plants. Genetic and biomolecular mechanisms largely contribute to this (Davies 2010). Microbes possess genes that are capable of resisting common antibiotics and are known to pass these genes to other species by horizontal gene transfer. Resistance genes have also been observed in ancient bacteria found in Siberian permafrost, thereby implying their occurrence since antiquity (D'Costa et al. 2011, Huddleston 2014). It has been observed in some cancers that tumors not only develop resistance toward drugs originally used to treat them but also cross-resistant toward other drugs with different mechanisms of action (Longley and Johnston 2005). All these factors have led to a push for more innovative medicines and drugs to control these infectious diseases.

The marine environment is a rich natural source with tremendous potential to find several biological entities with biotechnological significance. Oceans cover roughly 70% of the earth's surface and about 15% of biodiversity (Grosberg et al. 2012, James et al. 2012). Unexplored territories such as deep-sea vents and the abyss house a plethora of organisms. The oceans house several unculturable bacteria unknown to humans with several biological applications (Colwell 2002). Novel compounds of pharmaceutical, nutritional, cosmetic and agricultural relevance can be found here with tremendous potential for industrial application (König et al. 1994, Sharma et al. 2019). This has led to an investigation of marine organisms for their ability to synthesize bioactive compounds. Organisms such as microorganisms, algae, sponges, corals and ascidians have been explored to isolate biotherapeutic compounds (Fenical 1997, Blunt et al. 2003). It has been observed that the metabolites isolated from higher organisms such as invertebrates come from microbial origin. However, these microbes show symbiotic association with these invertebrates (Lindequist 2016).

Marine Organism-Derived Bio-Molecules

The marine ecosystem is one of the most under-utilized ecosystems in the world. This is largely due to the fact that most of it is still unexplored. Coastal communities worldwide rely on marine flora and fauna for their dietary requirements. Oceans provide fishes, seaweeds, crustaceans and mollusks which are consumed worldwide and are processed to make several nutritional and pharmaceutical products. Natural products have been a source of innovative drug discovery. Cellular pathways for the synthesis of these products have been extensively studied to understand molecular mechanisms leading to the formation of bioactive compounds and then replicated artificially to derive drugs and medicines. Thus, products of primary and secondary metabolism have found their industrial and pharmaceutical applications (James et al. 2012).

Marine organisms by their ability to survive extreme environmental conditions are able to synthesize novel metabolites which are not seen in territorial organisms. This results in the formation of metabolites which are not only unique in their structure but also possess unique chemical properties. Table 1 shows a list of some marine organisms-derived compounds and their corresponding biological activity.

Diverse sources such as marine bacteria, fungi, algae, sponges and cnidarians have been explored for their potential use in discovering bioactive compounds. Microbes and lower marine animals are known to lack an active means of defense such as a well-developed immune system. This has resulted in these organisms developing alternate chemical defense mechanisms to prevent predation. Marine invertebrates, such as sponges, tunicates, bivalves and others being filter feeders, tend to accumulate a significant bacterial and viral load. To prevent attacks from potentially harmful bacteria, they secrete antiviral and antibacterial compounds. As

Table 1. Bioactive compounds from marine resources.

Compound/ Metabolite	Source	Application	References
Euniatin B	*Fusarium* sp.	Antibacterial	Tsuda et al. 2003
Modiolides A-B	*Paraphaeospheria* sp.	Antibacterial	Okazaki et al. 1975
SS-228 Y	*Chania* sp.	Antibacterial	Christie et al. 1997
Byrostatin 1	*Bugula* sp.	Antibacterial	Fenical 2006
Xinghaiamine A	*Streptomycesxinghaiensis* NRRL	Antibacterial	Jiao et al. 2013
Merochlorins A–D	*Streptomyces* sp. strain CNH-189	Antibacterial	Sakoulas et al 2012, Kaysser et al. 2012
Ascochytain	*Ascochyta* sp.	Antibacterial	Kanoh et al. 2008
Brevianamides	*Aspergillusversicolor*	Antibacterial	Song et al. 2012
Penicillosides B	*Penicillium* sp.	Antibacterial	Murshid et al. 2016
Kocurin	*Kocuria palustris*	Antibacterial	Martín et al. 2013
Aurelin	*Aurelia* sp.	Antibacterial	Ovchinnikova et al. 2006
Desmethylisaridin C1	*Beauveriafelina*	Antibacterial	Du et al. 2014
Spongistatin 1	*Hyrtioserecta*	Antifungal	Cragg et al. 2006
Hypoxysordarin	*Hypoxyloncroceum*	Antifungal	Liu et al. 2003
Spongothymidine	*Tethyacrypta*	Antiviral	Bergmann and Feeney 1951
Mycalamide A	*Mycale* sp.	Antiviral	Perry et al. 1988
Microspinosamide	*Sidonopsmicrospinosa*	Antiviral	Rashid et al. 2001
Acyclovir	*Tethyacripta*	Antiviral	Elion et al. 1977
Griffithsin	*Griffithsia*sp	Antiviral	Mori et al. 2005

a result, such compounds have been receiving significant attention to develop novel drugs to combat infections (James et al. 2012).

The first drugs of marine origin to be commercialized were Ara-Aantiviral agent against Herpes simplex virus and Varicella zoster virus and Ara-C, a chemotherapeutic agent (Patrzykat and Douglas 2003). This has opened doors for several other drugs to undergo the screening and approval phase. In recent years, with the advent of structural elucidation of these compounds, their presumptive metabolic action has been explored to study the potential application of marine organism-derived bioactive compounds for disease control. Bivalves and mollusks are known to produce antiviral compounds, such as Ziconotide and Turbostatin 1.

Hence, it has also been proposed to explore the potential of these compounds against the novel Coronavirus leading to SARS COVID-19 (Yap 2020).

Challenges in the Development of Novel Drugs of Marine Origin

The marine environment is susceptible to damage due to unsustainable methods of fishing. Fragile coral reefs are already susceptible to damage due to ocean acidification and global warming (Hoegh-Guldberg 2007). This has resulted in a gradual push toward more sustainable methods to extract and develop these drugs. Marine organisms are particularly difficult to culture. Insufficient knowledge about the parameters and conditions has been deterring the commercialization of growing marine organisms to derive bioactive compounds (Yasuhara-Bell and Lu 2010). In past, research was restricted to near-shore samples and later with the advent of specialized submersibles, deeper frontiers of the sea have been explored in recent years. As a result, their progress has been rather slow (Tziveleka et al. 2003, Kanase et al. 2018).

Antibiotic resistance has been another significant challenge with the development of new drugs. Bacteria and viruses alter their genome to combat antibacterial and antiviral agents resulting in the development of resistance in them. This impairs the drug's efficiency in treating the infection rendering it inefficient (Yasuhara-Bell and Lu 2010). As a result, there has been a constant need to come up with creative solutions to tackle such problems. This has led several groups to investigate the shortcomings of conventional methods of drug discovery and solve roadblocks associated with them.

Microbes that show symbiotic associations with invertebrates tend to be fastidious and do not produce the metabolites in axenic cultures. It is speculated that their hosts play an important role in the production of these metabolites. Other biotechnologically important organisms are unable to grow under artificial conditions. This seriously affects the development of novel drugs (Lindequist 2016). Growing demand for large quantities of drugs has pressurized the pharmaceutical industry to come up with methods to upscale production. This is not always feasible with certain drugs as a result of which they are not favored for production. This has led to a push for the synthesis of these drugs. This may however not be always an economically viable option for all compounds (Bhadury et al. 2006).

Drug testing and approval is a time-consuming process, often taking up to 12–15 years to release a product into the market (Lauritano and Ianora 2018). Most companies may not commit to such a long timeline to develop new drugs to the market. This has led to general deterrence from developing new medicines, and the lengthy and expensive procedure to approve these drugs has also led to a significant increase in the price of these drugs.

Strategies to Develop Novel Drugs

Traditional methods of sampling that included time-consuming and inefficient extraction protocols, biochemical testing and gathering of massive amounts of biomass are rather cost-ineffective. Inconsistent results and lengthy protocols have led to the development of more effective molecular methods. The utilization of cell culturing techniques has led to a lesser dependence on marine biomass and has thus helped in the preservation of rare species. When it comes to microbial derivatives, traditional fermentation techniques are still preferable. For fungal derivatives, solid-state fermentation is preferred as it enables the formation of unbroken biofilms (Bhadury et al. 2006).

As we already know the majority of microorganisms are not cultivable and therefore are unknown. Such unidentified microorganisms from marine environments may hold molecular trove with therapeutic potential. There are multiple ways by which these underexplored treasures can be excavated. The first is through a screening of functional metagenomic libraries in expression vectors (Charles et al. 2015, Elrazak et al. 2015); the second is via metagenomic gene mining and expression in a surrogate host system (Ward et al. 2016); the third way is to extract draft genome and reconstruct their metabolic potential for targeted culturing in laboratory conditions and the last is biomolecules can be extracted from the marine environment to screen directly for therapeutic potential (Magdum et al. 2019). Once the therapeutically important molecule is successfully identified in laboratory conditions, it can be further characterized and pushed for clinical approval.

Gene manipulation, computer-based molecular modelling studies and semi-synthetic production methods are utilized in recent years. This has enabled the sustainable development of novel drugs and has led to a paradigm shift in approaches to drug development (Yasuhara-Bell and Lu 2010). Lesser dependence on biomass has prevented the risk of habitat destruction and species depletion. The chemical synthesis of these drugs has also enabled a fair amount of reduction in dependence on biomass for extraction.

Availability of better technology with respect to more efficient manned submersible vehicles, collaborative research worldwide and major government agencies funding research activities in the marine frontier have been some of the major advances in the field of research due to ease of access to valuable natural resources (Tziveleka et al. 2003, Kanase et al. 2018).

Future Prospects of Drug Discovery

Developing countries, such as India, Brazil, China and the African continent, have long been subjected to frequent epidemics of tropical diseases, such as malaria, Leishmaniasis and Chags' disease. With very limited drugs available for these diseases, treating infections is a difficult task. Costs of these treatments are also significant factors leading to high mortality in these countries. This has led to a push toward the

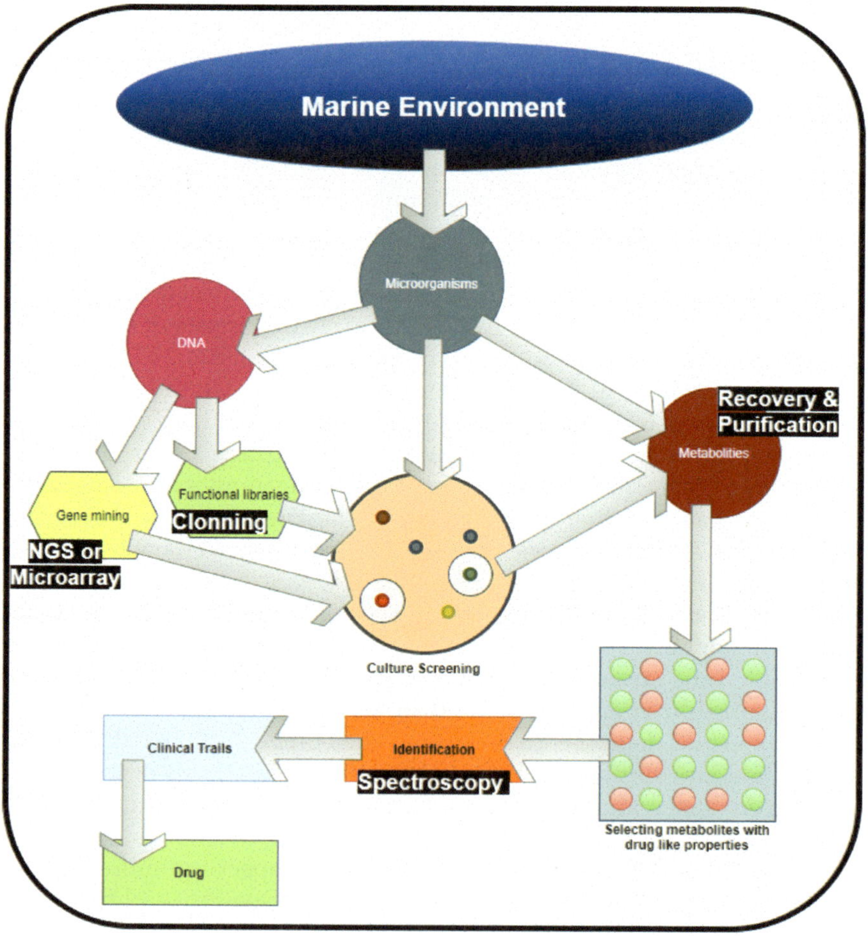

Figure 1. Strategies to develop novel drugs from the marine microbiome.

development of low-cost and effective drugs. The marine environment is an ideal candidate to explore these drugs as the rich species diversity can enable for development of several options for the treatment of these infectious diseases (Berlinck et al. 2004).

Designs for clinical trials need to be updated to suit current requirements. Typical drug testing takes about 12 years, which does not address matters of immediate concern. Clinical studies tend to be expensive owing to the high operation costs of phase II and phase III in drug trials. This has resulted in a push for cost-effective and efficient testing protocols. The development of modeling software which simulates *in vivo* drug testing can help in addressing this issue. This will not only significantly reduce the cost required to conduct the clinical trials but also fast-track the process significantly (Orloff et al. 2009).

The utilization of artificial intelligence (AI), machine learning and deep learning can be utilized to address several issues of drug development. AI can be used for various steps of drug development, such as identification of the drug target, drug designing, data management of all stages of drug development and identifying test subjects for clinical trials (Mak and Pichika 2019). Bayesian methodologies for drug development strategies can also significantly increase drug testing efficiency. By using adaptive dosing techniques, the dose-response curve can be utilized to accurately time doses to increase the drug efficiency, while simultaneously cutting costs of wasted doses in the lag and stationary phases (Orloff et al. 2009).

Conclusion

The marine environment serves as a habitat for a diversity of organisms which has tremendous application in the pharmaceutical industry. Most of the marine environment remains largely unexplored as a result of which there is a great scope for research in this field. This has resulted in research groups worldwide investigating this rich ecosystem and tapping into the rich biodiversity. The exploration of the oceans has led to the discovery of some of the most fascinating organisms in the world. This has also widened the scope of research in this field. Thus, bioprospection and mining for novel compounds have been lucrative fields for research.

There is a distinct lack of technical know-how as a result of which research in the past has been largely limited. The shortcomings in conventional experimentation and analytical methods have hindered progress in this area. However, with the advent of new technologies for sampling and production of drugs, more and more frontiers have been explored in recent years and have resulted in novel drugs coming into the market to treat diseases of infectious nature. Non-invasive methods of sampling have had a great impact on reducing sampling-based habitat disruption. The use of modern instrumentation technology has enabled the improvement of the efficiency of experimentation, resulting in high throughput screening. The combinational use of novel organisms and advanced technologies has significantly skewed the timeline for drug development.

Gene manipulation, tissue culturing techniques and recombinant DNA technology have facilitated a shift from the dependency on large volumes of biomass and thereby cutting down costs significantly. These techniques have improved the resultant output which has increased the success rates of experiments. The development of more effective extraction and elucidation methods has helped in improving net outcomes. Moreover, the chemical synthesis of several drugs has enabled the mass production of many drugs. This has pushed the frontiers further by detaching itself from the need for sophisticated bioreactors to more favorable chemical synthesis which has better controlling ability over conventional fermentation techniques. These in unison have opened new avenues for future research in the field and thus pushing the boundaries for more effective production techniques, innovations in problem-solving approaches and exploring new horizons of the abyss to search for the next wonder drug.

References

Akshada Amit Koparde, Rajendra Chandrashekar Doijad and Chandrakant Shripal Magdum (March 7th 2019). Natural Products in Drug Discovery, Pharmacognosy - Medicinal Plants, Shagufta Perveen and Areej Al-Taweel, IntechOpen, DOI: 10.5772/intechopen.82860. Available from: https://www.intechopen.com/chapters/65128.

Ali, M., Nelson, A. R., Lopez, A. L., and Sack, D. A. (2015). Updated global burden of cholera in endemic countries. PLoS Neglected Tropical Diseases, 9(6): p.e0003832.

Bergmann, W., and Feeney, R. J. (1951). Contributions to the study of marine products. XXXII. The nucleosides of sponges. I. The Journal of Organic Chemistry, 16(6): 981–987.

Berlinck, R. G., Hajdu, E., da Rocha, R. M., de Oliveira, J. H., Hernández, I. L., Seleghim, M. H., Granato, A. C., de Almeida, É. V., Nuñez, C. V., Muricy, G., and Peixinho, S. (2004). Challenges and rewards of research in marine natural products chemistry in Brazil. Journal of Natural Products, 67(3): 510–522.

Bhadury, P., Mohammad, B. T., and Wright, P. C. (2006). The current status of natural products from marine fungi and their potential as anti-infective agents. Journal of Industrial Microbiology and Biotechnology, 33(5): 325.

Blunt, J. W., Copp, B. R., Munro, M. H., Northcote, P. T., and Prinsep, M. R. (2003). Marine natural products. Natural Product Reports, 20(1): 1–48.

Chadburn, A. (2013 May). Immunodeficiency-associated lymphoid proliferations (ALPS, HIV, and KSHV/HHV8). Seminars in diagnostic pathology (Vol. 30, No. 2, pp. 113–129). WB Saunders.

Christie, S. N., McCaughey, C., McBride, M., and Coyle, P. V. (1997). Herpes simplex type 1 and genital herpes in Northern Ireland. International Journal of STD & AIDS, 8(1): 68.

Colwell, R. R. (2002). Fulfilling the promise of biotechnology. Biotechnology Advances, 20(3-4): 215–228.

Cragg, G. M., Newman, D. J., and Yang, S. S. (2006). Natural product extracts of plant and marine origin having antileukemia potential. The NCI experience. Journal of Natural Products, 69(3): 488–498.

Cross, K. L., Campbell, J. H., Balachandran, M. et al. (2019). Targeted isolation and cultivation of uncultivated bacteria by reverse genomics. Nat. Biotechnol. 37: 1314–1321. https://doi.org/10.1038/s41587-019-0260-6.

Davies, M. J. (2010). Myeloperoxidase-derived oxidation: mechanisms of biological damage and its prevention. Journal of Clinical Biochemistry and Nutrition, 48(1): 8–19.

D'Costa, V. M., King, C. E., Kalan, L., Morar, M., Sung, W. W., Schwarz, C., Froese, D., Zazula, G., Calmels, F., Debruyne, R., and Golding, G. B. (2011). Antibiotic resistance is ancient. Nature, 477(7365): 457–461.

Dimarco, G., Pareschi, L., Toscani, G., and Zanella, M. (2020). Wealth distribution under the spread of infectious diseases. arXiv preprint arXiv:2004.13620.

Du, F. Y., Zhang, P., Li, X. M., Li, C. S., Cui, C. M., and Wang, B. G. (2014). Cyclohexadepsipeptides of the isaridin class from the marine-derived fungus Beauveriafelina EN-135. Journal of Natural Products, 77(5): 1164–1169.

Fenical, W. (1997). New pharmaceuticals from marine organisms. Trends in Biotechnology, 15(9): 339–341.

Fenical, W. (2006). Marine pharmaceuticals: past, present, and future. Oceanography, 19: 111–119.

Edemekong, P. F., and Huang, B. (2017). Epidemiology of prevention of communicable diseases.

Elion, G. B., Furman, P. A., Fyfe, J. A., De Miranda, P., Beauchamp, L., and Schaeffer, H. J. (1977). Selectivity of action of an antiherpetic agent, 9-(2-hydroxyethoxymethyl) guanine. Proceedings of the National Academy of Sciences, 74(12): 5716–5720.

Grosberg, R. K., Vermeij, G. J., and Wainwright, P. C. (2012). Biodiversity in water and on land. Current Biology, 22(21): R900–R903.

Hoegh-Guldberg, O., Mumby, P. J., Hooten, A. J., Steneck, R. S., Greenfield, P., Gomez, E., Harvell, C. D., Sale, P. F., Edwards, A. J., Caldeira, K., and Knowlton, N. 2007. Coral reefs under rapid climate change and ocean acidification. Science, 318(5857): 1737–1742.

Huddleston, J. R. (2014). Horizontal gene transfer in the human gastrointestinal tract: potential spread of antibiotic resistance genes. Infection and Drug Resistance, 7: 167.

James, R. A., Vignesh, S., and Muthukumar, K. (2012). Marine drugs development and social implication. In Coastal Environments: Focus on Asian Regions (pp. 219–237). Springer, Dordrecht.

Jeffries, J. W. E., Dawson, N., Orengo, C., Moody, T. S., Quinn, D. J., Hailes, H. C., Ward, J. M. (2016). Metagenome Mining: A sequence directed strategy for the retrieval of enzymes for biocatalysis. ChemistrySelect , 1(10): 2217–2220. 10.1002/slct.201600515.

Jiao, W., Zhang, F., Zhao, X., Hu, J., and Suh, J. W. (2013). A novel alkaloid from marine-derived actinomycete Streptomyces xinghaiensis with broad-spectrum antibacterial and cytotoxic activities. PloS one, 8(10): e75994.

Kanase, H. R., and Singh, K. N. M. (2018). Marine pharmacology: Potential, challenges, and future in India. Journal of Medical Sciences, 38(2): 49.

Kanoh, K., Okada, A., Adachi, K., Imagawa, H., Nishizawa, M., Matsuda, S., Shizuri, Y., and Utsumi, R. (2008). Ascochytatin, a novel bioactive spirodioxynaphthalene metabolite produced by the marine-derived fungus, *Ascochyta* sp. NGB4. The Journal of Antibiotics, 61(3): 142–148.

Kaysser, L., Bernhardt, P., Nam, S. J., Loesgen, S., Ruby, J. G., Skewes-Cox, P., Jensen, P. R., Fenical, W., and Moore, B. S. (2012). Merochlorins A–D, cyclic meroterpenoid antibiotics biosynthesized in divergent pathways with vanadium-dependent chloroperoxidases. Journal of the American Chemical Society, 134(29): 11988–11991.

König, G. M., Wright, A. D., Sticher, O., Angerhofer, C. K., and Pezzuto, J. M. (1994). Biological activities of selected marine natural products. Plantamedica, 60(06): 532–537.

Lam, K. N., Cheng, J., Engel, K., Neufeld, J. D., and Charles, T. C. (2015) Current and future resources for functional metagenomics. Front. Microbiol., 6: 1196. doi: 10.3389/fmicb.2015.01196.

Lauritano, C., and Ianora, A. (2018). Grand challenges in marine biotechnology: overview of recent EU-funded projects. In Grand challenges in marine biotechnology (pp. 425–449). Springer, Cham.

Lindequist, U. (2016). Marine-derived pharmaceuticals–challenges and opportunities. Biomolecules & Therapeutics, 24(6): 561.

Liu, Z., Jensen, P. R., and Fenical, W. (2003). A cyclic carbonate and related polyketides from a marine-derived fungus of the genus Phoma. Phytochemistry, 64(2): 571–574.

Longley, D. B., and Johnston, P. G. (2005). Molecular mechanisms of drug resistance. The Journal of Pathology: A Journal of the Pathological Society of Great Britain and Ireland, 205(2): 275–292.

Mak, K. K., and Pichika, M. R. (2019). Artificial intelligence in drug development: present status and future prospects. Drug Discovery Today, 24(3): 773–780.

Martín, J., Sousa, D. S., Crespo, G., Palomo, S., González, I., Tormo, J. R., De la Cruz, M., Anderson, M., Hill, R. T., Vicente, F., and Genilloud, O. (2013). Kocurin, the true structure of PM181104, an anti-methicillin-resistant *Staphylococcus aureus* (MRSA) thiazolyl peptide from the marine-derived bacterium *Kocuria palustris*. Marine Drugs, 11(2): 387–398.

Mori, T., O'Keefe, B. R., Sowder, R. C., Bringans, S., Gardella, R., Berg, S., Cochran, P., Turpin, J. A., Buckheit, R. W., McMahon, J. B., and Boyd, M. R. (2005). Isolation and characterization of griffithsin, a novel HIV-inactivating protein, from the red alga *Griffithsia* sp. Journal of Biological Chemistry, 280(10): 9345–9353.

Mossong, J., Hens, N., Jit, M., Beutels, P., Auranen, K., Mikolajczyk, R., Massari, M., Salmaso, S., Tomba, G. S., Wallinga, J., and Heijne, J. (2008). Social contacts and mixing patterns relevant to the spread of infectious diseases. PLoS Med., 5(3): e74.

Murshid, S. S., Badr, J. M., and Youssef, D. T. (2016). Penicillosides A and B: New cerebrosides from the marine-derived fungus *Penicillium* species. Revista Brasileira de Farmacognosia, 26(1): 29–33.

Okazaki, T., Kitahara, T., and Okami, Y. (1975). Studies on marine microorganisms. IV. The Journal of Antibiotics, 28(3): 176–184.

Orloff, J., Douglas, F., Pinheiro, J., Levinson, S., Branson, M., Chaturvedi, P., Ette, E., Gallo, P., Hirsch, G., Mehta, C., and Patel, N. (2009). The future of drug development: advancing clinical trial design. Nature Reviews Drug Discovery, 8(12): 949–957.

Ovchinnikova, T. V., Balandin, S. V., Aleshina, G. M., Tagaev, A. A., Leonova, Y. F., Krasnodembsky, E. D., Men'shenin, A. V., and Kokryakov, V. N. (2006). Aurelin, a novel antimicrobial peptide from jellyfish *Aurelia aurita* with structural features of defensins and channel-blocking toxins. Biochemical and Biophysical Research Communications, 348(2): 514–523.

Patrzykat, A., and Douglas, S. E. (2003). Gone gene fishing: how to catch novel marine antimicrobials. Trends in Biotechnology, 21(8): 362–369.

Perry, N. B., Blunt, J. W., Munro, M. H., and Pannell, L. K. (1988). Mycalamide A, an antiviral compound from a New Zealand sponge of the genus Mycale. Journal of the American Chemical Society, 110(14): 4850–4851.

Pollock, G. ed. (2012). An epidemiological odyssey: the evolution of communicable disease control. Springer Science & Business Media.

Rashid, M. A., Gustafson, K. R., Cartner, L. K., Shigematsu, N., Pannell, L. K., and Boyd, M. R. (2001). Microspinosamide, a New HIV-Inhibitory Cyclic Depsipeptide from the marine sponge *Sidonops microspinosa*. Journal of Natural Products, 64(1): 117–121.

Trindade, M., van, Zyl L. J., Navarro-Fernández, J., and Abd Elrazak A. (2015). Targeted metagenomics as a tool to tap into marine natural product diversity for the discovery and production of drug candidates. Front. Microbiol. 6: 890. doi: 10.3389/fmicb.2015.00890.

Tsuda, M., Mugishima, T., Komatsu, K., Sone, T., Tanaka, M., Mikami, Y., and Kobayashi, J. I. (2003). Modiolides A and B, two new 10-membered macrolides from a marine-derived fungus. Journal of Natural Products, 66(3): 412–415.

Tziveleka, L. A., Vagias, C., and Roussis, V. (2003). Natural products with anti-HIV activity from marine organisms. Current Topics in Medicinal Chemistry, 3(13): 1512–1535.

Sakoulas, G., Nam, S. J., Loesgen, S., Fenical, W., Jensen, P. R., Nizet, V., and Hensler, M. (2012). Novel bacterial metabolite merochlorin A demonstrates in vitro activity against multi-drug resistant methicillin-resistant *Staphylococcus aureus*. PLoS One, 7(1): e29439.

Saxena, A. K. ed. (2018). Communicable Diseases of the Developing World (Vol. 29). Springer.

Sharma, S., Fulke, A. B., and Chaubey, A. (2019). Bioprospection of marine actinomycetes: recent advances, challenges and future perspectives. Acta Oceanologica Sinica, 38(6): 1–17.

Shaw, A., and Domachowske, J. (2019). Fever and pallor while living in, traveling to, or returning from just about anywhere in the tropics or subtropics. Introduction to Clinical Infectious Diseases: A Problem-Based Approach, p. 365.

Song, F., Liu, X., Guo, H., Ren, B., Chen, C., Piggott, A. M., Yu, K., Gao, H., Wang, Q., Liu, M., and Liu, X. (2012). Brevianamides with antitubercular potential from a marine-derived isolate of *Aspergillusversicolor*. Organic Letters, 14(18): 4770–4773.

Suryadevara, M. (2019). Human Papillomavirus Infection. Introduction to Clinical Infectious Diseases (pp. 181–190). Springer, Cham.

Yap, C. K. (2020). Antiviral compounds from marine bivalves for evaluation against SARS-CoV-2. Journal of PeerScientist, 2(2): e1000015.

Yasuhara-Bell, J., and Lu, Y. (2010). Marine compounds and their antiviral activities. Antiviral Research, 86(3): 231–240.

Chapter 6

Plant-Derived Microbial Bio-Similar for the Treatment of Tuberculosis

Rachana Khati,[1] *Alok K. Paul,*[2] *Maria de Lourdes Pereira,*[3] *Mohammed Rahmatullah,*[4] *Veeranoot Nissapatorn*[5] and *Karma G. Dolma*[1,*]

Introduction

Since the ancient period, the pathogenic bacterium *Mycobacterium tuberculosis* (*M. tuberculosis*) has been affecting mankind and has become a cause of contagious disease—tuberculosis (TB). According to the Global Tuberculosis Report 2020, TB infection has been considered the prominent cause of mortality from an infectious pathogen causing a serious burden on global public health. As per the WHO record in the year 2019, globally 10 million people were suffering from TB with 1.4 million deaths, including HIV-positive and HIV-negative patients (Chakaya et al. 2020). The entire nation throughout the world has adopted several measures to regulate the

[1] Department of Microbiology, Sikkim Manipal Institute of Medical Sciences, Sikkim Manipal University (SMU), Gangtok, Sikkim, India, Pin 737102.
[2] School of Pharmacy and Pharmacology, University of Tasmania, Hobart 7001, Australia.
[3] Department of Medical Sciences, CICECO-Aveiro Institute of Materials, University of Aveiro, Aveiro, Portugal.
[4] Department of Biotechnology & Genetic Engineering, University of Development Alternative, Lalmatia, Dhaka, Bangladesh.
[5] School of Allied Health Sciences and World Union for Herbal Drug Discovery (WUHeDD), Walailak University, Nakhon Si Thammarat, Thailand.
* Corresponding author: kgdolma@outlook.com

spread and control the rate of infection caused by tuberculosis. Despite these significant efforts, around one-third of the entire human population is still being affected by *M. tuberculosis* (Askun et al. 2013). The major factors that are responsible for the debilitating TB control measures are due to inadequate specific and sensitive diagnostic approaches, limited vaccine potency, and the development of antibiotic-resistant strains of *M. tuberculosis* due to co-infection with HIV (Igarashi et al. 2011). The regulation and management of TB remain a big hurdle because of the development of multidrug-resistant (MDR) strains and extensively drug-resistant strains (XDR). The enormous increase in antibiotic-resistant strains of TB needs quick actions to be taken with the development of advanced, more potent, shorter regimens, and safer drugs (Abuzeid et al. 2014). The urgency for the production of novel TB drugs has to overcome the current longer treatment regimen with shorter regimens and elevate the level of therapeutic treatment efficacy for MDR and XDR TB (Igarashi et al. 2011).

However, to globally overcome the health issues against MDR and XDR TB, there is an urgency to streamline more inputs for the development of novel anti-mycobacterial drugs derived from natural sources, including plants and available microorganisms. In order to overcome the battle against drug-resistant TB worldwide, more effort must be directed toward the discovery of innovative anti-tuberculosis medications derived from distinct products that are of natural origin to accomplish the goals of sustainable development for better health.

The natural product derived from the plants could be one of the prolific sources for the treatment of TB. Due to the several therapeutic effects and availability of diverse chemical and biological constituents, plants are the primary source of drug synthesis. The chemical and biological contents of the plants possess anti-inflammatory, antibiotic, immuno-modulating, and anti-cancer activities, which are explored for the treatment and therapeutic applications for humanity. Due to the limited outcome of novel drug research derived from plants, better innovative findings with efficient strategies are required. Various new methods have been improvised and implemented as advanced approaches for the effective formulation of novel drug and compound libraries via established high-throughput procedures. Approximately 60% of approved small molecule medications come from plants and 69% of all antibacterial drugs come from natural sources (Patridge et al. 2016, Matsumura et al. 2018). Several chemicals of natural origin can be employed as viable candidates for the creation of innovative drugs, but plant sources are restricted in nature, making large-scale, and cost-efficient drug discovery production problematic. A prominent indigenous technique was employed in bacterial and fungal hosts to express biosynthetic genes derived from natural sources (Song et al. 2014). Engineered microorganisms can create significant volumes of rare natural compounds, facilitating the production of the desired novel drug and its powerful analogs, including the authentication of their activities in a much easier way (Matsumura et al. 2018). The natural product industry is not the only one seen significant expansion or uses of medicinal items made by or from living organisms. An explosion of biologics has resulted from the development of unicellular and multicellular microbial cells as well as the progress of rDNA (Recombinant DNA) technology. Biologics are mainly a class of compounds whose potent medicinal constituents are taken from natural originslike plants. They are derived from living organisms including animals, human blood, and tissues and these are too complex to be produced naturally (Revers et al. 2010, Sanchez-Garcia et al. 2016). Engineered microorganisms can create significant volumes of scarcely available natural compounds. It enhances the production of the new-targeted antibiotics and their active products; besides, it validates their activities more conveniently. The natural product industry is booming with significant expansion by using medicinal items made by or from living organisms. Following the breakthrough production of Humulin R in the 1980s, human insulin was produced with the help of Recombinant DNA technology using the microorganism, *Escherichia coli*. *Escherichia coli* swiftly became the most widely used expression platform in the biopharmaceutical industry, followed by the yeast *Saccharomyces cerevisiae* (Sanchez-Garcia et al. 2016). Currently for the synthesis of approved recombinant products, mainly microorganisms, are utilised as the host due to their altered post-translational activities, unstable proteolysis, insolubility, and better initiation of cell stress feedback (Graumann et al. 2006). Despite various delays and hurdles, this shows that microbial hosts are a handy and reliable platform for the effective creation of recombinant proteins. To describe the biological constituents of natural products and biologics and further understand the mechanism of action, microbial systems are employed to manufacture these medicinal molecules. The new technological advancement in the synthesis of recombinant products using microbes and bioactive natural products has led to lucrative sources of medicines, which are currently under the process of development. The field of synthetic biology has enabled the large-scale production of natural products using microbes by decreasing the production cycle and eliminating the problem of product segregation (Yang et al. 2020). Figure 1 illustrates the schematic representation of the role of natural product-derived biosimilar for the treatment of tuberculosis.

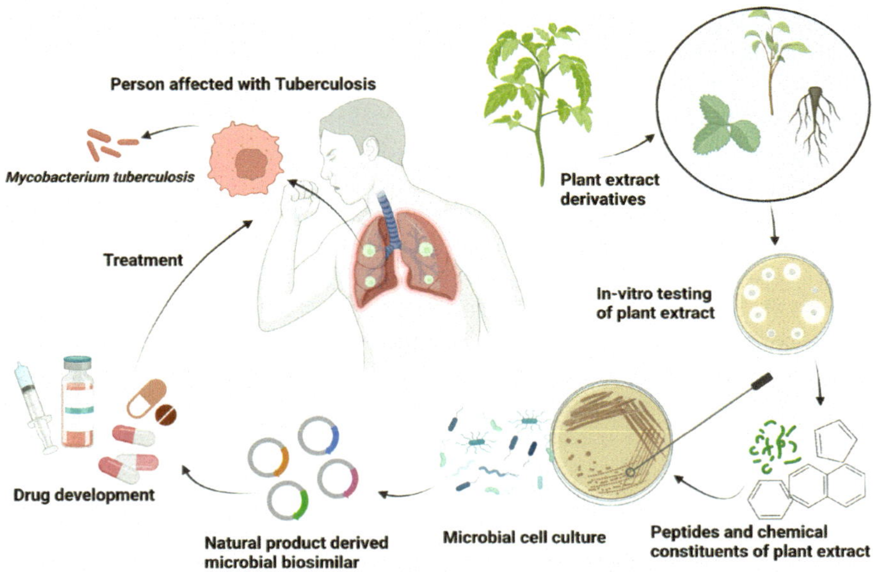

Figure 1. Schematic representation of the role of natural product-derived bio-similar for the treatment of tuberculosis. The Figure was made with Biorender.com.

Tuberculosis

Mycobacterium tuberculosis complex (*M. tuberculosis* complex) is the root cause of tuberculosis, which is a primaeval infectious illness in humans (Gagneux et al. 2018). The infection mainly arises due to the depleted immune system, inadequate diet, poverty, and overpopulation, which annually lead to the mortality of more than 1 million people worldwide (Orcau et al. 2011, Wang WF et al. 2020). The genus *Mycobacterium* comprises numerous species, among which few are prevalent in humans and several animals. The different species of (*M. tuberculosis* complex) are *Mycobacterium mungi*, *Mycobacterium tuberculosis variant pinnipedii*, *Mycobacterium orygis*, *Mycobacterium africanum*, *Mycobacterium tuberculosis sensu strict*, *Mycobacterium bovis Mycobacterium microti*, and *Mycobacterium caprae*. The species of (*M. tuberculosis* complex) obligate human pathogen that are responsible for infection in humans, which are *Mycobacterium tuberculosis sensu strict* and *Mycobacterium africanum*. *Mycobacterium bovis* and *Mycobacterium caprae* are also seen in some animals and humans, and the chance of transmission is sparse (Gagneux et al. 2018). *Mycobacterium tuberculosis* is a slow-growing bacillus with a doubling time of 12–24 hours under ideal circumstances. The aberrant cell wall structure of *M. tuberculosis* is the main characteristic feature that creates an exceptionally strong non-porous barrier to toxic chemicals and medicines that significantly contributes to its virulence (Delogu et al. 2013). The size of the bacterium is approximately 0.8–4 mm in diameter, and it is an acid-fast aerobic bacillus and non-motile in nature (Wang et al. 2020). A significant group of the population are affected by latent tuberculosis, which in some cases will be effectively contained and will not proceed to TB. However, 5–15% of people with latent tuberculosis will develop active tuberculosis (Möller et al. 2018, Behr et al. 2019). Tuberculosis is disseminated by inhaling infected droplet nuclei containing viable *Mycobacterium tuberculosis* (Möller et al. 2018, Behr et al. 2019). Tuberculosis usually spread from person to person through air droplets with tubercle bacilli infecting lung alveoli as the primary site of infection. Bacilli can travel and affect other organ systems in the body, including abdominal organs, lymphatic system, musculoskeletal system, and central nervous system causing extra-pulmonary tuberculosis. This process includes a pinpoint source of tuberculosis, infectious particle formation, particle survival in the air, inhalation by susceptible persons, and finally the development of tuberculosis (Churchyard et al. 2017).

The Clinical Manifestations of Tuberculosis

As with primary tuberculosis, which is usually self-resolving, TB might present with common mild and general symptoms. Patients with pulmonary tuberculosis experience high fever, chills, haemoptysis, night sweats, and weight loss and nearly 95% of cases have a persistent cough with rigorous chest pain accompanied by dyspnoea or subpleural involvement (Churchyard et al. 2017). Extrapulmonary tuberculosis (EPTB) is tuberculosis affecting all the organs excluding the lungs, and it occurs in addition to pulmonary tuberculosis. It can cause TB in the lymphatic system affecting regional lymph nodes; it also affects the meninges causing meningeal TB. The preliminary symptoms of meningeal TB are headache and mental instability or neurological abnormalities. TB affects the bones and the joints of individuals who show symptoms of swollen joints with persistent localised discomfort (Sanford et al. 2017). Due to a shortage of accurate and precise diagnostic testing kits, the advent of MDR and XDR TB, low vaccination efficacy and co-infection with HIV, and the paucibacillary nature of EPTB have aggravated the occurrences of TB worldwide, and this factor tends to impede TB control programmes (Venketaraman et al. 2015). Figure 2 shows the illustration of infection caused by *M. tuberculosis*.

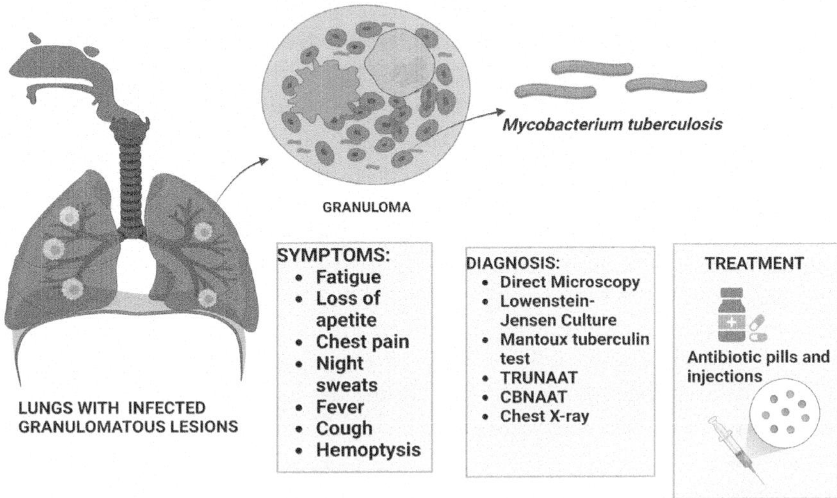

Figure 2. *Mycobacterium tuberculosis* infection. The Figure was made with Biorender.com.

Current Treatment Strategies of Tuberculosis

The current anti-tuberculosis drugs that are under the developmental stages are divided into three different categories below and are listed in the Table 1:

1. First-line medications that are under extensive re-evaluation to improve their efficacy;
2. Current anti-tuberculosis drugs that have been changed or used as anti-tuberculosis drugs; and
3. Novel therapies with new ideal structures and different modes of action.

Table 1. List of anti-tuberculosis drugs.

Isoniazid
Rifampicin
Pyrazinamide
Ethambutol
Streptomycin
Fluoroquinolones
p-aminosalicyclic acid
Ethionamide

Many manufactured compounds, such as fluoroquinolones, moxifloxacin, gatifloxacin, and DC-159a oxazolidinone compounds [Linezolid, Sutezolid, and AZD-584712 and diamine compounds (SQ109)], can be enhanced with simple modifications in categories 1 and 2 (Alcalá et al. 2003, Nuermberger et al. 2004, Cynamon et al. 2007, Nikonenko et al. 2007, Rustomjee et al. 2008, Williams et al. 2009, Disratthakit et al. 2010, Balasubramanian et al. 2014). In addition to tuberculosis, fluoroquinolone and oxazolidinone drugs are efficient against a wide spectrum of microorganisms. Cross-tolerance concerns must be avoided when using compounds like these. The majority of the compounds in Category 3 are compounds with unique morphology and different effects, which are derived from natural sources. One of the few recent instances of synthetic chemicals is bedaquiline (TMC207). Bedaquiline is a kind of Diarylquinoline (Andries et al. 2005, Diacon et al. 2009).

Plant—The Source of Natural Product Synthesis and its Importance

The majority of natural product collections begin with fresh extracts or shade dried materials made by dissolving in several kinds of organic solvents. The extracts are mixed combinations of hundreds of different chemicals. Usually, the techniques adopting traditional bioassay-guided fractionation are considered to be lagging behind matching the level of high throughput screening; in an intensive screening campaign, the experiment is conducted only for a period of a few months, so within the given period the active compounds may not be purified (Harvey et al. 2007). Since the birth of medicine, many substances derived from living sources including plants, animals, and microorganisms have been employed to cure illnesses among mankind. Natural product research as a source of innovative human treatments peaked in the Western pharmaceutical industry between 1970 and 1980, resulting in a pharma-dominated landscape where there is the use of non-synthetic compounds. The products derived from nature, that is mainly plants, natural product analogues that are semi-synthetic in nature, or synthetic compounds based on natural-product make up approximately half (49%) of the 877 small-molecule. The new chemical entities (NCEs) were introduced between 1981 and 2002 (Butler et al. 2005). A total of 70 natural product-related compounds were in clinical trials in 2004, in addition to the released products. The research on the potent bioactive role of natural products forms a basis for innovative chemical scaffolds for future therapeutic inventions (Koehn et al. 2005, Chin et al. 2006). The use of natural products success in drug development has been attributed to a variety of factors, including their enormous chemical diversity, evolutionary drive to form biologically potent compounds, the structural analogy of protein targets across numerous species, and so on (Harvey et al. 2005). Natural products have gotten a lot of attention as possible anti-TB agents because they are proven templates for developing novel pharmacological scaffolds (Pauli et al. 2005). Antimycobacterial active chemicals have been discovered in a variety of skeleton types, primarily from plants, but also from fungi and marine organisms (Copp et al. 2007).

Plant Extract with Anti-Mycobacterial Activity

Inhibitory activity against M. tuberculosis has been found in a variety of natural substances, including extracts from all parts of plants, microbes, and aquatic organisms. Researchers interested in phytochemical biodiversity with the help of preliminary functional assays have supplied many substances. Due to the unavailability of data, the potential pharmacological value of these chemicals derived from natural products is generally unknown to prove that these chemicals are harmful to mycobacterial survival mechanisms in humans (Okunade et al. 2004). Plant-based remedies have long been utilised in traditional medicine around the world to treat a variety of ailments. Medicinal herbs are still used by over 60% of the world's population for illness at a primary level. The different species of plants continue to be a potent source of new biologically active chemicals, despite the fact that only a few have been thoroughly explored for their therapeutic properties (Heinrich et al. 2001). Thus, throughout the last decade, there has been a renewed interest in phytomedicine, and numerous plants with medicinal values are now being explored to evaluate their pharmacological and therapeutic activities (Gautam et al. 2007). To develop novel medications, there are possibilities to explore ethnobotanical and anti-bacterial research. Plants were generally acknowledged as a source of antiseptic materials before antibiotic therapy with raw plants releasing a variety of natural compounds having antibacterial action (Table 2).

In rare cases, anti-mycobacterial chemicals have been discovered, such as (E)- and (Z)-phytol and phytanol, which were obtained from *Leucas volkensii* and pentacyclic triterpenoids from *Lantana hispida* (Rajab et al. 1998, Jimenez-Arellanes et al. 2007). The two crucial plants possessing chemicals diterpenes, such as mulinane with anti-tuberculosis activity are obtained from *Azorellama dreporica* Clos and calanolide A which is a native plant common in the tropical rain forest of Sarawak, Malaysia, and was first found by researchers at the National Cancer Institute (Xu et al. 2004, Wachter et al. 1998). Tryptanthrin, an alkaloid derived

Table 2. List of plants with active compounds.

Name of plants	References
1. Garlic *Allium sativum* (Fam. Liliaceae)	Jain et al. 1998
2. Bushy seaside oxeye *Borrichia frutescens* (Fam. Asteraceae)	Cantrell et al. 1996
3. Giant fennel *Ferula communis* (Fam. Umbelliferae)	Appendino et al. 2004
4. Cow parsnip *Heracleum maximum* (Fam. Umbelliferae),	Newton et al. 2000
5. Coyotillo *Karwinskia humboldtiana* (Fam. Rhamnaceae)	
6. Gurke *Leucas volkensii* (Fam. Labiatae)	Rajab et al. 1998
7. Wood nymph *Moneses uniflora* (Fam. Ericacaeae),	Newton et al. 2000
8. Devils walking stick *Oplopanax horridus* (Fam. Araliaceae)	
9. Stemmed sage *Salvia multicaulis* (Fam. Labiatae)	Ulubelen et al. 1997
10. Kuntze *Strobilanthus cusia* (Fam. Acanthaceae)	Mitscher et al. 1998
11. Shan-shan *Senna silvestris* (Fam. Leguminosae)	Graham et al. 2003
12. *Sommeras abiceoides* (Fam. Rubiaceae)	
13. Sweetwood *Nectan drahihua* (Fam. Lauraceae)	
14. *Senna obliqua* (Fam. Leguminosae)	
15. *Heisteria accuminata* (Fam. Olacaceae)	
16. *Zanthoxylum sprucei* (Fam. Rutaceae)	
17. *Lantana hispida* (Verbenaceae)	Jimenez-Arellanes et al. 2007
18. *Citrus aurantifolia* (Fam. Rutaceae)	Camacho-Corona et al. 2008
19. *Citrus sinensis* (Fam. Rutaceae)	
20. *Olea europaea* (Fam. Oleaceae)	

from the Chinese herb *Strobilanthes cusia*, is an example of an active plant with promising activity. After oral administration to mice, tryptanthrin and its analogues are effective against multi-resistant tuberculosis strains and are non-toxic (Mitscher et al. 1998). Furthermore, three molecules that are extracted from the plant *Indigofera longeracemosa* (Fabaceae) with the MIC value of 0.38 and MIC value of the (24R)-isomer of the triterpene Saringosterol 0.13 and are derived from *Lessonia nigrescens* (Lessoniaceae). The compound difenilalkyl ether ketone Engelhardione extracted from *Engelhardia roxburghiana* (Juglandaceae) with the MIC value of 0.21 mg/L can be considered a promising anti-TB compound based on the *in vitro* effect performed on H37R strains of MTB (Wächter et al. 2001, Thangadurai et al. 2002, Lin et al. 2005). The developmental phases of the novel anti-tuberculosis drug are a gradual process of development. The prominent and frequently used drug for the treatment of TB in humans was approved lastly in 1963; the drug was Rifampicin. Since then only a substantial number of drugs have undergone human trials. The two such drugs are bedaquiline and delamanid, and these drugs are currently being authorised primarily for the treatment of MDR-TB (Brigden et al. 2015). Both medications, however, have negative effects and are only indicated for patients who have exhausted all other therapy choices. For successful eradication of TB, more effective drugs and their analogues are produced as MDR. XDR TB remains untreatable with the current treatment strategy. The potent anti-mycobacterial antibiotics must show the characteristics of high efficacy, mainly against both MDR and XDR strains and should be safe for use. In addition to this, the compound should also be effective against latent and multiplying forms of *M. tuberculosis* with restricted compound or drug interactions mainly with anti-retroviral agents. In order to develop new effective antibiotics, the field of drug discovery is utilising target-based and genomic approaches. But these approaches are not yet relevant due to the restriction of enzyme activity as it does not relate to bacterial death (Payne et al. 2007). Advanced screening programs with large high-throughput (HTS) are being used to evaluate and discover useful molecules. Furthermore, clinically successful antibacterial are produced without following the 'rule of five' mentioned by Lipinski for the drug likeliness (Payne et al. 2007). According to the analysis of HTS programmes by Novartis natural products are tested vigorously as they possess the most varied classes of compounds with much better hit rates than synthetic compounds, and the molecules produced from different libraries (Sukuru et al. 2009). The utilisation of natural products has greatly enhanced over the past few years due to its elevated stereo-chemistry, complex pharma constituents, and precise three-dimensional structure (Harvey et al. 2015). The process of identifying bioactive compounds from sources of natural origin entails a set of cascades to characterise and manufacture the desired products. Furthermore, natural products are generally bioactive molecules with high bioavailability, enhancing their ability to reach their site of action within target cells (Quana et al. 2017).

Plant-Based Secondary Metabolites with Anti-Mycobacterial Properties

Plant-based anti-mycobacterial therapeutics are well versed in traditional medicines in many parts of the world. Plants are the main valuable origin of natural molecules with diverse bioactivities and therapeutic effects (Gupta et al. 2017). Plant extracts produce a structurally distinct bioactive chemical that has sparked a huge amount of drug development and findings based on the used value index by local healers for the treatment of illnesses (Khazir et al. 2013). In Mozambique, leaf extracts of the aromatic plant *Lippia javanica* (Verbenaceae) are used as a therapy for a variety of illnesses, including flu symptoms, cough, gastrointestinal illness, and more (Mitra et al. 2012). This herb yielded the triterpene euscaphic acid with the MIC value of 50 microgram/mL and showed a great anti-TB effect against *M. tuberculosis* (Mohamad et al. 2011). Diospyrin, shinanolone, and 7-methyljuglone were extracted from the roots of the *Euclea natalensis* (Ebenaceae) plant. However, it showed a substantial effect against drug-resistant and drug-sensitive *M. tuberculosis* strains (Zofou et al. 2013). This herb has been used to cure infections related to chest diseases, toothaches, and other ailments in the past, which supports its ethnomedicinal use (Anochie et al. 2018). *In vitro Artemesia afra* was also found to be effective for the treatment of depression, it had strong radical scavenging activity with improved efficacy for cardiac health. It also showed a strong anti-mycobacterial effect (Anochie et al. 2018, Sieniawska et al. 2017). This plant possesses a therapeutic effect against respiratory and oro-pharyngeal ailments, such as colds, coughs, and asthmatic problems and is generally found in Eastern and Southern Africa (Sánchez et al. 2010).

Ursolic acid and hydroquinone were extracted from *Artemisia capillaries* Thunb. (Asteraceae), a plan t historically used to treat malaria was also tested for anti-TB activity with potential effect against drug-resistant strains of *M. tuberculosis* and in *in vitro* it showed MIC values of 0.0125 to 0.025 mg/mL, respectively (Jyoti et al. 2008). *Curcuma longa* L. (Zingiberaceae), a herb historically used to treat whooping cough, was also tested for anti-mycobacterial properties. This plant yielded isoxazole curcuminoids, which were reported to exhibit anti-mycobacterial activity against MDR-TB with MIC values of 0.0019 to 0.00312 mg/mL (Changtam et al. 2010). In South Africa, the plant *Knowltonia vesicatoria* (Ranunculaceae), has been used to treat tuberculosis infection. The isolation of two compounds, 5-(hydroxymethyl) dihydrofuran-2(3H)-one and 5-(hydroxymethyl)furan-2(5H)-one, with a MIC value of 50.0 mg/mL, resulted in the discovery of the anti-mycobacterial drug. Isoniazid (INH) was combined with the raw herbal extract to observe if there was an increased anti-mycobacterial action due to a possible synergistic effect, confirming its historic use in tuberculosis treatment (Efange et al. 2002, Anochie et al. 2018).

Microbial Cell Factories and Biologics

The soil has yielded a large number of important antibiotics and metabolites from microbial sources (Zahner et al. 1995). In the past few years, it has been documented that approximately 23,000 bioactive molecules were synthesized by bacteria, including actinomycetes producing around 10,000 of these metabolic derivatives (McCarthy et al. 1992, Berdy et al. 2007). Due to the regular occurrence of *M. tuberculosis* resistance to several chemotherapeutics, the need for novel chemotherapeutics remains the main hurdle. Thus, there is an urgency to keep searching for novel natural compounds originating from plants and bacteria in understudied ecosystems, such as aquatic and terrestrial ecosystems. Scientists theorised the capability of the bacteria to develop unique and novel bioactive substances that may have therapeutic potential (Harvey et al. 2000). Microorganisms are also one of the most promising sources of drugs which potentially acts as major antibiotics, immunosuppressant and statins that significantly decreases lipid synthesis. Moreover, these are produced from a minimum range of microbial sources available globally (Tripathi et al. 2005). The discovery of streptomycin reaffirmed the commitment of Selman Waksman's for deriving small bioactive compounds from soil microbes (Harvey et al. 2000). Aminoglycoside-aminocyclitol antibiotics have been found in a number of bacterial species. *Streptomyces, Micromonospora, Bacillus,* and other bacteria are among them. Only *Streptomyces*-derived substances are called '- mycins' (e.g., tobramycin), whereas others are called '-micins', '-osins', '-asins', or '-acins'. Kanamycin, amikacin, and capreomycin are three more aminoglycosides that used to treat tuberculosis. Rifamycins (RMP) are a class of semisynthetic antibiotics derived from *Streptomyces mediterrani* that are also effective against tuberculosis. Thiolactomycin [(4R)(2E,5E)-2,4,6-trimethyl-3-hydroxy-2,5,7-octatriene-4-hiolide] is another microorganism-derived chemical. It is a novel thiolactone antibiotic that inhibits mycolic acid production and has anti-TB activity. It was originally isolated from a soil *Nocardia* species (Davies et al. 2007).

Microbial Cell Factories

One of the most critical components of designing natural products and recombinant protein bioprocesses is choosing the right host strain Gram-negative. *Escherichia coli* is a type of bacteria. *E. coli* has been considered a suitable tool for the synthesis of natural products due to its easy accessibility, well-understood physiology high productivity, and well-understood genetic mechanism (Abdin et al. 2003). The natural product synthesis is facilitated by using microbes to enhance the manufacturing of secondary metabolites. The natural product with enhanced secondary metabolites manufactured with the use of microbes can generally be synthesized by two systems, including Gene shuffling mechanism and ribosome engineering. Moreover, the integration of information from 'omics' has played a significant role in the drug discovery of natural products. It can precisely predict the changes occurring biochemically and describe the pathways of different metabolic. Advances in metagenomics have helped researchers better comprehend a wide range of microbes and their source of origin including water bodies and harsh environments, like seabed and frozen glaciers (Mohana et al. 2018). Overall, the diversified and vital significance of microbial natural products and biologics in human life will continue to grow. The potential for recombinant pharmaceuticals is growing as it extensively uses novel protein production platforms for product enhancement. Despite the numerous hurdles, the synthesis of natural products from microbes using rDNA technology is now booming because of their genuine and inexpensive features (Pham et al. 2019). Continued efforts in the production of natural product analogues will pave the way for the extraction of molecules with improved biological constituents. With the help of today's improved methodologies and techniques, natural products derived from microbes can continue to be a reliable source of novel molecules in drug discovery (Koyama et al. 2010).

Antimycobacterial properties have also been discovered in fungi. The terrestrial landscape of Japan reported the presence of *Mortierella alpine* FKI-4905 fungus, which was reported to exhibit anti-TB effect against *M. tuberculosis* with the MIC value of 12.5 and *M. smegmatis* with the MIC values of 0.78 g/mL (Koyama et al. 2010). Another natural molecule anthraquinone, 4-deoxybostrycin was derived from South China from the fungus *Nigrospora* spp. and the endophytes found in mangrove. The H37RV strain of *M. tuberculosis* revealed its effect on gene expression involved in nucleotides, proteins, coenzymes, lipids, or energy (Wang et al. 2013).

Other studies investigated the anti-TB effect of the microbes actinobacteria derived from the water sources, mainly the lake situated in Michigan by observing the chemical profile of aquatic bacteria (Mullowney et al. 2015). In comparison to clinically utilised anti-TB drugs, a small portion of *in vitro* activity of the microorganisms (*Micromonospora* sp.) showed a satisfactory inhibitory effect. Two novel secondary metabolites, diazaquinomycins H (DAQH) with MIC values of 0.04 and J (DAQJ) with MIC value of 0.07 g/mL isolated from the microorganisms, also showed the effect against *M. tuberculosis* (Manikkam et al. 2014). At least 1% of microbes, leaving the vast amount of microbes and their biological metabolic pathway inaccessible (Koyama et al. 2010, Mohana et al. 2018, Pham et al. 2019). The metagenomics method for drug discovery from uncultured microbes provides a promising platform for drug discovery. Through a study of their DNA recovered from environmental samples, these microbes have proven to be a valuable source of new antibiotics.

As a result of the introduction of metagenomic methods as a culture-independent methodology, multiple techniques are now accessible to assess the complete range of uncultured arrays of microorganisms which make it easier to understand the physiological and biochemical scenarios of these uncultured microbes (Wang et al. 2013, Manikkam et al. 2014, Mullowney et al. 2015). Different investigations have established the presence of indigenous actinobacteria using a culture-independent molecular method. These methods concentrate on nucleic acids collected purely from the samples and amplified DNA derived from RNA by PCR reaction isolated from the samples collected from the environment (Monciardini et al. 2002, Mincer et al. 2005, Das et al. 2007, Sun et al. 2010, Zhang et al. 2010). Alternatively, these amplicons might be multiplied and replicated, and the unique constituent present in the bacterial samples will then be determined and counted (Stach et al. 2003, Riedlinger et al. 2004). The shorter fragments of DNA with non-identical sequences but of similar length can be evaluated by observing the band formed by the electrophoresis technique (Riedlinger et al. 2004). The abundance actinomycetes and diversity in environmental samples have been determined using the PCR-DGGE/TGGE method (Rath et al. 2011). Newer strategies for gene, genome, protein, and metabolic pathway identification are enabled by metaproteomic and metagenomics technologies (Wawrik et al. 2004).

The production of drugs and commercial medicines with a wide genetic variety possessing untraced chemical resources has gradually declined due to uncultured microorganisms from aquatic sources which have the potential to synthesise secondary metabolites (Fenical et al. 2006). These rare biochemical and complex metabolic arrays will contribute to the basic idea related to current drug discovery for the treatment of a variety of disorders. Products that are derived naturally are of great importance as it provides the basic foundation for the synthesis of chemotherapeutic drugs for tuberculosis treatment (Rath et al. 2011).

Researchers recently found a source of various chemical morphology whose structural constituents might be employed for the synthesis of compounds using synthetic chemistry. The low efficacy in the drug development from natural sources necessitates a need for the discovery of alternative improved drug-producing strategies. In order to produce a pure novel drug, advanced techniques with modified and enhanced screening are to be employed which ultimately can provide the libraries of different novel biochemical compounds derived from natural sources (Manivasagan et al. 2013).

Plant Extract-Based Microbial Bio-Similar with Anti-Mycobacterial Activity

The products that are of natural and synthetic origin play a profound role in the large-scale production of medicine and treatment therapies as it uses a chemical probe for identifying the protein targets to understand the phenomenon of its action. However, it is worth noting that newer techniques (e.g., genetics involving chemical sciences) have been used significantly more in studies of unicellular and multicellular organisms where the relative genome of prokaryotes can be modified and altered (Gutierrez-Lugo 2008). Aside from target recognition and authoriSation, current inputs have concentrated on the development of novel antibiotics and the improvement of the pharmacological effects of those that are already under application in order to reduce treatment exposure and enhance the effect of drugs against drug-resistant tuberculosis. In order to identify new medication classes, a variety of methodologies have been used; the successful outcomes of some of these investigations have led to the findings in *M. tuberculosis* studies where fresh and improved chemical probes are developed (Maria-Teresa Gutierrez-Lugo 2008). The nitroimidazoles 18 (PA-824) (Stover et al. 2000) and 19 (OPC-67863) (Matsumoto et al. 2006), which are linked, are one of the trusted compounds with a profound effect against *M. tuberculosis*. Both substances are prodrugs that need ignition by a similar type of enzyme that is dependent on F420 (Rv3547) and affect *M. tuberculosis* and MDR-TB growth by blocking the production of cellular mycolic acid and protein translation (Manjunatha et al. 2006, Matsumoto et al. 2006). The relevant information deduced from these two observations related to the action of the inhibitor is appreciable. The animal model study closely deciphering the latent phase *M. tuberculosis* showed an effect of the first 18 molecules against non-replicating *Bacillus* thriving in microaerophilic settings. In comparison to the treatment regimens of the combined therapy of 18 or 19 with RIF and PZA, the effective action was seen within a shorter duration as compared to conventional therapy. The evaluation of molecule 18 and 19 in phase-II clinical trials are being conducted by Global Alliance for TB Drug Development and Otsuka Pharmaceutical. The recent discovery of diarylquinoline 20 (TMC207) showed that it is a powerful inhibitor of *M. tuberculosis* species including drug-resistant strains with the ability to identify an unanticipated target; ATP synthase which is necessary for mycobacterial growth (Andries et al. 2005). Compound 20, discovered by a whole-cell screening of *M. smegmatis* from a library of diarylquinolines, are stronger than both isoniazid and rifampicin and it shows a non-cross-resistance against other anti-TB drugs. The *in vivo* study demonstrated that the combination of compound 20 with either of two medications,

INH, RIF, or PZA, showed greater efficacy compared to conventional combination therapy of INH, RIF, and PZA, implying that substituting either any of the two drugs with 20 could shorten the current longer TB treatment process.

Diarylquinoline 20 is unique as it has greater efficacy against all the species of *M. tuberculosis* (*M. smegmatis* and *Mycobacterium tuberculosis*) (Andries et al. 2005). The mice study showed its potent action irrespective of stages of infection. The inadequacy of overt toxicity of the chemical is perplexing as ATP synthase is available only in the mitochondria of the host cell but not in the mycobacteria. Compound 20 is now through phase II testing (Laurenzi et al. 2007, http://www.tballiance.org/newscenter/view-brief.php?id=726). Lupin Limited is developing a new class of anti-TB drugs called Compound 17 (LL-3858 or Sudoterb) (India) (Arora et al. 2004). The 17 drugs were found to be active against all forms of *M. tuberculosis* infection, demonstrating the novel route of its effect. It is associated with molecules of pyrrole alkaloid mostly isolated from *Lupinus* plants (Arora et al. 2004).

Compound 17 is currently being tested in a multi-dose phase I clinical trial (Ginsberg et al. 2007). The new class of fluoroquinolone that inhibit DNA gyrase are Gatifloxacin (GAT, 21) and Moxifloxacin (MXF, 22), and it is more effective Thanofloxacin and Ciprofloxacin the second-line fluoroquinolones that are already in use. A mice model showed the efficacy of compound 22, which exhibited an anti-mycobacterial effect similar to INH and the study reported that the compound destroyed the rifampicin-resistant *M. tuberculosis* and its activity was seen more enhanced when used in combination with INH, RIF, and PZA compared to the three-drug treatment alone (Zhang et al. 2006).

All semisynthetic versions of the substances of natural origin including Rifamycin, Rifapentin, Rifabutin, Rifalazil, and Rifametane, are in multi-stages of experimental tests, exhibiting better efficacy against *M. tuberculosis* with refined pharmacologic properties (Rothstein et al. 2006, Portero et al. 2007). The 16 (SQ109) is an ethambutol derivative that has been shown to behave collectively with INH and RIF with enhanced pharmacological characteristics (Protopopova et al. 2005). Despite the fact that 16 is a second-generation molecule derived from EMB, it relatively does not hinder cell wall formation in the same way as its parent molecule does. Compound 16 is now undergoing phase I testing (Laurenzi et al. 2007). Pre-clinical testing is now underway for new anti-TB drugs such as molecules derived from the natural sources of Capuramycin, Oxazolidinones, and Sulfonylcarboxamides. Capuramycins act as an anti-TB drug by inhibiting the enzyme Translocase-I which synthesises peptidoglycan as crucial components of the cell wall.

Also, 13 (RS-118641) is the most active Capuramycin derivative discovered so far (Kogal et al. 2004). Linezolid (14) is a synthetic oxazolidinone that inhibits the synthesis of proteins (Sood et al. 2006). The drug was authorised by the FDA for the treatment of MDR-TB patients (Protopopova et al. 2005). A number of oxazolidinone compounds are currently being researched (Arora et al. 2004). For the effective treatment of MDR and latent TB, Sulfonylcarboxamide 15 (FAS20013) compounds are used which is similar to the enzyme ketacyl synthase; the enzyme required for the synthesis of fatty acid (Parrish et al. 2001). The compounds cerulenin and thiolactomycin of natural origin hinder the two-carbon homologation catalysed by ketoacyl synthase, which prompted the synthesis. Cerulenin restricts the enzyme permanently, whereas Thiolactomycin disrupts the acetyl-coenzyme A/ACP transacylase and the ketoacyl carrier protein (ACP) synthase (Parrish et al. 2001). According to Copp and Pearce, the identification of the natural goods extracts with anti-mycobacterial activity has explored eight natural items that could be helpful foundations for anti-TB drug development in the near future. Pleuromutilin and erythromycin are the two compounds of natural origin that are currently in the developmental stage and are produced by TB Global Alliance (Thomford et al. 2018).

Recent Advancements in the Synthesis of a Plant-Based Microbial Bio-Similar as Anti-TB Drugs

Natural products face numerous hurdles in the therapeutic development pipeline due to (1) insufficient knowledge of their biological mechanisms, (2) problems related to the isolation of pure chemical components, (3) restricted standardisation procedures, and (4) documented clinical trials (Özdemir et al. 2018). To utilise the therapeutic qualities of natural products, a combinatorial approach has been used with the utilisation of advanced technologies including computational biological approaches, metadata, sophisticated profiling tools, and artificial intelligence (Gupta et al. 2017, Özdemir et al. 2018). Due to the presence of several mixtures of raw metabolic extracts of natural products, it is critical to employ other effective ways of inducing a non-replicating state in *M. tuberculosis* strains by incubating them at sub-optimal pH. The examination of luminescence is used to determine anti-microbial susceptibility testing of bacteria against chemicals. This simple assay has the advantages of not requiring an outgrowth phase and allowing results to be extrapolated with minimum adjustments. It could be used to screen natural products for bactericidal action against non-dividing *M. tuberculosis* bacteria. The anti-TB effects of test chemicals are commonly measured by lysing macrophages with hypotonic buffer and measuring the bacterial load within them (1.5 mM $MgCl_2$, 10 mM HEPES, and 10 mM KCl) (Gupta et al. 2017). Post incubation after a 21-days, the bacterial load is counted at different time intervals by placing the samples in Dubos agar plates (Khan et al. 2008). The number of colony-forming units (CFUs) determines the viability of bacteria (Jiménez-Arellanes et al. 2013). The drug development process has been assisted by these extraction and fractionation procedures, which have resulted in the highest yield for hit drug leads (Luo et al. 2013). The different technologies that are used for the synthesis of drugs are semi-bionic extraction, microwave-assisted, and uses molecular distillation methods like ultrasonic and enzyme-assisted extraction and supercritical fluid extraction (Gan et al. 2016, He et al. 2016, Thomford et al. 2018).

Conclusion

Due to the advent of drug-resistant strains of *Mycobacterium tuberculosis*, a prolonged treatment strategy is needed for the discovery of alternative therapeutic options. Multi-functionality of the drugs is required to overcome the emergence of drug-resistant strains with better efficacy, potency, and safety. Drugs with improved half-life and least toxicity are necessary for the development of anti-tuberculosis drugs. The plant's natural products since the ancient period have proved to be effective against various illnesses with the least side effects in mankind. As natural products are a rich source of various bioactive compounds, it is the primary source of several compounds used for the synthesis of drugs. Due to the scarce availability of plant material with anti-tuberculosis activity, the cell factories with greater production of drugs are required to enhance the large production with the help of recombinant DNA technology. In this technology, several easy-to-cultivate and readily available microorganisms are utilised. The natural product of plants possessing multiple therapeutic activities are incorporated into these microorganisms, and this facilitates the production of highly target-specific and potent effective anti-mycobacterial compounds with a shorter treatment regimen. These compounds are currently in preliminary phases with promising therapeutic effects. The use of several advanced technologies including recombinant DNA technology with green chemistry could pave a path for battling the infections caused by *Mycobacterium tuberculosis*. The plant extract-derived microbial bio-similar could be one of the specific and target-oriented drugs with a shorter treatment regimen and safer for the treatment of *Mycobacterium tuberculosis* infection.

Acknowledgments

Project CICECO-Aveiro Institute of Materials, UIDB/50011/2020, UIDP/50011/2020 & LA/P/0006/2020, financed by national funds through the FCT/MEC (PIDDAC).

References

Abdin, M. Z., Israr, M., Rehman, R. U., and Jain, S. K. (2003). Artemisinin, a novel antimalarial drug: biochemical and molecular approaches for enhanced production. Planta Medica, 69: 289–299. doi: 10.1055/s-2003-38871.

Abuzeid, N., Kalsum, S., Larsson, M., Glader, M., Andersson, H., Raffetseder, J., Pienaar, E., Eklund, D., Muddathir, S. Alhassan, Haidar A. AlGadir, Waleed S. Koko, Schön, T., Mesaik, M. A., Abdalla, O. M., Khalid, A., and Lerm, M. (2014). Antimycobacterial activity of selected medicinal plants traditionally used in Sudan to treat infectious diseases. Journal of Ethnopharmacology, 157: 134–139, ISSN 0378-8741,https://doi.org/10.1016/j.jep.2014.09.020.

Adelina Jiménez-Arellanes, Julieta Luna-Herrera, Jorge Cornejo-Garrido, Sonia López-García, María Eugenia Castro-Mussot, Mariana Meckes-Fischer, Dulce Mata-Espinosa, Brenda Marquina, Javier Torres, and Rogelio Hernández-Pando. (2013). BMC Complementary and Alternative Medicine volume 13, Article number: 258.

Alcalá, L., Ruiz-Serrano, M. J., Pérez-Fernández Turégano, C., Darío García de Viedma, D. C., Infantes, M. D., Mercedes, Arriaza, M., and Bouza, E. (2003). *In Vitro* Activities of Linezolid against Clinical Isolates of *Mycobacterium tuberculosis* That Are Susceptible or Resistant to First-Line Antituberculous Drugs Antimicrobial Agents and Chemotherapy, 47(1). doi: https://doi.org/10.1128/AAC.47.1.416-417.2003.

Andreas, H. Diacon, Alexander Pym, Martin Grobusch, D. T. M. & H., Ramonde Patientia., Roxana Rustomjee, Liesl Page-Shipp, Christoffel Pistorius, Rene Krause, Mampedi Bogoshi, Gavin Churchyard, Amour Venter, Nat Dip, Jenny Allen, Juan Carlos Palomino, Tine De Marez, Rolf, P. G. van Heeswijk, NacerLounis, Paul Meyvisch, Johan Verbeeck, Wim Parys, Karel de Beule, Koen Andries, and David F. Mc Neeley. (2009). The diarylquinoline TMC207 for multidrug-resistant tuberculosis. The New England Journal of Medicinevol. 360 no. 23360:2397-405.

Andries, K., Verhasselt, P., Guillemont, J., Göhlmann, H. W., Neefs, J. M., Winkler, H., Van Gestel, J., Timmerman, P., Zhu, M., Lee, E., Williams, P., de Chaffoy, D., Huitric, E., Hoffner, S., Cambau, E., Truffot-Pernot, C., Lounis, N., and Jarlier, V. (2005). A diarylquinoline drug active on the ATP synthase of *Mycobacterium tuberculosis*. Science, 14; 307(5707): 223–7. doi: 10.1126/science.1106753. PMID: 15591164.

Anochie, P. I., Ndingkokhar, B., Bueno, J., Anyiam, F. E., Ossai-Chidi, L. N. et al. (2018). African medicinal plants that can control or cure tuberculosis. International Journal of Pharmaceutical Sciences and Developmental Research, 4(1): 001–008. DOI: 10.17352/ijpsdr.000016.

Appendino, G., Mercalli, E., Fuzzati, N., Arnoldi, L., Stavri, M., Gibbons, S., Ballero, M., and Maxia, A. (2004). Antimycobacterial coumarins from the sardinian giant fennel (*Ferula communis*). Journal of Natural Product; 67(12): 2108–10. doi: 10.1021/np049706n.

Arora, S. K., Sinha, N., Jain, S., Upadhayaya, R. S., Jana, G., Shankar, A., and Sinha, R. K. (2004). Preparation of pyrrole derivatives as antimycobacterial compounds. PTC/IN2002/000189 WO2004026828NO.A1.

Askun, T., Tekwu, E. M., Satil, F., Modanlioglu, S., and Aydeniz, H. (2013). Preliminary antimycobacterial study on selected Turkish plants (Lamiaceae) against *Mycobacterium tuberculosis* and search for some phenolic constituents. Bio-Med Central Complementary Medicine and Therapies 13: 365. DOI: https://doi.org/10.1186/1472-6882-13-365.

Balasubramanian, V., Solapure, S., Iyer, H., Ghosh, A., Sharma, S., Kaur, P., Deepthi, R., Subbulakshmi, V., Ramya, V., Ramachandran, V., Balganesh, M., Wright, L., Melnick, D., Butler, S. L., and Sambandamurthy, V. K. (2014). Bactericidal activity and mechanism of action of AZD5847, a novel oxazolidinone for treatment of tuberculosis. Antimicrobial Agents and Chemotherapy, 58: (1): 495–502, doi:10.1128/AAC.01903-13.

Behr, M. A., Edelstein, P. H., and Ramakrishnan, L. (2019). Is *Mycobacterium tuberculosis* infection life long? British Medical Journal doi: 10.1136/bmj.l5770.

Berdy, J. (2005). Bioactive microbial metabolites. Journal of Antibiotics (Tokyo), 58(1): 1–26. doi: 10.1038/ja.2005.1.

Bo Yang, Xudong Feng, and Chun Li. (2020). Microbial cell factory for efficiently synthesizing plant natural products via optimizing the location and adaptation of pathway on genome scale. Frontiers in Bioengineering and Biotechnology, VOL:8, DOI:10.3389/fbioe.2020.00969.

Brigden, G., Hewison, C., and Varaine, F. (2015). New developments in the treatment of drug-resistant tuberculosis: clinical utility of bedaquiline and delamanid. Infection and Drug Resistance, 8: 367–378, DOI: https://doi.org/10.2147/IDR.S68351.

Butler, M. S. (2005). Natural products to drugs: natural product derived compounds in clinical trials. Natural Product Reports Journal; 22(2): 162–95. doi: 10.1039/b402985m.

Camacho-Corona, M. del R., Ramirez-Cabrera, M. A., Santiago, O. G., Garza-Gonzalez, E., Palacios, I. de P., and Luna-Herrera, J. (2008). Activity against drug resistant-tuberculosis strains of plants used in Mexican traditional medicine to treat tuberculosis and other respiratory diseases. Phytotherapy Research, 22(1): 82–5. doi: 10.1002/ptr.2269.

Cantrell, C. L., Lu, T., Fronczek, F. R., Fischer, N. H., Adams, L. B., and Franzblau, S. G. (1996). Antimycobacterial cycloartanes from *Borrichia frutescens*. Journal of Natural Products, 59(12): 1131–6. doi: 10.1021/np960551w.

Chakaya, J., Khan, M., Ntoumi, F., Aklillu, E., Fatima, R., Mwaba, P., Kapata, N., Mfinanga, S., Hasnain, S. E., Katoto, N. H., Bulabula, A., Sam-Agudu, N. A., Nachega, J. B., Tiberi, S., McHugh, T. D., Abubakar, I., and Zumla, A. (2021). Global Tuberculosis Report 2020—Reflections on the Global TB burden, treatment and prevention efforts International Journal of Infectious Diseases. Volume 113, Supplement 1,S7-S12 doi:https://doi.org/10.1016/j.ijid.2021.02.107.

Chandra Mohana, N., Yashavantha Rao, H. C., Rakshith, D., Mithun, P. R., Nuthan, B. R., and Satish, S. (2018). Omics based approach for biodiscovery of microbial natural products in antibiotic resistance era. Journal of Genetic Engineering and Biotechnology, 16: 1–8. doi: 10.1016/j.jgeb.2018.01.006.

Chang-Blanc, D., and Nunn, P. (1999). Incentives and disincentives for new anti-tuberculosis drug development. STOP TB Initiative and the special programme for research and training in tropical diseases World Health Organization, Geneva, Switzerland.

Changtam, C., Hongmanee, P., and Suksamrarn, A. (2010). Isoxazole analogs of curcuminoids with highly potent multidrug-resistant antimycobacterial activity. European Journal of Medicinal Chemistry, 45(10): 4446–4457.

Chin, Y. W., Balunas, M. J., Chai, H. B., and Kinghorn, A. D. (2006). Drug discovery from natural sources. American Association of Pharmaceutical Scientists Journal 8: E239-53.

Churchyard, G., Kim, P., Shah, N. S., Rustomjee, R., Gandhi, N., Mathema, B., Dowdy, D., Kasmar, A., and Cardenas, V. (2017). What we know about tuberculosis transmission: an overview. The Journal of Infectious Diseases, 216(6): S629–S635, https://doi.org/10.1093/infdis/jix362.

Churchyard, G., Kim, P., Shah, N., Rustomjee, R., Gandhi, N., Mathema, B. and others. (2017). An outline of what we know about tuberculosis transmission. J. Infect. Dis., 216(S6): S629–35.

Copp, B. R., and Pearce, A. N. (2007). Natural product growth inhibitors of *Mycobacterium tuberculosis*. Natural Product Report, 24: 278–97.

Cynamon, M., Sklaney, M. R., and Shoen, C. (2007). Gatifloxacin in combination with rifampicin in a murine tuberculosis model. Journal of Antimicrobial Chemotherapy, 60: 429–432.

Das, M., Royer, T. V., and Leff, L. G. (2007). Diversity of fungi, bacteria, and actinomycetes on leaves decomposing in a stream. Appl. Environ. Microbiol., 73(3): 756–67.

Davies, J. (2007). In the Beginning there was streptomycin. Wiley Library https://doi.org/10.1002/9780470149676.ch1.

Delogu, G., Sali, M., and Fadda, G. (2013). The biology of *Mycobacterium tuberculosis* infection. The Mediterranean Journal of Hematology and Infectious Diseases, 5(1): e2013070.

Disratthakit, A., and Doi, N. (2010). *In vitro* activities of DC-159a, a novel fluoroquinolone, against *Mycobacterium* species. Angewandte Chemie (a journal of the German Chemical Society) 54: 2684–2686.

Fenical, W., and Jensen, P. R. (2006). Developing a new resource for drug discovery: marine actinomycete bacteria. Natural Chemical Biology, 2(12): 666.

Efange, S. M. N. (2002). Natural products: a continuing source of inspiration for the medicinal chemist. Advances in phytomedicine, Ethnomedicine and Drug Discovery 2002 pp.61-69 ref.38.

Eitaro, M., Akira, N., Yusuke, T., Shinichi, I., Toshiyuki, S., Takane, K., Kenji, Y., Hidehiko, K., Fumihiko, S., and Hiromichi, M. (2018). Microbial production of novel sulphated alkaloids for drug discovery. Scientific Reports volume 8, Article number: 7980.

Gagneux, S. (2018), Ecology and evolution of *Mycobacterium tuberculosis*. Nature Review Microbiology, 16(4): 202.

Gautam, R., Saklani, A., and Jachak, S. M. (2007). Indian medicinal plants as a source of antimycobacterial agents. The Journal of Ethnopharmacology, 110: 200–34.

Ginsberg, A. M., and Spigelman, M. (2007). Challenges in tuberculosis drug research and development. Nature. Medicine 13: 290–294. [PubMed: 17342142].

Graham, J. G., Pendland, S. L., Prause, J. L., Danzinger, L. H., Schunke, Vigo, J., Cabieses, F., and Farnsworth, N. R. (2003). Antimycobacterial evaluation of Peruvian plants. Phytomedicine, 10: 528–35.

Graumann, K., and Premstaller, A. (2006). Manufacturing of recombinant therapeutic proteins in microbial systems. Biotechnology Journal (BTJ), 1: 164–186. doi: 10. 1002/biot.200500051.

Gupta, V. K., Kumar, M. M., Bishta, D., and Kaushik, A. (2017). Plants in our combating strategies against *Mycobacterium tuberculosis*: progress made and obstacles met. Pharmaceutical Biology, 55(1): 1536–44.

Harvey, A. (2000). Strategies for discovering drugs from previously unexplored natural products. Drug Discovery Today, 5: 294–300.

Harvey, A. L. (2007). Natural products as a screening resource. Current Opinion in Chemical Biology, 11: 480–4.

Harvey, A. L., Edrada-Ebel, R., and Quinn, R. J. (2015). The re-emergence of natural products for drug discovery in the genomics era. Nature Reviews Drug Discovery, pp. 111–29.

He, G., Yin, Y., Yan, X., and Wang, Y. (2016). Semi-bionic extraction of effective ingredient from fishbone by high intensity pulsed electric fields. Journal of Food Process Engineering, 40: 1–9. doi: 10.1111/jfpe.12392.

Heemskerk, D., Caws, M., Marais, B., and Farrar, J. (2015). Clinical manifestations. Tuberculosis in adults and children. Springer International Publishing AG., pp. 17–20. DOI: 10.1007/978-3-319-19132-4.

Heinrich, M., and Gibbons, S. (2001). Ethnopharmacology in drug discovery: an analysis of its role and potential contribution. Journal of Pharmacy and Pharmacology, 53: 425–32. http://www.tballiance.org/newscenter/view-brief.php?id=726.

Igarashi, M., Takahashi, Y., and Nihon Rinsho. (2011). Journal of Clinical Medicine. 2011 Aug, 69(8): 1482–1488. PMID: 21838051.

Jain, R. C. (1998). Anti tubercular activity of garlic oil. Indian Journal of Pathology Microbiology, 41: 131. Vol 65 | Issue 2.

Janette, V. Pham, Mariamawit, A. Yilma, Adriana Feliz, Murtadha, T. Majid, Nicholas Maffetone, Jorge, R. Walker, Eunji Kim, Hyo Je Cho, Jared M. Reynolds, Myoung Chong Song, Sung Ryeol Park, and Yeo Joon Yoon, A. (2019). Review of the microbial production of bioactive natural products and biologics. Frontiers in Microbiology., https://doi.org/10.3389/fmicb.2019.01404.

Jimenez-Arellanes, A., Meckes, M., Torres, J., and Luna-Herrera, J. (2007). Antimycobacterial triterpenoids from *Lantana hispida* (Verbenaceae). Journal of Ethnopharmacology, 111: 202–5.

Jyoti, M. A., Nam, K. W., Jang, W. S., Kim, Y. H., Kim, S. K., Lee, B. E., and Song, H. Y. (2016). Antimycobacterial activity of methanolic plant extract of *Artemisia capillaris* containing ursolic acid and hydroquinone against *Mycobacterium tuberculosis*. Journal of Infection and Chemotherapy, (4): 200–8. doi: 10.1016/j.jiac.2015.11.014. PMID: 26867795.

Khan, A., and Sakhar, D. A. (2008). Simple whole cell based high throughput screening protocol using M. bovis BCG for inhibitors against dormant and active tubercle bacilli. The Journal of Microbiological Methods, 73(1): 62–8.

Khazir, J., Mir, B. A., Mir, S. A., and Cowan, D. (2013). Natural products as lead compounds in drug discovery. Journal of Asian Natural Products Research. 15(7): 764–88.

Koehn, F. E., and Carter, G. T. (2005). The evolving role of natural products in drug discovery. Nature Reviews Drug Discovery, 4: 206–20.

Kogal, T., Fukuoka, T., Doi, N., Harasaki, T., Inoue, H., Hotoda, h., Kakuta, M., Muramatsu, Y., Yamamura, N., Hoshi, M., and Hirota, T. (2004). Activity of capuramycin analogues against *Mycobacterium tuberculosis*, *Mycobacterium avium*, and *Mycobacterium intracellulare in vitro* and *in vivo*. Journal of Antimicrobial Chemotherapy, 54: 755–760. [PubMed: 15347635].

Koyama, N., Kojima, S., Fukuda, T., Nagamitsu, T., Yasuhara, T., Omura, S., and Tomoda, H. (2010). Structure and total synthesis of fungal calpinactam, a new antimycobacterial agent. Organic Letters, 12(3): 432–5. doi: 10.1021/ol902553z. PMID: 20030344.

Laurenzi, M., Ginsberg, A., and Spigelman, M. (2007). Challenges associated with current and future TB treatment. Infectious Disorders - Drug Targetsrgets, 7: 105–119. [PubMed: 17970222].

Lin, W.-Y., Peng, C.-F., Tsai, I.-L., Chen, J.-J., Cheng, M.-J., and Chen, I.-S. (2005). Antitubercular constituents from the roots of *Engelhardia roxburghiana*. Planta Medica, 71: 171–5.

Luo, X., Pires, D., Aínsa, J. A., Gracia, B., Duarte, N., Mulhovo, S., Anes, E., and Ferreira, M. J. (2013). *Zanthoxylum capense* constituents with antimycobacterial activity against *Mycobacterium tuberculosis in vitro* and *ex vivo* within human macrophages. Journal of Ethnopharmacology. 2013 Mar 7; 146(1): 417–22. doi: 10.1016/j.jep.2013.01.013.PMID: 23337743.

Manikkam, R., Venugopal, G., Subramaniam, B., Ramasamy, B., and Kumar, V. (2014). Bioactive potential of *Actinomycetes* from less explored ecosystems against *Mycobacterium tuberculosis* and other nonmycobacterial pathogens. International Scholarly Research Notices. 1–9. 10.1155/2014/812974.

Manivasagan, P., Venkatesan, J., Sivakumar, K., and Se-Kwon, K. (2013). Marine actinobacterial metabolites: current status and future perspectives. Microbiology Research, 168(6): 311–32.

Manjunatha, U. H., Boshoff, H., Dowd, C. S., Zhang, L., Albert, T. J., Norton, J. E., Daniels, L., Dick, T., Pang, S. S., and Barry, C. E. (2006). Identification of a nitroimidazooxazine-specific protein involved in PA-824 resistance in *Mycobacterium tuberculosis*. Proceedings of the National Academy of Sciences, 103: 431–436. [PubMed: 16387854].

Maria-Teresa Gutierrez-Lugo, and Carole A. Bewley. (2008). Natural products, small molecules, and genetics in tuberculosis drug development. Journal of Medicinal Chemistry, 51(9): 2606–2612. doi:10.1021/jm070719i.

Matsumoto, M., Hashizume, H., Tomishige, T., Kawasaki, M., Tsubouchi, H., Sasaki, H., Shimokawa, Y., and Komatsu, Y. (2006). OPC-67683, a nitro-dihydroimidazooxazole derivative with promising action against tuberculosis *in vitro* and in mice. Public Library of Science Medicine., 3, e466. [PubMed: 17132069].

McCarthy, A. J., and Williams, S. T. (1992). *Actinomycetes* as agents of biodegradation in the environment-a review. Gene, 115: 189–92.

Michael, W. Mullowney, M. W., Hwang, C. H., Newsome, A. G., Wei, X., Tanouye, U., Wan, B., Carlson, S., Barranis, N. J., Óh Ainmhire, E., Chen, W. L., Krishnamoorthy, K., White, J., Blairr, Lee, H., Burdette, J. E., Rathod, P. K., Parish, T., Cho, S., Scott, G. Franzblau, and Murphy, B. T. (2015). Diaza-anthracene antibiotics from a freshwater-derived actinomycete with selective antibacterial activity toward *Mycobacterium tuberculosis*. ACS Infectious Diseases, 1(4): 168–74.

Mincer, T. J., Fenical, W., and Jensen, P. R. (2005). Culture-dependent and culture-independent diversity within the obligate marine actinomycete genus *Salinispora*. Applied Environmental Microbiology, 71(11): 7019–28.

Mitra, P. P. (2012). Drug discovery in tuberculosis: a molecular approach. Indian Journal of Tuberculosis, 59: 194–206.

Mitscher, L. A., and Baker, W. (1998). Tuberculosis: a search for novel therapy starting with natural products. Medicinal Research Reviews, 18: 363–74.

Mohamad, S., Zin, N. M., Wahab, H. A., Ibrahim, P., Sulaiman, S. F. et al. (2011). Antituberculosis potential of some ethnobotanically selected Malaysian plants. Journal of Ethnopharmacology, 133: 1021–6.

Möller Marlo, Kinnear Craig J., Orlova Marianna, Kroon Elouise E., van Helden Paul D., Schurr Erwin, and Hoal Eileen, G. (2018). Genetic resistance to *Mycobacterium tuberculosis* infection and disease. Frontiers in Immunology. DOI = 10.3389/fimmu.2018.02219 ISSN=1664-3224.

Monciardini, P., Sosio, M., Cavaletti, L., Chiocchini, C., and Donadio, S. (2002). New PCR primers for the selective amplification of 16S rDNA from different groups of *Actinomycetes*. Federation of European Microbiological Societies Microbiology Ecology, 42: 419–29.

Newton, S. M., Lau, C., and Wright, C. W. (2000). A review of antimycobacterial natural products. Phytotherapy Research, 14: 303–22.

Nikonenko, B. V., Protopopova, M., Samala, R., Einck, L., and Nacy, C. A. (2007). Drug therapy of experimental tuberculosis (TB): improved outcome by combining SQ109, a new diamine antibiotic, with existing TB drugs. Antimicrobial a Agents Chemotherapy. 51: 1563–1565.

Nuermberger, E. L., Yoshimatsu, T., Tyagi, S., O'Brien, R. J., Vernon, A. N., Chaisson, R. E., Bishai, W. R., and Grosset, J. H. (2004) Moxifloxacin-containing regimen greatly reduces time to culture conversion in murine tuberculosis. American Journal of Respiratory and Critical Care Medicine, 169(3): 421–6. doi: 10.1164/rccm.200310-1380OC. PMID: 14578218.

Okunade, A. L., Elvin-Lewis, M. P., and Lewis, W. H. (2004). Natural antimycobacterial metabolites: current status. Phytochemistry, (8): 1017–32. doi: 10.1016/j.phytochem.2004.02.013. PMID: 15110681.Enfermedades Infecciosas y MicrobiologíaClínica., 29(1): 2–7.

Özdemir, V., and Hekim, N. (2018). Birth of industry 5.0: making sense of big data with artificial intelligence, "the internet of things" and next-generation technology policy. Omics. 22: 65–76.

Parrish, N. M., Houston, T., Jones, P. B., Townsend, C., and Dick, J. D. (2001). *In vitro* activity of a novel antimycobacterial compound, N-octane-sulfonylacetamide, and its effects on lipid and mycolic acid synthesis. Antimicrobial Agents Chemotherapy, 45: 1143–1150. [PubMed: 11257028].

Patridge, E., Gareiss, P., Kinch, M. S., and Hoyer, D. (2016). An analysis of FDA approved drugs: natural products and their derivatives. Drug Discovery Today, 21: 204–207. doi: 10.1016/j.drudis.

Pauli, G. F., Case, R. J., Inui, T., Wang, Y., Cho, S., Fischer, N. H., and Franzblau, S. G. (2005). New perspectives on natural products in TB drug research. Life Science, 78: 485–94.

Payne, D. J., Gwynn, M. N., Holmes, D. J., and Pompliano, D. L. (2007). Drugs for bad bugs: confronting the challenges of antibacterial discovery. Nature Reviews Drug Discovery, 6: 29–40.

Portero, J.-L., and Rubio, M. (2006). *New antituberculisis* therapies. Expert Opinion on Therapeutic Patents, 2007, 17: 617–637.

Protopopova, M., Hanrahan, C., Nikonenko, B., Samala, R., Chen, P., Gearhart, J., Einck, L., and Nacy, C. A. (2005). Identification of a new antitubercular drug candidate, SQ109, from a combinatorial library of 1,2- ethylenediamines. J. Antimicrobial. Chemotherapy 56: 968–974. [PubMed: 16172107].

Quana, D., Nagalingama, G., Paynec, G. R., and Triccas, J.A. (2017). New tuberculosis drug leads from naturally occurring compounds. International Journal of Infectious Diseases, 56: 212–220.

Rajab, M. S., Cantrell, C. L., Franzblau, S. G., and Fischer, N. H. (1998). Antimycobacterial activity of (E)-phytol and derivatives: a preliminary structure-activity study. Planta Medica, 64: 2–4.

Rath, C. M., Janto, B., Earl, J., Ahmed, A., Hu, F. Z., Hiller, L., Dahlgren, M., Kreft, R., Yu, F., Wolff, J. J., Kweon, H. K., Christiansen, M. A., Håkansson, K., Williams, R. M., Ehrlich, G. D., and Sherman, D. H. (2011). Meta-omic characterization of the marine invertebrate microbial consortium that produces the chemotherapeutic natural product ET-743. American Chemical Society Chemical Biology, 6(11): 1244–56. doi: 10.1021/cb200244t. PMID: 21875091; PMCID: PMC3220770.R2011.

Revers, L., and Furczon, E. (2010). An introduction to biologics and biosimilars. Part II: subsequent entry biologics: biosame or biodifferent? Can. Pharmaceutical. Journal, 143: 184–191. doi: 10.3821/1913-701x-143.4.184.

Riedlinger, J., Reicke, A., Zähner, H., Krismer, B., Bull, A. T., Maldonado, L. A., Ward, A. C., Goodfellow, M., Bister, B., Bischoff, D., Süssmuth, R. D., and Fiedler, H. P. (2004). Abyssomicins, inhibitors of the para-aminobenzoic acid pathway produced by the marine Verrucosispora strain AB-18-032. Journal Antibiotics (Tokyo), (4): 271–9. doi: 10.7164/antibiotics.57.271. PMID: 15217192.

Rothstein, D. M., Shalish, C., Murphy, C. K., Sternlicht, A., and Campbell, L. A. (2006). Development potential of rifalazil and other benzoxazinorifamycins. Expert Opinion on Investigational Drugs Volume 15,2006-Issues 6 15,603–623.

Rustomjee, R., Lienhardt, C., Kanyok, T., Davies, G. R., Levin, J., Mthiyane, T., Reddy, C., Sturm, A. W., Sirgel, F. A., Allen, J., Coleman, D. J., Fourie, B., and Mitchison, D. A. (2008). Gatifloxacin for TB (OFLOTUB) study team. A Phase II study of the sterilising activities of ofloxacin, gatifloxacin and moxifloxacin in pulmonary tuberculosis. International Journal of Tuberculosis and Lung Disease, (2): 128–38. PMID: 18230244.

Sánchez, J. G. B., and Kouznetsov, V. V. (2010). Antimycobacterial susceptibility testing methods for natural products research. Brazilian Journal of Microbiolog, 41: 270–7.

Sanchez-Garcia, L., Martin, L., Mangues, R., Ferrer-Miralles, N., Vazquez, E., and Villaverde, A. (2015). Recombinant pharmaceuticals from microbial cells: Microbial Cell Factories, 15: 33. doi: 10.1186/s12934-016-0437-3.

Sanford, C. A., Pottinger, P. S., and Jong, E. C. (2017). Tuberculosis in travelers and immigrants. The travel and tropical medicine manual. 5th edn. Philadelphia: Elsevier Inc;. pp. 356–9.

Sieniawska, E., Swatko-Ossor, M., Sawicki, R., SkalickaWoźniak, K., and Ginalska, G. (2017). Natural terpenes influence the activity of antibiotics against isolated *Mycobacterium tuberculosis*. Medical Principal and Practices, 26: 108–12.

Song, M. C., Kim, E. J., Kim, E., Rathwell, K., Nam, S. J., and Yoon, Y. J. (2014). Microbial biosynthesis of medicinally important plant secondary metabolites. Natural Product Reports, 31: 1497–1509. doi: 10.1039/c4np00057a.

Sood, R., Bhadauriya, R., Rao, M., Gautam, R., Malhotra, S., Barman, T. K., Upadhyay, D. J., and Rattan, A. (2006). Antimycobacterial activities of oxazolidinones: a review. Infectious Disorders Drug Targets, 6: 343–354. [PubMed: 17168800].

Stach, J. E. M., Maldonado, L. A., Ward, A. C., Goodfellow, M., and Bull, A. T. (2003). New primers for the class actinobacteria: application to marine and terrestrial environments. Environmental Microbiology, 5(10): 828–41.

Stover, C. K., Warrener, P., Van Devanter, D. R., Sherman, D. R., Arain, T. M., Langhorne, M. H., Anderson, S. W., Towell, J. A., Yuan, Y., McMurray, D. N., Kreiswirth, B. N., Barry, C. E., and Baker, W. R. (2000). A small-molecule nitroimidazopyran drug candidate for the treatment of tuberculosis. Nature 405: 962–966. [PubMed: 10879539] (53).

Sukuru, S. C., Jenkins, J.L., Beckwith, R.E., Scheiber, J., Bender, A., Mikhailov, D., Davies, J. W., and Glick, M. (2009). Plant-based diversity selection based on empirical HTS data to enhance the number of hits and their chemical diversity. Journal of Biomolecular Screening, (6): 690–9. doi: 10.1177/1087057109335678. PMID: 19531667.

Sun, W., Zhang, F., He, L., Karthik, L., and Li, Z. (2010). *Actinomycetes* from the South China Sea sponges: isolation, diversity and potential for aromatic polyketides discovery. Frontiers Microbiology, 6:1048.

Thangadurai, D., Viswanathan, M. B., and Ramesh, N. (2002). Indigofera bietone, a novel abietane diterpenoid from *Indigofera longeracemosa* with potential antituberculous and antibacterial activity. Pharmazie, 57: 714–15.

Thomford, N. E., Senthebane, D. A., Rowe, A., Munro, D., Seele, P., Maroyi, A., and Dzobo, K. (2018). Natural products for drug discovery in the 21st century: innovations for novel drug discovery. International Journal of Molecular Sciences, 19(6): 1578. doi: 10.3390/ijms19061578. PMID: 29799486; PMCID: PMC6032166.

Tripathi, R. P., Tewari, N., Dwivedi, N., and Tiwari, V. K. (2005). Fighting tuberculosis: an old disease with new challenges. Medicinal Research Reviews, 25: 93–131.

Ulubelen, A., Topcu, G., and Johansson, C. B. (1997). Norditerpenoids and diterpenoids from *Salvia multicaulis* with antituberculosis activity. Journal of Natural Products, 60: 1275–80.

Venketaraman, V., Kaushal, D., and Saviola, B. (2015). *Mycobacterium tuberculosis*. Journal of Immunology Research; 857598: 1–2. doi: 10.1155/ 2015/857598.

Ventura, M., Canchaya, C., Tauch, A., Chandra, G., Fitzgerald, G. F., Chater, K. F., and van Sinderen, D. (2007). Genomics of Actinobacteria: tracing the evolutionary history of an ancient phylum. Microbiology and Molecular Biology Reviews, 71(3): 495–548. doi: 10.1128/MMBR.00005-07. PMID: 17804669; PMCID: PMC2168647.

Wächter, G. A., Franzblau, S. G., Montenegro, G., Hoffmann, J. J., Maiese, W. M., and Timmermann, B. N. (2001). Inhibition of *Mycobacterium tuberculosis* growth by saringosterol from *Lessonia nigrescens*. Journal of Natural Products, 64(11): 1463–4. doi: 10.1021/np010101q. PMID: 11720535.

Wachter, G. A., Franzblau, S. G., Montenegro, G., Suarez, E., Fortunato, R. H., Saavedra, E., and Timmermann, B. N. (1998). A new antitubercular mulinane diterpenoid from *Azorella madreporica*. Journal of Natural Products, 61: 965–8.

Wang, C., Wang, J., Huang, Y., Chen, H., Li, Y., Zhong, L., Chen, Y., Chen, S., Wang, J., Kang, J., Peng, Y., Yang, B., Lin, Y., She, Z., and Lai, X. (2013). Anti-mycobacterial activity of marine fungus-derived 4-deoxybostrycin and nigrosporin. Molecules, 18(2): 1728–40. doi: 10.3390/molecules18021728. PMID: 23434859; PMCID: PMC6269944.

Wang, W. F., Lu, M. J., Cheng, T. R., Tang, Y. C., Teng, Y. C., Hwa, T. Y., Chen, Y. H., Li, M. Y., Wu, M. H., Chuang, P. C., Jou, R., Wong, C. H., and Li, W. H. (2020). Genomic analysis of *Mycobacterium tuberculosis* isolates and construction of a beijing lineage reference genome. Genome Biology and Evolution, 1; 12(2): 3890–3905. doi: 10.1093/gbe/evaa009. PMID: 31971587; PMCID: PMC7058165.

Wawrik, B., Kerkhof, L., Zylstra, G. J., and Kukor, J. J. (2005). Identification of unique type II polyketide synthase genes in soil. Applied and Environmental Microbiology Journal. 71(5): 2232–8.

Williams, K. N., Stover, C. K., Zhu, T., Tasneen, R., Tyagi, S., Grosset, J. H., and Neurmberg (2009). Promising antituberculosis activity of the oxazolidinone PNU-100480 relative to that of linezolid in a murine model. Antimicrobial Agents and Chemotherapy. 53: 1314–1319.

Xu, Z. Q., Barrow, W. W., Suling, W. J., Westbrook, L., Barrow, E., Lin, Y. M., and Flavin, M. T. (2004). Anti-HIV natural product (+)-calanolide A is active against both drug-susceptible and drug-resistant strains of *Mycobacterium tuberculosis*. Bioorganic & Medicinal Chemistry, 12: 1199–207.

Zahner, H., and Fiedler, H. P. (1995). Fifty years of antimicrobials: past perspectives and future trends. SGM Symposium 53. Cambridge: Cambridge University Press. pp. 67–85.

Zhang, Y., Post-Martens, K., and Denkin, S. (2006). New drug candidates and therapeutic targets for tuberculosis therapy. Drug Discovery Today, 11: 21–27. [PubMed: 16478687].

Zhilin Gan., Zheng Liang., Xiaosong Chen., XinWen., Yuxiao Wang., MoLi., and Yuanying Ni. (2016). Separation and preparation of 6-gingerol from molecular distillation residue of Yunnan ginger rhizomes by high-speed counter-current chromatography and the antioxidant activity of ginger oils *in vitro*. Journal of Chromatography B Volume 1011, Pages 99-107https://doi.org/10.1016/j.jchromb.2015.12.051.

Zofou, D., Ntie-Kang, F., Sippl, W., and Efange, S. M. (2013). Bioactive natural products derived from the Central African flora against neglected tropical diseases and HIV. Natural Product Reports, 30: 1098–120.

CHAPTER 7

Effects of Antimicrobial Peptides on Bacteria and Viruses

Tânia D. Tavares,[1] *Marta O. Teixeira,* [1] *Marta A. Teixeira,*[1] *Joana C. Antunes*[1,2] and *Helena P. Felgueiras*[1,*]

Introduction

For many years, antibiotics have been the most widely used antimicrobial agents to fight infections. However, their excessive consumption has led to an alarmingly high development of resistance by microbial pathogens, raising a serious global, public-health problem (Felgueiras 2021, Lewies et al. 2019, Rončević et al. 2019). Hence, the growing search for alternatives to these agents. In recent years, antimicrobial peptides (AMPs) have been the focus of great interest since they are the most widespread and evolutionarily conserved components of the innate immune system, acting as a primary line of host defense against microbial infections (Magana et al. 2020). They can be of natural origin obtained from all organisms ranging from bacteria to plants, vertebrates, and invertebrates or of synthetic origin and synthesized as analogs of those naturally occurring (Felgueiras and Amorim 2017, Kumar et al. 2018, Rima et al. 2021). AMPs exhibit several advantages over conventional antibiotics, including lower risk of developing microbial resistance and rapid killing (antibiotics are slower in action than AMPs which allow bacteria, for instance, to

[1] Centre for Textile Science and Technology (2C2T), University of Minho, Campus de Azurém, 4800-058 Guimarães, Portugal.
[2] Fibrenamics, Institute of Innovation on Fiber-based Materials and Composites, University of Minho, Guimarães, 4800-058, Portugal.
* Corresponding author: helena.felgueiras@2c2t.uminho.pt

complete their replication cycle, passing the developed resisting tools to other cells), and lower propensity to induce host toxicity (the human body is also a producer of AMPs, being an important effector in the innate immune response) (Annunziato and Costantino 2020, Cabrele et al. 2014). These peptides also possess a broad spectrum of activity against a great variety of microorganisms, including Gram-positive and Gram-negative bacteria, viruses, fungi, and parasites. In addition, AMPs are also involved in a diversity of biological functions, such as immune regulation, angiogenesis, tissue regeneration, etc. (Moravej et al. 2018, Rončević et al. 2019). One of the crucial reasons that explain the low development of microbes' resistance to AMPs is their biological mechanisms of action and their adaptation to different pathogens. The structural variety that characterizes these peptides contributes to multiple bioactive roles against microbial, ranging from membrane bonding to permeabilization and the interaction with an array of intracellular target molecules. On the contrary, antibiotics tend to target specific molecular receptors of pathogens, often being exclusive to those (Divyashree et al. 2020, Huan et al. 2020, Mahlapuu et al. 2020).

In this chapter, different sources and types of AMPs are briefly introduced, as well as their inherent mechanisms of action against bacteria and viruses. It should be noted, however, that the exact nature of these mechanisms is not yet completely understood, particularly regarding viruses.

Antimicrobial Peptides

AMPs, also designated as host defense peptides (HDPs), are an integral part of non-specific defense systems and the innate immunity of all organisms against exogenous pathogens (De Mandal et al. 2021, Lima et al. 2021). Currently, the AMP database (Data Repository of Antimicrobial Peptides—DRAMP, http://dramp.cpu-bioinfor.org/) accounts for 5,891 natural and synthetic AMPs, which comprises 2,519 from animals (including humans), 824 from plants, 431 from bacteria, 7 from protozoal, 6 from fungal, and 4 from archaea. These peptides are small molecules typically formed of 12 to 50 amino acids and have a low molecular weight, which is less than 10 kDa. Despite the high variability of their amino acid sequences, most AMPs share a cationic character with a net charge of +2 to +9 (Felgueiras 2021, Felgueiras and Amorim 2017, Hassan et al. 2021, Tornesello et al. 2020). One of the most important and widely explored features of AMPs is their amphipathic structure due to the positioning of their amino acid residues. On the hydrophilic side, the arginine, lysine, and histidine residues prevail, while on the hydrophobic opposite side, the tryptophan, phenylalanine, isoleucine, and leucine residues can be found

(He et al. 2020, Pandit et al. 2021). Peptides are frequently organized into four main classes of amphipathic structures; linear α-helical peptides (e.g., human cathelicidin LL-37, cecropins, and magainins), β-sheet peptides (e.g., α-defensin peptides), peptides containing both α and β structures (e.g., vertebrate β-defensin, invertebrate big defensin, and *cys* defensin that come from invertebrates and plants), and peptides with an extended linear structure in which both α and β elements are absent and glycine, proline, tryptophan, or histidine amino acids are present (e.g., bovine indolicidin, human histatins, and hymenoptaecins from various insects) (Liu et al. 2018, Moretta et al. 2021). The properties of α-helical peptides are influenced by the amino acid sequence, helix length, orientation of hydrophobic, and charged residues. β-sheet peptides are characterized by their rigid structure, guaranteed by one or more disulfide bonds which makes them more stable in aqueous solution than the α-helical (Koga et al. 2021). More recently, new classes of AMPs have been proposed in response to an increase of studies about cyclic and disulfide-rich AMPs (e.g., cyclotides and θ-defensins) and other more complex conformations, including lasso-peptides and thioether-bridged structures (Bakare et al. 2020, Koehbach and Craik 2019, Moretta et al. 2021).

Sources and Types

Natural AMPs are preserved in the genome of all life forms, that is from bacteria to mammals and display extraordinary structural and functional variability (Gan et al. 2021, Zhang et al. 2021). These peptides are produced either by ribosomal translation of messenger RNA, genetically encoded within most species, or by non-ribosomal peptide synthesis exclusively synthetized by bacteria and fungi (Mahlapuu et al. 2016, Rima et al. 2021). Ribosomally synthesized AMPs are derived from relatively short precursor peptide sequences and are translated as inactive pro-peptides, requiring at least one proteolytic cleavage for their activation such as the defensins and the cathelicidins (Mahlapuu et al. 2016). Recently, they have been recognized for their therapeutic potential because of their intervention in innate immunity (Martínez-Núñez 2016). On another hand, the non-ribosomally synthesized AMPs are produced by large assembly-line enzymes, named non-ribosomal peptide synthetases. They have been known for various decades and consist of many front-line antibiotics used in clinical (e.g., polymyxins, gramicidins, bacitracins, and glycopeptides such as vancomycin or teicoplanin) (Koehbach and Craik 2019, Martínez-Núñez 2016). Non-ribosomal peptides possess two main structural characteristics that distinguish them from ribosomally synthesized AMPs: (1) their primary structure is more regularly cyclic, branched, or polycyclic instead of linear and (2) their biodiversity of monomers containing amino acids, such as imino acids or ornithine, is superior (Awan et al. 2017, Caboche et al. 2010).

In mammals, the main identified families of AMPs are the defensins and the cathelicidins, predominantly found in secretions from epithelial cells, covering mucosal surfaces and skin and on granules of neutrophils (Moretta et al. 2021). In its turn, magainin is the most popular and well-studied family of amphibian AMPs, which can be extracted from skin secretions of frogs from the genera *Hymenochirus, Silurana, Xenopus,* and *Pseudhymenochirus* (McMillan and Coombs 2020). In insects, the most famous family of AMPs is the cecropin and can be isolated from *Drosophila,* bees, and guppy silkworms (Peng et al. 2019). Plants are renowned as a great source of AMPs and are found and extracted from their leaves, stems, and seeds, and each is classified into several groups including thionins and snakins (Höng et al. 2021). Various AMPs have also been obtained from microorganisms, such as bacteria and fungi; nisin and gramicidin are synthetized from *Lactococcus lactis, Bacillus brevis,* and *Bacillus subtilis* (Yazici et al. 2018). In the following sub-chapters, a detailed characterization of the main classes of AMPs derived from mammals, plants, insects, and microorganisms will be provided.

Defensins

Defensins are cysteine-rich AMPs described as ancient innate immunity molecules, which can be found in animals, plants, and fungi as well as in mollusks, cnidarians, annelids, and nematodes (Gerdol et al. 2020, Stambuk et al. 2021). Structurally, they are composed of a triple-stranded antiparallel β-sheet motif, usually packed against an α-helix which is stabilized by three to six disulfide bonds between the six cysteine residues that are commonly present (Solanki et al. 2021). Defensins have been classified as *cis-* and *trans*-defensins (Omidvar et al. 2021, Shafee et al. 2016). The *cis*-defensins are the largest superfamily, containing sequences that possess two parallel disulfide bonds that link the final β-strand to the α-helix. Examples of *cis*-defensins include the cysteine-stabilized αβ proteins from fungi, invertebrates, and plants. Regarding *trans*-defensins, these possess two disulfide bonds oriented in opposite directions from the final β-strand, connecting them to different secondary structure elements. This superfamily includes the vertebrate defensins as well as the invertebrate big defensins, the lobster β-defensin-like peptides, and several cnidarian defensin-like sequences (Santana et al. 2021, Shafee et al. 2017).

In mammals, defensins are composed of 18 to 45 amino acid residues and are divided into three subgroups: α, β, and θ. This subdivision is related to the spatial arrangement and pairing of cysteine residues and hence by the posterior generation of disulfide bonds (Gao et al. 2021).

From an evolutionary perspective, α-defensins are restricted to mammals and seem to have evolved from ancestral β-defensin genes. Interestingly, some mammalian species (e.g., cattle) have lost this sub-family during evolution, although the reason for this loss is not understood (Santana et al. 2021). In mammalian, α-defensins (despite the six characteristic cysteine residues), arginine, glutamic acid, and glycine residues are also frequently found. Arginine and glutamic acid have been extremely important in forming a salt bridge critical for the *in vivo* stability of defensins, while the glycine amino acid has been essential for specific defensin functions and to ensure its folding abilities (Gao et al. 2021). In contrast with α-defensins, β-defensins have been identified in a large variety of groups, including fish, amphibians, reptiles, birds, and mammals. There are many that defend that β-defensins evolved from the C-terminal domain of the invertebrate big defensins because of their similar cysteine disulfide motif, sites of expression as well as comparable biological functions (Tassanakajon et al. 2015). More than 30 different β-defensin genes are present within the human genome. Yet, research has focused mostly on human β-defensin-1 to 4, which are expressed by epithelial cells (Liu et al. 2021). The θ-defensin is the least explored and for that they are considered the youngest defensins, being only expressed in primates. They are produced from mutated α-defensin genes that contain a premature stop codon in the defensin domain. The most remarkable feature of θ-defensins is their organized pair ladder pattern of six internal cysteine residues that cross-connect two antiparallel β-strands, forming a circular octa-deca-peptide that is stabilized structurally by three disulfide bonds (Montero-Alejo et al. 2016, Santana et al. 2021).

Cathelicidins

Cathelicidins are an important family of AMPs, found mostly in vertebrates. The name of this family is based on its limited homology to cathelin, a protein acronym for cathepsin-L-inhibitor and a member of the cystatin family of cysteine protease inhibitors (Wang et al. 2020). Generally, cathelicidins consist of peptides with a highly conserved N-terminal region, containing a signal peptide and a domain with homology to the cathelin protein and a C-terminal region with wide variability in length and sequence, which is where the active portion of the peptide is located (Luo et al. 2021, Oliveira et al. 2020, Scheenstra et al. 2019). The release of the active portion of the peptide depends upon proteolytic processing which occurs at a site located between the cathelin and the antimicrobial peptide domains. For most bovine and porcine cathelicidins, this separation is accomplished by elastase and a serine protease present in neutrophils, which is specific to neutral side chains, such as glycine, alanine, serine, and valine (Shinnar et al. 2003).

The hCAP-18 (human cationic protein 18 kDa) is the only cathelicidin identified in humans. Here, however, elastase is not effective in cleaving the active peptide since threonine is positioned at the separation site. Instead, neutrophil proteinase 3 and epithelial kallikreins (i.e., kallikrein gene 5) are the ones responsible for such a task, liberating the LL-37, the active peptide, from hCAP-18. LL-37 expression occurs in neutrophils, mast cells, and epithelial cells. The designation of LL-37 arises from the fact that the peptide contains 37 amino acids and that its chain is initiated with two leucines (Engelberg and Landau 2020). LL-37 exhibits a broad spectrum of antimicrobial activity, being particularly effective against bacteria and viruses (Felgueiras et al. 2020, Kulkarni et al. 2021). Still, various studies have shown that human immune-mediated inflammatory diseases, like rosacea or psoriasis, are promoted by the presence of excess LL-37 within the skin (Takahashi et al. 2018).

Cecropins

The cecropins family is a diverse class of AMPs naturally present in living organisms, namely in insects (primary source), bacteria *Helicobacter pylori*, tunicates, ascarid nematodes and various other mammals (Mandel et al. 2021, Peng et al. 2019). This family is subdivided into five sub-types, the cecropins A to E (Peng et al. 2019). Cecropin B, secreted by silkworms, is one of the most antimicrobial peptides in the cecropin family (Timur and Gürsoy 2021). In general, cecropins are small peptides of 29 to 42 amino acids with an α-helical structure. They present strongly basic, amphipathic N-terminal regions and neutral-charged, hydrophobic C-terminal parts (Mandel et al. 2021, Ziaja et al. 2020).

Thionins

Thionins are plant-specific AMPs, composed of 45 to 48 amino acid residues rich in lysine, arginine, and cysteine (Barashkova and Rogozhin 2020, Li et al. 2021). In general, this family possess a three-dimensional structure consisting of antiparallel α-helices and a double-stranded β-sheet with three to four disulfide bridges. Thionins have been characterized according to the number of cysteine residues and their organization in the peptide sequence. Thionins class 1 contains eight cysteine residues (e.g., HTH1, α-hordothionin, Barley leaf-thionin DB4, and Pyrularia thionin); class 2 comprises six cysteine residues (e.g., viscotoxin A3; crambin 2, and Arabidopsis thaliana thionin 2.1); class 3 is composed of six cysteine residues without the second and eighth cysteine residue in the default positions three and four (e.g., pTTH20 Neutral wheat thionin); and finally, class 4 encompasses eight cysteine residues, without the second cysteine and a new cysteine residue between the fourth and fifth position (e.g., Tulipa gesneriana thionin) (Höng et al. 2021, Parthasarathy et al. 2021).

Bacteriocins

Bacteriocins are an abundant and heterogeneous class of AMPs made of 20 to 60 amino acids, produced predominantly by bacteria to eradicate other bacterial species in a competitive environment, particularly species phylogenetically close to the producing strain. Bacteriocins have also been reported in some archaea. Based on their general mode of production, bacteriocins can be classified into two main classes (Kranjec et al. 2021, Soltani et al. 2020). Class I includes peptides subjected to post-translational modifications ensured by enzymes encoded in the bacteriocin gene cluster. This class encompasses all the peptides that undergo enzymatic modification during biosynthesis. Some examples are the lanthipeptides, circular peptides, sactipeptides, and glycocins from Gram-positive bacteria, lasso peptides, and linear azole(ine)-containing peptides as well as from both Gram-negative and Gram-positive bacteria and siderophore peptides and nucleotide peptides, which are exclusive from Gram-negative bacteria. Additionally, linaridins and thiopeptides produced by *Actinobacteria* and cyanobactins obtained from various cyanobacteria belong to this class (El-Gendy et al. 2021, Heilbronner et al. 2021, Kaur et al. 2013). Class II integrates bacteriocins that do not undergo post-translational modifications. Instead, they are synthesized with or without disulfide bridges as prebacteriocin peptides with an N-terminal leader, which is removed during secretion. This confers these peptides with high stability against proteolytic enzyme attacks, extreme pH alterations, and significant temperature rises (Chen et al. 2021, Ongey et al. 2018). Some examples of Class II bacteriocins are the pediocin PA-1 from *Pediococcus acidilactici;* the bacteroidetocins produced by *Bacteroidetes* species and the enterobacterial microcins S, L, and V (with disulfide bridges) or the epidermicin NI01, aureocin A53, and capidermicin bacteriocins all produced by *Staphylococcus* spp. (lacking disulfide bridges) (Kumariya et al. 2019). Many researchers also consider a third class (Class III), which encompasses unmodified peptides with a molecular weight greater than 10 kDa that show bacteriolytic or non-lytic mechanisms of action (e.g., carnobacteriocin, piscicolin, curvaticin, ruminococcin, subtilin, lichenicidin, cinnamycin, actagardine, lacticin, divergicin, mutacin, mundticin, duramycin, sakacin, mesenterocin, curvacin, enterocin, leucosin, enterocin, lysostaphin, brevinine, and columbicin) (Hassan et al. 2012, Kranjec et al. 2021).

Peptaibiotics

Peptaibiotics are a group of linear and more rarely cyclic peptides composed of 4 to 21 amino acid residues, produced mainly by fungi (Das et al. 2018). They contain in their sequence a high content of α,α-dialkylated amino acids [such as α-aminoisobutyric acid (Aib)], isovaline, non-proteinogenic amino acids and/or lipoamino acids, and acylated N-terminus. The linear peptides have a C-terminal residue that entails a free or acetylated amide-bonded β-amino alcohol. Because of its heterogenous nature, the C-terminus might also include free amino acid, amine, amide, 2,5-diketopiperazine, or sugar alcohol (Liu et al. 2020, Mukherjee et al. 2011). The presence of these unusual amino acids is a result of the non-ribosomal synthesis that peptaibiotics undergo via large multifunctional enzymes, the peptide synthetases, which by the multiple carrier thiotemplate mechanism assemble these molecules from a notable range of precursors (Mukherjee et al. 2011, Szekeres et al. 2005). More than 1,000 different peptaibiotics have been identified and categorized based on their chemical structure in peptaibols, lipopeptaibols, lipoaminopeptides, and cyclic peptaibiotics. Peptaibols constitute the largest and the most studied group. *Trichoderma* species are the highest sources of peptaibols, and their Aib-containing peptides are characterized by acylated N-terminus and C-terminus β-amino alcohol, such as phenylalaninol, leucinol, valinol, tryptophanol, and isoleucinol which are the most prevalent. Peptaibols have shown effective action against several Gram-positive bacteria, fungi, and viruses (Ray et al. 2017).

Action Against Bacteria: Mechanisms and Effects

Antibacterial activity is one of the most basic functions of AMPs, and their mechanisms of action have been extensively researched. The difference in cell wall composition of Gram-positive and Gram-negative bacteria can affect the AMPs' mode of action. Some AMPs have shown selectivity toward specific classes of bacteria (e.g., bacteriocins), while others exhibit activity against both (e.g., LL-37, indolicidin, magainins, melittin, and cecropins) (Wang et al. 2019). The architecture and molecular components of the peripheral cell wall differ between these two classes of bacteria, particularly in what concerns membrane and cell wall structure and disposition (Bahar and Ren 2013, Felgueiras 2021, Tavares et al. 2020). Gram-negative bacteria are more complex, containing two distinct lipid membranes, the cytoplasmic cell membrane, and the outer membrane with a thin layer of peptidoglycans and a periplasm space in between.

The peptidoglycans are formed of linear polysaccharide chains cross-linked by short peptides that generate a three-dimensional rigid structure that protects bacteria; the periplasmic space contains enzymes capable of degrading molecules introduced from the extracellular medium. The outer membrane acts as an additional compound-selective barrier protecting the cell from the environment, containing lipopolysaccharides (LPS) which is a large glycolipid complex responsible for low membrane permeability (Figure 1a). Conversely, in Gram-positive bacteria, the cytoplasm is only enveloped by a lipid bilayer membrane, the cytoplasmic membrane, attached to a much thicker layer of peptidoglycans and a periplasmic space. This layer contains glycopolymers, such as lipoteichoic acids, which act on membrane function and stability and in intercellular interactions (Figure 1b) (Bahar and Ren 2013, Drayton et al. 2021).

The cationic AMPs possess a strong affinity toward the anionic components present in the bacterial membrane, such as the phospholipids in both Gram-positive and Gram-negative bacteria, the LPS present in Gram-negative bacteria, and the lipoteichoic acids found in Gram-positive bacteria. In Gram-negative bacteria, the initial action involves the AMPs binding to the lipid A core of LPS, replacing LPS-associated divalent cations, such as Ca^{2+} and Mg^{2+}. In this way, peptides destabilize the outer membrane and can interact with the cytoplasmic membrane. For Gram-positive bacteria, AMPs cross capsular polysaccharides and other components of the cell wall before interacting with the cytoplasmic membrane (Liu et al. 2021, Moravej et al. 2018). These interactions can cause membrane disruption leading to cell lysis or can generate pores that penetrate the membrane and interact with intracellular molecules, including nucleic acids, proteins, enzymes, and ribosomes interfering with their functions (Bahar and Ren 2013). Thus, the AMPs' mode of action against bacteria can be divided into two different mechanisms, which are

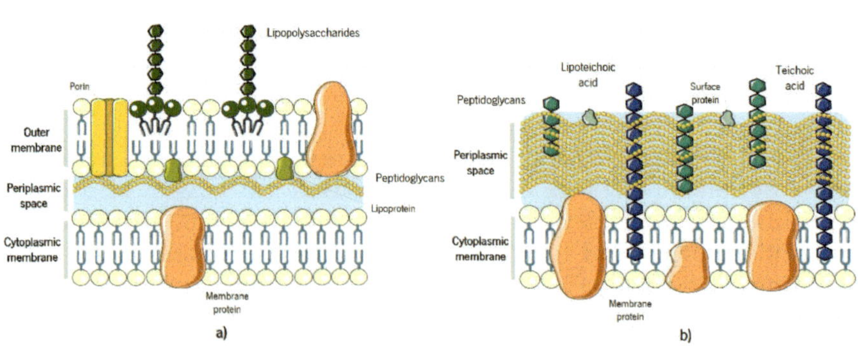

Figure 1. Schematic presentation of the cell membranes of (a) Gram-negative bacteria and (b) Gram-positive bacteria. The figure was produced using elements from Servier Medical Art. Adapted from (Huan et al. 2020) with CC BY 4.0 permission.

the membrane-targeting mechanism that impairs the structural integrity of the cell membrane and the non-membrane targeting mechanism that inhibits the biosynthesis of the cell wall, nucleic acids, essential enzymes, and other functional proteins (Moravej et al. 2018). Some AMPs can act using both mechanisms and sometimes switch from one to the other, depending on the peptide concentration, the membrane characteristics of a particular bacterial species, or its growth phase (Annunziato and Costantino 2020, Rončević et al. 2019). In addition to these two mechanisms of direct killing, AMPs can also act through immune modulation by means of pro- and anti-inflammatory effects instigated via their binding abilities with chemokine receptors (many AMPs are potent chemoattractants for several immune cell types, including neutrophils, monocytes, dendritic cells, and T cells) or by triggering differentiation of specific cell types (i.e., macrophages) and controlling their functions (i.e., AMPs can regulate the apoptosis of neutrophils, for instance, by prolonging their life after activation) (Kumar et al. 2018). The rapid killing, which can occur in seconds after the initial contact with the cell membrane, and the ability of AMPs to act at multiple targets within the microbial cell reduces the risk of bacteria developing resistance (Zhang et al. 2021).

Membrane-Targeting Mechanism

The cell membrane plays a crucial role in bacterial survival. In fact, one-third of its proteins are membrane-associated and related to multiple functions such as nutrient transportation, respiration, proton motive force, adenosine triphosphate (ATP) generation, and intercellular communication. A dysfunction in the membrane, such as leakage of ions and intracellular metabolites, leads to depolarization, impairment of membrane functions (i.e., osmotic regulation), and loss of pH gradient and respiration, which eventually causes direct or indirect cell death (Bahar and Ren 2013, Liu et al. 2021). Thus, the bacterial membrane is the most important target of AMPs and even though intracellular targets are involved, initial interaction with it is imperative. This interaction can be receptor-mediated or non-receptor-mediated (Kumar et al. 2018). The receptor-mediated pathway mostly includes bacteriocins, such as nisin, which act against Gram-positive bacteria by specifically binding to lipid II, a molecule required for cell wall biosynthesis, blocking peptidoglycan biosynthesis and leading to pore formation that results in cell lysis (Breukink et al. 1999). However, most vertebrate and invertebrate AMPs establish the initial interaction with general targets present along the cell surface without any specific receptor (Ryu et al. 2021). The peptide-membrane interaction can be affected by several factors including physicochemical properties of AMPs, such as cationicity and amphipathicity, which are essential for the initial attraction and other factors like peptide/lipid ratio and secondary structure of AMPs that are important for peptide permeation (Moravej et al. 2018). The primary contact with target cells occurs through electrostatic and/or hydrophobic interactions (Figure 2a,b). Electrostatic binding of hydrophilic/positively charged regions of peptides to negatively charged phospholipid head groups in the outer leaflet of the membrane, which is the main driving force in peptide-membrane interaction since bacterial cell membranes are characterized by a high content of anionic lipids highly attractive to cationic AMPs (Bahar and Ren 2013). In contrast, mammalian cell membranes possess mainly zwitterionic phospholipids in the outer leaflet of the membrane and negatively charged phospholipids in their inner leaflet. Thus, this affinity is weaker and the difference between bacterial and mammalian cells imparts the selectivity characteristic of the AMPs (Erdem Büyükkiraz and Kesmen 2021). The hydrophobic regions of AMPs interact with the fatty acid tails of bacterial membrane lipids, which aid in the peptide insertion into the cell membrane (Lima et al. 2021). The percentage of hydrophobic residues within a peptide sequence is proportional to its antimicrobial performance. In general, moderately hydrophobic peptides display optimal antimicrobial activity, while highly hydrophobic peptides are less effective, exhibiting a stronger hemolytic profile (Erdem Büyükkiraz and Kesmen 2021). The AMPs antibacterial effect is also related to the peptide/lipid ratio. At a low concentration, AMPs are in an orientation parallel to the lipid membrane surface; with an increase in concentration, peptides aggregate and acquire a perpendicular orientation and reach a critical concentration responsible for inducing lipids disruption and membrane structure modifications (Figure 2c). Some of these changes are electrostatics modifications, membrane

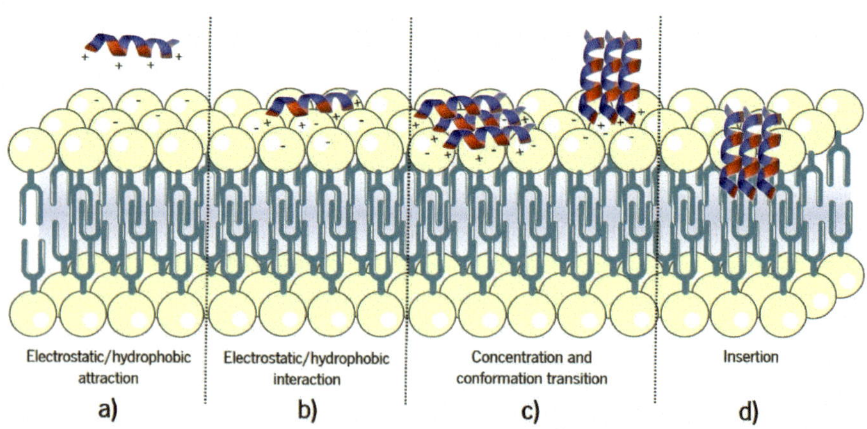

Figure 2. Interactions between AMPs and bacterial membrane. The red color represents the hydrophilic portions of AMPs, while the blue color represents the hydrophobic portions. The figure was produced using elements from Servier Medical Art. Adapted from (Kumar et al. 2018) with CC BY 4.0 permission.

thinning, alterations in membrane curvature, pore formation, and disturbance of the bilayer permeability barrier. These changes allow for a reorientation of peptides in the membrane and their translocation to the cytoplasm (Figure 2d) (Lima et al. 2021, Liu et al. 2021). Once bound to the membrane, AMPs can change their conformations, which is considered one of their most important traits. Most α-helical AMPs are unstructured in aqueous solution; however, in contact with the lipid membrane, they can form highly structured amphipathic conformations that facilitate their interaction. On the other hand, β-sheet AMPs are more structured in solution due to the presence of disulfide bridges and thus leading to smaller conformational changes upon interaction with lipid membranes (Gong et al. 2021, Zhang et al. 2021).

Various models have been proposed to explain the membrane disruption process, that is membrane permeability through pore formation (barrel-stave and toroidal-pore models), micelles formation via detergent-like effects (carpet model), or through aggregate channel models (Figure 3). The barrel-stave model was the first mechanism suggested and proposed that AMPs penetrate perpendicularly into the membrane, forming a transmembrane pore. The amphipathic structure of peptides is essential for pore formation as its hydrophobic regions interact with membrane lipids, while its hydrophilic regions form the lumen of the barrel-shaped channels and promote lateral peptide-peptide interactions (Figure 3a) (Moravej et al. 2018, Rončević et al. 2019). In order to span the lipid bilayer, AMPs require a minimum length of ≈ 22 residues for

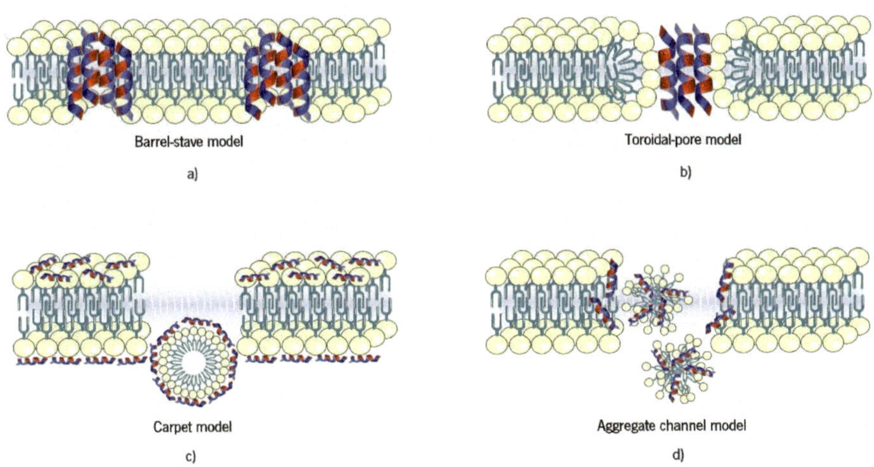

Figure 3. Proposed action models of membrane disruption process by AMPs (a) barrel-stave model, (b) toroidal-pore model, (c) carpet model, and (d) aggregate channel model. The red color represents the hydrophilic portions of AMPs, while the blue color represents the hydrophobic. The figure was produced using elements from Servier Medical Art. Adapted from (Bahar and Ren 2013) with CC BY 4.0 permission.

α-helical AMPs or ≈ 8 residues for β-sheet AMPs (Kumar et al. 2018). Ceratotoxins, an α-helical medfly AMP, display a well-defined voltage-dependent ion channel according to the barrel-stave model. Bessin et al. correlated their strong antibacterial activity against both Gram-positive and Gram-negative bacteria with their pore-forming properties (Bessin et al. 2004). This mechanism, however, has been shown to be rare and applies to a limited number of AMPs (Table 1). In the toroidal-pore model, peptides are also inserted perpendicularly into the membrane; however, a peptide-lipid complex is formed instead of a peptide-peptide interaction, making it much more permissive for diverse primary structures than the barrel-stave model (Rončević et al. 2019). Hence, a transmembrane pore is composed partly of AMPs and partly of phospholipids head groups of the lipid membrane. The locally accumulated peptides induce deformation in lipid molecules, causing them to bend continuously acquiring a membrane curvature that allows a bond between hydrophilic regions of peptides and lipids and form a toroidal-pore complex (Figure 3b) (Kumar et al. 2018, Lima et al. 2021). The formation of these pores can be mutually influenced by the peptide and the lipids, while the peptide promotes the acquisition of curvature and the organization of the lipids modulates peptide conformation (Yanmei et al. 2012). In this model, pores

Table 1. Examples of AMPs for the different mechanisms of membrane interactions.

Antibacterial Mechanism	**Representative AMPs**	**References**
Barrel-stave model	Alamethicin, Pardaxin, Dermcidin, Ceratotoxins	(Bessin et al. 2004, Chen et al. 2002, Lee et al. 2004, Rapaport and Shai 1991, Song et al. 2013)
Toroidal-pore model	Magainin 2, Aurein, Melittin, Protegrin-1, Lacticin Q, Indolicidin, Arenicin, LL-37, PG-1, Maculatin 1.1	(Henzler Wildman et al. 2003, Kim et al. 2009, Lee et al. 2004, Pan et al. 2007, Sani et al. 2013, Schibli et al. 2002, Sengupta et al. 2008, Shenkarev et al. 2011, Tang and Hong 2009, Yang et al. 2000, Yoneyama et al. 2009)
Carpet model	Plantaricin-149, Cecropin P1, Dermaseptin S, Caerin 1.1, Apomyoglobin 56-131, LL-37, Aurein 1.2, Melittin, NaD1, Ovispirin, BP100	(Fernandez et al. 2012, Gazit et al. 1996, Ghosh et al. 1997, Järvå et al. 2018, Mak et al. 2001, Manzini et al. 2014, Müller et al. 2007, Naito et al. 2000, Pandidan and Mechler 2019, Porcelli et al. 2008, Pouny et al. 1992, Rasmusson and Moller 1991, Wong et al. 1997, Yamaguchi et al. 2001)
Aggregate channel model	BP100, Maculatin 1.1	(Bond et al. 2008, Manzini et al. 2014)

have relatively short lifetimes and can collapse, which allows AMPs to extend and combine leading to membrane micellization or translocating to the cytoplasm and potentially targeting intracellular components. One of the characteristics of this model is the discrete pore size (Kumar et al. 2018, Rončević et al. 2019). Yet, a few years back, Yoneyama et al. reported a novel lacticin Q-mediated antimicrobial mechanism called the 'huge toroidal-pore' which was the first report of a bacteriocin from Gram-positive bacteria that formed a toroidal pore. This mechanism allows macromolecules to leak through the peptide-lipid complex formed by the AMP and shows that when pores close some lacticin Q molecules migrate from the outer to the inner membrane leaflets (Yoneyama et al. 2009) (AMPs capable of forming pores using the toroidal-pore model are listed in Table 1).

AMPs can also act in bacterial membranes without specific pore formation. In the carpet model, peptides bind electrostatically to the membrane in a parallel orientation, accumulating and forming a carpet-like structure. Once a critical concentration of AMPs is reached, they exert a detergent-like effect that disintegrates the membrane by forming micelles with a hydrophobic core (Figure 3c) (Erdem Büyükkiraz and Kesmen 2021, Yanmei et al. 2012). The carpet model does not require peptide-peptide interactions or peptide insertion into the hydrophobic membrane core to form transmembrane channels or even any specific peptide structures. Thus, this non-specific mechanism can explain why many AMPs act as antimicrobial agents regardless of their sequence length and their specific amino acid composition (Table 1) (Zhang et al. 2021). Lastly, in the aggregate channel model, AMPs insert into the membrane in a perpendicular orientation, forcing lipids to cluster into unstructured peptide-lipid aggregates (Figure 3d). Unlike the carpet model, these aggregates possess water molecules associated, providing channels for ions and other intracellular components leakage across the membrane. Furthermore, this mechanism may also aid the translocation of AMPs into the cytoplasm to exert functions on intracellular substances (Yanmei et al. 2012, Zhang et al. 2021).

In addition to the mechanisms described above, other related models have been proposed, such as electroporation (Jean-François et al. 2008), interfacial activity (Wimley 2010), and floodgate (Azad et al. 2011) models. In any case, they are all attempts to simplify mechanisms that are extremely complex and dependent on a number of variables. Thus far, there is no one model that fits all AMPs and the precise process of peptide uptake and channel formation remains unknown and a subject of controversy. However, it is safe to say that AMPs act *in vivo* using the possibly widest variety of membrane-disrupting mechanisms. Ion channels, transmembrane pores, and extensive membrane disruption do not represent completely different modes of action, but instead are a continuous gradient between them. For instance, toroidal-pore and carpet models are not necessarily mutually exclusive but are probably two facets of the permeabilization process (Kumar et al. 2018). For instance, the AMP aurein can act in different ways depending on lipid composition. Cheng et al. showed that this peptide in different membrane models for Gram-positive bacteria, such as *Staphylococcus aureus*, *Staphylococcus epidermidis*, and *Bacillus cereus*, disrupts the lipid bilayer by forming toroidal pores or works in a detergent-like way, indicating that the extent of the membrane perturbation is strongly concentration-dependent and partly dependent on peptide nature (Cheng et al. 2011). Other AMPs such as LL-37 (Henzler Wildman et al. 2003, Porcelli et al. 2008) and melittin (Naito et al. 2000, Tosteson and Tosteson 1981) can also disrupt bacterial membranes via different pathways and switch between mechanisms depending on their concentration in the site. BP100, a cecropin-melittin AMP hybrid, exhibits selectivity for Gram-negative bacteria and also shows that its mechanism of action depends on conditions, such as the peptide-lipid ratio and the proportion of phosphatidylglycerol in the lipid membrane. The peptide promoted lipid clustering that in turn promotes peptide aggregation on the membrane surface, giving rise to peptide-lipid patches that will eventually leave the membrane in a carpet-like mechanism (Manzini et al. 2014).

Non-Membrane Targeting Mechanism

Even though membrane disruption is the primary and the most important mode of action of AMPs, many AMPs can act on extracellular targets by inhibiting cell wall biosynthesis or need to reach intracellular targets to exert their bactericidal effects. Some peptides act externally by interacting with various precursor molecules required for cell wall biosynthesis, which weakens the structural integrity and survival of the bacterial cell and impairs the subsequent bacterial replication. The highly conserved lipid II molecule is one of the major targets of this type of AMPs (Kumar et al. 2018, Rončević et al. 2019). Leeuw et al. proposed a novel antibacterial mechanism for human defensins through their binding to lipid II, reducing its levels in the bacterial membrane and this way inhibiting peptidoglycan biosynthesis (Leeuw et al. 2010). They provided evidence that this important class of AMPs may be involved in direct defensin-lipid interactions in addition to their primary mode of action and cell membrane disruption. Manabe et al. found that D-form KLKLLLLLKLK-NH2 AMP enhanced membrane permeability through specific interaction with cell wall components, including peptidoglycans (Manabe and Kawasaki 2017). Lysostaphin is an AMP capable of cleaving pentaglycine crosslinks unique to the *Staphylococci* cell wall and hydrolyzing them (Chen et al. 2014).

For the inhibition of intracellular targets, AMPs need to interact with the cell membrane by penetrating it independently or in synergy with membrane permeabilization. Therefore, AMPs can enter using specific membrane transport proteins or by direct translocation, which leads to pore formation in the membrane without causing cell lysis (Bahar and Ren 2013, Rončević et al. 2019). Several mechanisms involving intracellular targets have been discovered, including inhibition of nucleic acids and protein biosynthesis and disruption of enzymatic activity, which can lead to the collapse of the metabolic pathways and result in cell death (Kumar et al. 2018). Some AMPs can induce the degradation of nucleic acid molecules to inhibit their biosynthesis. Buforin II is an AMP homologous to the N-terminal fragment of histone H2A, a protein that interacts directly with nucleic acids. This peptide penetrates the membrane without affecting its permeability and binds to DNA molecules (Uyterhoeven et al. 2008). Hao et al. demonstrated that designed analogues of buforin II may possess a greater affinity to RNA (Hao et al. 2013). Indolicidin acts by both disrupting the bacterial membrane and inhibiting DNA synthesis by binding to the abasic site of nucleic acids to crosslink single- or double-stranded DNA (Schibli et al. 2002). The DNA topoisomerase I is inactivated, and DNA replication and transcription are inhibited (Subbalakshmi and Sitaram 1998). Some AMPs can inhibit both DNA and protein syntheses, such as the tPMP-1 and the aHNP-1 peptides derived from the human immune system which exert their function within the first hour after entry into the cells (Xiong et al. 1999). PR-39, a proline- and arginine-rich peptide, also acts as a proteolytic agent, causing the degradation of proteins required for DNA replication (Shi et al. 1996). Typically, proline-rich AMPs interfere with protein synthesis by binding to ribosomes affecting the translation. For instance, apidaecin is an AMP blocking protein synthesis that is only effective against Gram-negative bacteria, which is actively transported by a non-specific binding action on an outer membrane component followed by specific binding to a component of probably a permease-type transporter system on the cytoplasmic membrane (Castle et al. 1999). Once in the cytoplasm, apidaecin targets the ribosome 50S large sub-unit blocking its assembly, which inhibits mRNA translation (Schmidt et al. 2017). Another study showed that the N-terminal fragments (1–25) and (1–31) of Bac5 bind to the tunnel of the ribosome, preventing the transition from the initial phase to the elongation phase of translation (Mardirossian and Barrière et al. 2018). Similarly, Bac7 (1–35) and Tur1A inhibit protein synthesis, binding to ribosomes and interacting with the ribosomal peptide exit tunnel in different ways (Mardirossian and Pérébaskine et al. 2018). Proline-rich AMPs are also reported to exert antibacterial activity by interfering with chaperones, key proteins for the proper folding, and assembly of newly synthesized proteins. Pyrrhocoricin, drosocin, and apidaecin AMPs bind specifically to *Escherichia coli* DnaK, a 70-kDa heat shock protein, and non-specifically to GroEL, a 60-kDa bacterial chaperone, blocking the protein folding pathway. These AMPs prevent refolding of misfolded proteins by inducing a permanent closure of the DnaK peptide-binding cavity (Otvos et al. 2000) and in some cases inhibit its ATPase activity (Kragol et al. 2001). Considering that AMPs involved in intracellular mechanisms target ATP-dependent processes, it is suggested that these cationic molecules interact with the highly negatively charged ATP (Hilpert et al. 2010). Bacterial protease activity can also be suppressed by some AMPs. Human histatin 5, a salivary AMP, has shown a strong inhibitory effect on both host-secreted and bacterial-secreted proteases that are associated with oral diseases, such as periodontitis (Clements et al. 2001). Couto et al. reported that equine neutrophil AMP (eNAP-2) selectively forms a non-covalent complex with bacterial serin proteases, inactivating them as a result (Couto et al. 1993). It has been proposed that AMPs can also interfere with cell division processes by inhibiting DNA replication and DNA response (SOS induction), blocking the septation process or causing the failure of chromosome segregation (Huan et al. 2020). For example, human α-defensin-5 can provoke extensive cell elongation resulting in a filamentous morphology in addition to targeting other intracellular mechanisms that cause multiple cellular damages (Chileveru et al. 2015). Similarly, the bacterial AMP microcin J25 also interferes with this process, acting via a non-SOS-dependent mechanism (Salomón and Farías 1992).

Immunomodulatory Activity

As mentioned earlier, aside from the direct killing mechanisms of action, many AMPs have been shown to recruit and activate host immune cells, promoting resistance to bacterial infection (Drayton et al. 2021). The immune responses produced by AMPs include reduction of inflammation by suppression of pro-inflammatory cytokines and preventing excessive and harmful inflammatory conditions that can result in organ failure and sepsis. Other functions include modulation of chemokine expression, stimulation of angiogenesis, promotion of cellular differentiation of leukocytes and dendritic cells, and control of the excessive release of reactive oxygen/nitrogen species (Kumar et al. 2018, Mwangi et al. 2019). Furthermore, AMPs are also capable of enhancing wound healing by stimulating epithelial cell migration and proliferation (Lima et al. 2021). The immunomodulatory properties of these peptides have been intensively studied in cathelicidin and defensin families. Human LL-37 has the ability to chemoattract and activate various immune cells, such as neutrophils, mast cells, and monocytes through formyl peptide receptor-like 1 (Niyonsaba et al. 2002, Yang et al. 2000). Similarly, β-defensin 2 and

3 can induce chemotaxis by interacting with chemokine receptor type 6 (Röhrl et al. 2010) or toll-like receptors (Funderburg et al. 2007). Defensins can also enhance the production of inflammatory cytokines, such as interleukin-8 in bronchial epithelial cells, inducing degranulation and activation of mast cells to recruit neutrophils (Van Wetering 1997).

Action on Biofilms

A biofilm is a highly organized arrangement of bacterial communities attached to surfaces and protected by an extracellular polymeric matrix (Lima et al. 2021). Many bacterial species develop these structures with the intention of becoming considerably more resistant to antibiotics. However, some AMPs have demonstrated the ability to prevent their formation or disrupt mature biofilms (Erdem Büyükkiraz and Kesmen 2021). They can hamper different stages of biofilm formation, even though their mode of action is not yet fully understood. It has been shown that some AMPs can inhibit the adhesion of bacteria to surfaces by decreasing swarming and swimming motilities and increasing twitching motility. These peptides may also act as inhibitors of quorum sensing by down-regulating crucial genes involved in this process, which are implicated in biofilm formation and/or the organization and communication of bacteria within the biofilm (Divyashree et al. 2020, Rima et al. 2021). LL-37 and derivatives of this AMP have been shown to prevent biofilm formation of different bacteria species through these mechanisms (Breij et al. 2018, Fuente-Núñez et al. 2012, Habets and Brockhurst 2012). Reffuveille et al. demonstrated that IDR-1018 is a broad-spectrum anti-biofilm peptide, acting by bonding to the bacterial guanosine pentaphosphate (p)ppGpp, which is involved in stress responses and biofilm development (Reffuveille et al. 2014). Membrane-targeted AMPs are more advantageous for treating biofilms since they are capable to interfere with their metabolic processes, including slow-growing cells such as persister cells which attribute to these structures (Bahar and Ren 2013).

Action Against Viruses: Mechanisms and Effects

Viruses are non-living entities that have their genome surrounded by a protein capsid with different architectures (e.g., helical or icosahedral). In some viruses, the nucleocapsid is surrounded by a lipid bilayer membrane called an envelope, which incorporates different glycoproteins that facilitate receptor binding and cell fusion (Figure 4) (Chaitanya 2019a, Hagan 2014). This structure protects the genomic material and prevents recognition by the immune system (Buchmann and Holmes 2015). Viruses can be classified according to their genome chemical composition in RNA

Viral genome	Non-enveloped-virus	Enveloped-virus
DNA	Human adenovirus, Porcine circovirus 2, BK polyomavirus, Human papillomavirus 16, Human parvovirus B19	Herpes simplex virus-1, Hepatitis C virus, Vaccinia virus, Pseudorabies virus
RNA	Rhinovirus, Norovirus, Poliovirus, Norwalk virus, Hepatitis E virus, Rotavirus	West Nile virus, Dengue virus , Yellow fever virus, Zika virus, Influenza A Virus, Human immunodeficiency virus type 1, SARS-CoV-2, Porcine Reproductive and Respiratory Syndrome virus, Ebola virus

Figure 4. Schematic representation of the enveloped and non-enveloped virus structure and identification of viruses with those specific characteristics. The figure was produced using elements from Servier Medical Art. Information to construct this figure was collected from (Dimitrov 2004, Heidary and Gharebaghi 2020, Sherman et al. 2020, Sutter 2020).

and DNA viruses with the genome in single-stranded (ss) or double-stranded (ds) forms (Domingo 2020). Because they lack the machinery necessary for their replication, viruses use the genetic and metabolic components of the host cell for their intracellular growth, which makes them intracellular parasites (Chaitanya 2019b). Through this, the virus introduces its genome into the host cell and thus carries out its expression, which allows viral replication (Levinson and Jawetz 1996). The mechanism of viral entry depends on the type of virus and may occur by direct penetration into the membrane, pore formation, lipid modification, or fusion (enveloped-virus) (Daussy and Wodrich 2020). Viruses are known to have different targets and can infect and cause numerous diseases in almost all species. Over the years, new viral diseases have appeared, leading to high mortality rates in humans, such as the current case of the SARS-CoV-2 virus (or COVID-19). The severity of the diseases triggered by the COVID-19, influenza, and Ebola viruses as well as the increasing resistance to antiviral drugs emphasizes the need to develop new antiviral compounds to combat the conventional viruses and the new and emerging threats (Furuse and Oshitani 2020, Luteijn et al. 2020, Nath 2021, Smyth 2020).

Recent reports have shed a light on the potential of AMPs or antiviral peptides (AVPs) to mitigate viral infection and propagation because of their broad spectrum of antimicrobial activities (Luteijn et al. 2020, Yu et al. 2020). AVPs have low molecular weight, low toxicity, high specificity and efficacy, low side effects, and can perform both against non-enveloped and enveloped viruses (Chowdhury et al. 2020, Feng et al.

2020, Li et al. 2019, Maiti 2020). At the structural level, it is known that these peptides are mainly composed of cysteine, glycine, leucine, lysine, valine, and isoleucine amino acid residues (Mousavi Maleki et al. 2021). The activity of AVPs is dependent on several factors, such as the total charge of the peptide, its concentration as well as the electrostatic affinity between its cationic charges and anionic charges of the microorganism's membrane (Gan et al. 2021, Jung et al. 2019, Kumar et al. 2018, Lei et al. 2019). Consequently, modifications to increase the positive charge of peptides have been reported to improve their binding to negatively charged membranes (e.g., virus envelope) (Jung et al. 2019, Li et al. 2011). Kozhikhova et al. demonstrated that in the dendrimer peptides, KK-16, KK-18, and TA-11, an increase in the positive charge allowed a greater destabilization of the virus and consequently a reduction in the viability of the respiratory syncytial virus (RSV) in 5 to 10 times (Kozhikhova et al. 2020). Accordingly, Jung et al. showed that increasing the positive charge of the M2 MH peptide on the polar face resulted in a 16.3-fold improvement in antiviral activity (Jung et al. 2019). Furthermore, increasing bulkier hydrophobic residues on the non-polar face may improve peptide insertion through hydrophobic membrane interactions. Yet, one peptide can mediate different antimicrobial activities in distinct ways. In the case of the chicken peptide Gga-AvBD11, the antibacterial activity is predominantly mediated by the N-terminal domain; however, the antiviral activity requires the entire protein (Guyot et al. 2020).

Virus entry is a key early stage of viral infection and some AVPs can act at this stage by interacting with the envelope and/or the capsid of the virus (Moreno-Habel et al. 2012, Vilas Boas et al. 2019) or preventing binding/fusion between the virus and the host cell by blocking of glycoproteins and/or cellular receptors (Chen et al. 2017, Leikina et al. 2005). Additionally, AVPs can also interfere in different stages of the viral replication cycle (Xu and Lu 2020) (Ruzsics et al. 2020) or induce immune responses in the host (Figure 5) (Feng et al. 2020). To this day, there is still no consensus regarding the AVPs' mechanisms of action.

AVPs can interfere with viral infection outside (1–4) and/or inside (5–13) the cell or through the induction of immunomodulatory responses by the host cell (14); (1) AVPs inactivate a virus by destroying the viral envelope, forming pores, or deforming the membrane; (2) AVPs prevent virus entry into the host cell by inducing viral extracellular aggregation; (3) AVPs interact with different glycoproteins present in the viral envelope; (4) AVPs interact with specific cell receptors present on the host cell membrane; (5) AVPs inhibit endosomal escape which prevents virus entry/fusion process; (6) AVPs prevent viral capsid from being removed; (7) AVPs prevent the viral genome from being released; (8) AVPs block the import of the viral genome into the host cell nucleus; (9) AVPs prevent

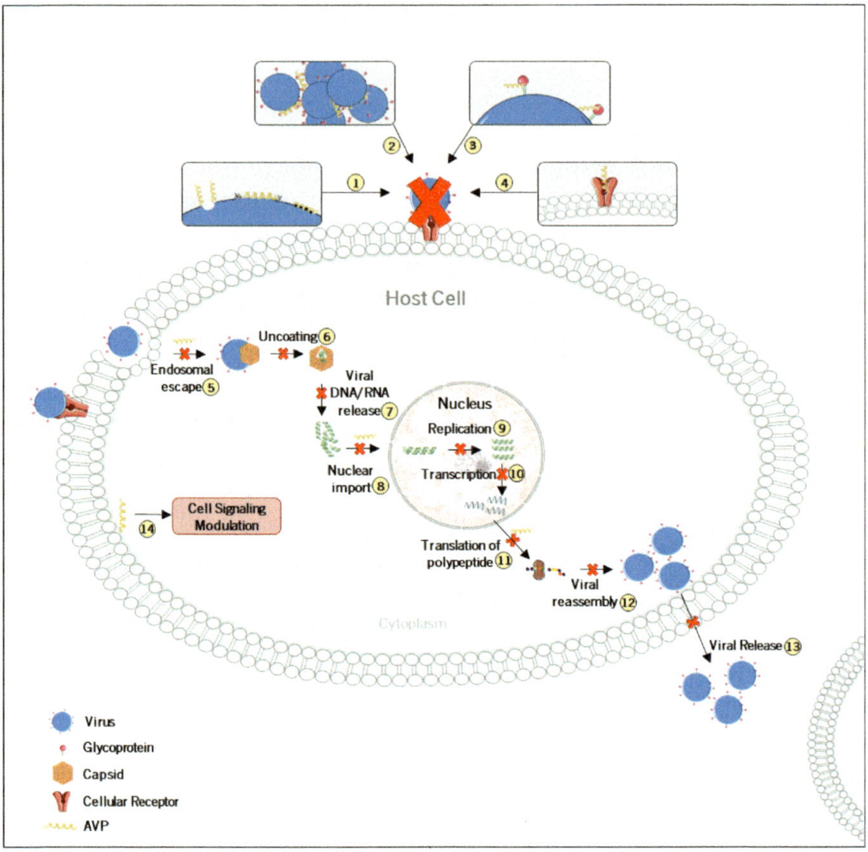

Figure 5. Mechanism of action of antiviral peptides (APVs). The figure was produced using elements from Servier Medical Art. Adapted from (Pen et al. 2020) with CC BY 4.0 permission.

viral replication by binding to DNA or to essential replication enzymes; (10) AVPs can interfere with transcription and/or block the necessary elements for this process; (11) AVPs can inhibit translation, hence the formation of functional proteins; (12) AVPs hijack the essential elements for the assembly of new viral particles and thus prevent the formation of new viruses; (13) AVPs prevent the release of new viruses into the extracellular medium and thus viral propagation is contained; (14) AVPs induce the activation of different immune modulatory pathways.

Direct Virus Inactivation

AMPs, due to their amphiphilic-cationic organization, can interact with lipid membranes. Some peptides can insert themselves into viral membranes through hydrophobic interactions between the lipid tails and the different hydrophobic residues of the peptide, causing a deformation/morphological change in the lipid membrane (Jung et al. 2019). Due to the lipidic nature of the viral envelope, AVPs can selectively associate with the viral envelopes and cause their rupture or deformation (without rupture) (Figure 5; 1). Furthermore, AVPs can induce the formation of pores in the membrane as well as the aggregation of viruses (Figure 5; 1 and 2) (Gordon et al. 2009, Luteijn et al. 2020, Mammari et al. 2021, Moreno-Habel et al. 2012). The binding of AVP to viral envelopes by surface charge interactions is one way to decrease their infectious profile. This mechanism is observed in mucroporin-M1 (scorpion venom peptide variants) which, given its strong electrostatic affinity, interacts/interrupts the viral envelope causing a direct virucidal effect against measles viruses, SARS-CoV, and influenza A virus H5N1 (Li et al. 2011). Recently, Sala et al. showed that the presence of hydrophobic amino acids and the total positive charge of the killer peptide (KP) allow its interaction with the lipid bilayer of the viral envelope and the neutralization of the viral infectious load. However, the authors observed that KP causes different levels of inhibition in herpes simplex virus type 1 (HSV-1) and vesicular stomatitis virus. This can be explained by the variability in the composition and organization of virus membranes, which is a consequence of the different origins of formation, suggesting that AVP may interact differently across the lipid layer (Sala et al. 2018). Marcocci et al. showed that temporin-B can inactivate the HSV-1 virus by disrupting the outer viral layer, following electrostatic interactions with the viral capsid (virucidal effect) (Marcocci et al. 2018). Cyclotides can also cause the rupture of lipid membranes, namely the cyclotide kalata B1 that breaks the membranes of human immunodeficiency virus (HIV) particles (Henriques et al. 2011). Dermaseptin is another example of AVP with similar activity, which by compromising the virus integrity inhibits HIV-1 (Lorin et al. 2005). In its turn, melittin (positively charged C-terminus) disturbs the viral envelopes, which can cause direct virolysis of several viruses (Memariani et al. 2020) by interacting with the phospholipid bilayer and establishing electrostatic interactions (Hall et al. 2011, Memariani et al. 2020).

One of the most important mechanisms of action of defensins and cathelicidins against enveloped viruses resorts to destabilizing the viral envelope and inhibiting their infectious nature (Mookherjee et al. 2020). Interactions of defensins with the viral lipid bilayer, encouraged by the presence of negatively charged phospholipids, can trigger membrane depolarization (Solanki et al. 2021, Wilson et al. 2013). Cathelicidin-derived peptides are reported to disrupt the viral envelope of Kaposi's sarcoma-associated herpes virus. In the case of cathelicidin LL-37, it causes the destruction of the HIV viral envelope by binding to the anionic lipids (e.g., phosphatidylserine) present along the envelope (Angel 2021). Furthermore, this peptide can inhibit the Vaccinia virus and RSV by the same mechanism of action (Barlow et al. 2014). A study by Quintero-Gil et al. based on docking suggested that porcine β-defensin-2 has the potential to inhibit the hepatitis E virus (non-enveloped RNA virus) by establishing hydrogen bonds with the capsid proteins (Quintero-Gil et al. 2017). Gulati et al. showed that α-defensin HD5 inhibits the human papillomavirus 16 (HPV-16; non-enveloped viruses) through its interaction with the virus capsid. The binding of this AVP to disordered or flexible regions of the capsid L1 and L2 proteins resulted in an alteration of its structure (Gulati et al. 2019). Moreover, HD5 can inhibit human adenovirus (HAdV) by binding to its capsid, stabilizing it, and restricting the release of endosomal protein VI (mediates membrane disruption), which prevents the dissociation of the capsid from the viral genome and prevents the viral genome from reaching the target cell nucleus (Nguyen et al. 2010, Smith et al. 2010). Gordon et al. showed that CAP37 is also capable of disrupting the capsid and/or envelope of AdV and HSV-1 (Gordon et al. 2009). Moreno-Habel et al. reported that gloverin can fight viral infection by disrupting the viral envelope through the formation of pores. The accumulation of this peptide on the surface of budded viruses can cause membrane tension or the formation of pores that breaks their envelope (Moreno-Habel et al. 2012). Additionally, defensin peptides can also suppress viral infection of non-enveloped viruses by aggregating viral plaques or by preventing the viral genome from entering the host cell. When bound to viruses, defensins can establish links between them, forming multimeric structures that allow neighboring viruses to aggregate. This aggregation is facilitated by the neutralization of the capsid charge and consecutive reduction of the repulsion effect between the viruses. Through this process, the virus binding/entry into host cells are reduced since they are in the form of clusters. This mechanism is verified in human α-defensins that prevent infection by the neutralization of the BK virus (Dugan et al. 2008, Gounder et al. 2012). Furthermore, these binding to the capsid avoids the removal of the viral coat, which inhibits the release of the viral genome into the host cell (Mookherjee et al. 2020). Jung et al. and other authors reinforce the idea that viral envelope shape is closely related to viral activity, and membrane deformation is an antiviral mechanism with a high level of selectivity (Jung et al. 2019).

Interaction with Viral Glycoproteins

The specific binding between the virus and the host cell depends on the glycan pattern of the viral envelope and the respective cell-binding receptors. Each enveloped virus exhibits specialized glycoproteins responsible for binding to receptors and merging with the host cells (Li et al. 2021). In the case of Coronavirus, the N-terminal and C-terminal domains (receptor binding domain; RBD) of the spike (S) glycoprotein recognize and bind to host membrane receptors (angiotensin-converting enzyme 2; ACE2), thus starting the infection process (Wielgat et al. 2020). Some AVPs can bind to glycoproteins present on the surface of the viral envelope (specific viral targets) and consequently prevent viruses from attaching to potential host cells (Figure 5; 3) (Mertinková et al. 2021). Electrostatic charge differences improve peptide-glycoproteins interactions and the more negatively charged the glycoprotein domains are, the greater the contribution to non-specific hydrogen bonds and electrostatic interactions (Vedadghavami et al. 2020). Glycoproteins are composed of several domains that are formed by different amino acid residues. Thus, there are domains that have a greater or lesser affinity for binding with AVPs; for instance, glycoprotein E present in the West Nile virus (WNV) is composed of three domains and yet only domain III mediates binding with cell receptors (Mertinková et al. 2021). Lokhande et al. and Zhang et al. also reported that inhibition of the SARS-CoV-2 virus can occur through the binding of LL-37 and human β-defensin-2 to the S-protein, respectively (Lokhande et al. 2021, Zhang et al. 2021). LL-37 prevents the interaction between S-protein and ACE2 (primary receptor) by establishing intermolecular interactions (salt bonds, hydrogen bonds, and hydrophobic interactions) with the RBD of the S-protein. These interactions allow this AVP to bind to various residues of the RBD (e.g., LYS417, GLN493, and HR500) (Wang and Wang et al. 2021). This binding appeared to be specific and thermostable, thus preventing the S-protein from being free to interact and form a stable complex with ACE2 (Lokhande et al. 2021).

Retrocyclin AVP shows a strong affinity for binding to glycosylated proteins by exhibiting three disulfide bonds and four positively charged residues (Jia et al. 2021). Specifically, retrocyclin-2 binds to glycoprotein-B of HSV-2 with high affinity, which prevents further attachment to the host cells' surface (Wong et al. 2014). Also, HNP1-3, HD5 and HBD3 defensins can prevent HSV-1/HSV-2 adhesion to host cells by binding to viral glycoproteins (Hazrati et al. 2006). Similarly, influenza virus infection can be inhibited by retrocyclin-2 and HBD3 as these interfere with viral fusion, which is mediated by hemagglutinin (Leikina et al. 2005) or may be inhibited by the binding of urumin to hemagglutinin H1 (Holthausen et al. 2017). For the inhibition of other viruses, such as HIV, the retrocycline peptide binds to glycoprotein 120 (gp120) of the viral envelope to prevent the virus from binding to the cluster of differentiation 4 (CD4) cell receptor (Teixeira et al. 2011). By preventing the binding of gp120 to CD4, the conformational change that exposes the binding site of co-receptors (CCR5 or CXCR4) on gp120 does not occur and therefore fusion between the viral envelope and the cell membrane is not possible (Su et al. 2020). Lactoferrin (Lf) can also inhibit HIV-1 adsorption by binding to the positively charged oV3 loop of the gp120 (Groot et al. 2005) as well as LL-37 peptide by competing with gp120 for co-receptors, which prevents gp120-CD4 binding in both cases (Angel 2021). Additionally, enfuvirtide, a fusion inhibitor peptide, suppresses HIV by blocking the conformational change of gp41, which prevents it from binding to the membrane fusion domain (Matthews et al. 2004). The viral peptide-glycoprotein interaction appears to play a major role in inhibiting viral infection not only by preventing viruses from binding to cell receptors but also by preventing their fusion with the cell membrane.

Interaction with Specific Cellular Receptors

As mentioned earlier, cell receptors are an important element in viral infection. These can have different functions, such as directing viruses to endocytosis, activating specific signaling pathways for virus entry, inducing fusion/penetration events on the surface of the target cell, and/or inducing conformational changes in the virus surface structure (Grove and Marsh 2011, Maginnis 2018). There are numerous cellular receptors, namely heparan sulfate (HS), CD4, CD81, CD46, ACE2, and sialic acid, which allow the virus-target cell interaction (Furci et al. 2012, Hoffmann et al. 2020, Kumar et al. 2021). The binding of AVPs to cellular receptors prevents them from interacting with viral plaques, making them disruptors of viral binding, entry, and infection (Figure 5; 4). HS is a very important and well-known cell surface molecule, detrimental to viral plaque binding for HSV, HIV, and SARS-CoV viruses (Kell et al. 2020, O'Donnell and Shukla 2008). This receptor has a negative charge which facilitates the binding to positively charged molecules, such as Lf, via the N-terminal (Madavaraju et al. 2021). Electrostatic binding between Lf-HS prevents HS-HVS binding from occurring, which makes viral infection less likely to take place (O'Donnell and Shukla 2008). Furthermore, the α-defensins HNP4 and HD6 can bind to the HS receptor to inhibit HSV-1 and HSV-2 infections (Hazrati et al. 2006). HNP1-2 and HD5 defensins are associated with blocking HIV-1 entry into cells by competitively binding to the CD4 receptor, preventing HIV-CD4 binding (Furci et al. 2012). It is in the N-terminal helix of the ACE2 cell receptor that most of the residues responsible for the interaction with the RBD of SARS-CoV-2 are found (Lokhande et al. 2021). Recently, it was reported that LL-37 has a strong inhibitory effect on the viral entry of SARS-CoV-2 due to its possible binding to ACE2 receptors. This inhibitory effect is due to the different interactions (salt bridges, hydrogen bonds, and hydrophobic interaction pairs) with the ligand-binding domain (LBD) of ACE2 and its respective binding to the residues of the LBD, including TYR41 (Wang et al. 2021). Chen et al. reported that Lf can protect the host cell against dengue virus

(DENV-2) infection by binding to different cellular receptors, including HS, lipoprotein receptors low-density, and dendritic cell-specific intercellular adhesion molecule 3-grabbing non-integrin, demonstrating the versatility of Lf binding to different receptors which can be an asset in inhibiting viral infection (Chen et al. 2017).

The antiviral action of AMPs can also result from the interaction with ATPases. For instance, melittin by inhibiting Na^+ K^+ ATPase, a cellular enzyme involved in the membrane fusion process, avoids the fusion of HSV-1 with the target cell (Matanic and Castilla 2006). It should be noted that some AVPs (e.g., HBD3, Lf, and LL-37) are able to bind to both cell receptors and viral glycoproteins, allowing them to inhibit viral infection through a dual mechanism (Groot et al. 2005, Hazrati et al. 2006, Madavaraju et al. 2021, Wang et al. 2021).

Intracellular Targets and Stimulation of Immunomodulatory Responses

In addition to the mechanisms of action discussed earlier, some AVPs can act intracellularly and inhibit viral dissemination. Cellular internalization of AVPs can result in suppression of viral gene/protein expression at various levels, namely replication, transcription, and translation (Figure 5; 9–11); It can block the release of viral particles to the outside (Figure 5; 13) or induce immunomodulatory activities (Figure 5; 14) (Jenssen et al. 2006). AVPs can also inhibit viral replication indirectly, such as defensin-derived peptide P9R, which due to its alkaline nature (presence of several basic amino acid residues in its structure), has been shown to inhibit infection of coronavirus, influenza virus, and rhinovirus (pH-dependent viruses) by preventing endosomal acidification. Thus, the viral infection does not occur since the low endosomal acidification blocks the virus entry/fusion process (Figure 5; 5) (Zhao et al. 2020). The venom peptide Ev37 is reported to inhibit DENV-2, HCV, Zika virus, and HSV-1 (low pH-dependent fusion peptides) by the same mechanism of action (Li et al. 2019). For viral replication to occur, the normal functioning of nuclear import is necessary (Figure 5; 8); however, it has been described that human α-defensins can act at this stage by inhibiting nuclear import of HIV-1 viral RNA into the host cell nucleus (Xu and Lu 2020). Other sets of AVPs can directly interact with the viral genome. Ruzsics et al. showed that TAT-I24 peptide can inhibit the replication of several double-stranded DNA viruses (HSV, cytomegalovirus, AdV type 5, polyomavirus SV40, and VV) by linking to the viral DNA (Ruzsics et al. 2020). Indeed, this APV binds directly and with great affinity to the viral DNA, which prevents polymerase binding and consequently the synthesis of a new nucleic acid (Harant et al. 2021). LL-37 is a good example of an AVP with an anti-HIV activity that interferes with viral replication by inhibiting reverse transcriptase that blocks DNA synthesis (Wong et al. 2011). Furthermore, this is associated with a decrease in the metabolic activity of cells infected with human rhinovirus 1B, directly affecting the virus without compromising the target cells (Sousa et al. 2017).

After the translation of the genomic RNA into a polyprotein, the intervention of several proteases is necessary for the processing of this viral polyprotein to occur. Of all the viral serines, protease (NS2B-NS3 protease) is responsible for cleaving the polyprotein on the cytoplasmic side into individual proteins and thus ensuring the normal process of replication and maturation (Chew et al. 2017). Jia et al. reported that rectocycline-101 can inhibit the replication of flaviviruses (e.g., DENV and WNV) by inhibiting the activity of the serine protease NS2B-NS3, which possesses negatively charged surfaces (Jia et al. 2021). This AVP blocks the catalytic motif of the NS3 proteolytic domain, preventing the binding of the substrate to the domain, and the binding of NS2B and NS3. Similarly, in COVID-19 and influenza respiratory viruses, the serine protease is a potential viral target since these viruses hijack their host proteases to increase viral spread in the host cell (Rahbar Saadat et al. 2021). In the case of LL-37, this peptide inhibits HIV protease, which makes it impossible to convert polyproteins into functional proteins (Angel 2021).

Beyond the mechanisms discussed so far, AVPs can also stimulate the host cells' antiviral immune system and thus control viral replication/ propagation (Chang et al. 2020, Takiguchi et al. 2014). In the case of LL-37, its immunomodulatory role is associated with the upregulation of type 1 interferon (IFN) (Pahar et al. 2020, Takiguchi et al. 2014). These types of cytokines are known to play an important role in the innate response of cells, activating a set of proteins that interfere with viruses' replication to restrict spreading between cells (Lin and Young 2014, Soper et al. 2018). Lf also exhibits an immunomodulatory role, reducing cell damage through the modulation of cytokines, chemokines, and receptors involved in cascades of signaling pathways. At the level of inhibition of SARS-CoV-2 infection, Lf upon binding to the receptor increases IFN-α and IFN-β response and consequently viral replication is inhibited (Chang et al. 2020, Mirabelli et al. 2021). It has been described that protein kinase C (PKC) is a key element in HIV-1 replication and for replication to occur, this protein must be phosphorylated. In this sense, α-defensin-1 seemed to mediate the inhibition of HIV-1 in primary $CD4^+$ T cells by interfering with PKC signaling pathways, reducing its phosphorylation (Chang et al. 2005). On the other hand, influenza A virus infection can be controlled by the immunomodulatory activity of Alloferon 1 and 2 peptides, which influence the activation of natural killer cells and the release of IFN (Chernysh et al. 2002). At the same time, the action of certain AVPs can only be observed at a later stage of the viral infection; cecropin D, for instance, attenuates cellular apoptosis induced by porcine reproductive and respiratory syndrome virus by suppressing virus release to the extracellular medium (Liu et al. 2015).

Currently, of all characterized AMPs, only about 4–6% exhibit activity against viruses. Consequently, the percentage of viral AMPs approved by the U.S. Food and Drug Administration (FDA) is still very low (Kurpe et al. 2020).

Conclusions and Future Perspectives

AMPs are unique biomolecules with great potential to fight microbial infections. The broad antimicrobial activity and immunomodulatory action associated with these peptides are attractive features for their application in the prevention of infectious and non-infectious diseases. To this end, a thorough understanding of their underlying mechanism of action is essential for future clinical development. Strong evidence suggests that there is a close correlation between the mechanism of action and the physical characteristics or primary structure of AMPs. However, to date, the exact mechanism of each AMPs has not been well studied, especially with respect to viruses. Although some information about their mechanism of action already exists, further investigations are needed for a better characterization that will allow their greater applicability as antiviral therapies. It is urgent to counteract the intriguing way viruses replicate, adapt, and evolve so quickly. Furthermore, there are still some challenges that require a detailed and careful analysis of the applicability of AMPs in clinical use, such as potential toxicity to humans, sensitivity to harsh environmental conditions (i.e., susceptibility to proteases and extreme pH), lack of selectivity against specific strains, high production cost, folding issues of some large peptides, and reduced activity when used for surface coating. To overcome these limitations, several approaches have already been considered, such as peptides designed to enhance their antimicrobial activities and minimize susceptibility to proteolytic degradation and cytotoxicity. The other way is the development of formulations to protect and allow the control of peptides release without causing cytotoxicity, such as their encapsulation, attachment to the surface of materials, and formulations by self-assembly and by covalent conjugations (Schito et al. 2021, Wang and Hong et al. 2021), but there is still much to be uncovered.

Acknowledgements

The authors acknowledge the Portuguese Foundation for Science and Technology (FCT), FEDER funds by means of Portugal 2020 Competitive Factors Operational Program (POCI) and the Portuguese Government (OE) for funding the project PEPTEX with reference PTDC/

CTMTEX/28074/2017 (POCI-01-0145-FEDER-028074). The authors also acknowledge project UID/CTM/00264/2021 of 2C2T funded by national funds through FCT/MCTES. T.D.T, M.O.T. and M.A.T. acknowledge FCT for PhD funding via scholarships 2020.06046.BD, 2021.06906.BD and SFRH/BD/148930/2019, respectively.

References

Angel, J. B. (2021). Antimicrobial peptide, LL-37, and its potential as an anti-HIV agent. Clinical and Investigative Medicine (Online), 44(3): E64–E71.

Annunziato, G., and Costantino, G. (2020). Antimicrobial peptides (AMPs): a patent review (2015–2020). Expert Opinion on Therapeutic Patents, 30(12): 931–947.

Awan, A. R., Blount, B. A., Bell, D. J., Shaw, W. M., Ho, J. C. H., McKiernan, R. M., and Ellis, T. (2017). Biosynthesis of the antibiotic nonribosomal peptide penicillin in baker's yeast. Nature Communications, 8(1): 15202.

Azad, M. A., Huttunen-Hennelly, H. E. K., and Friedman, C. R. (2011). Bioactivity and the First Transmission Electron Microscopy Immunogold Studies of Short <i>*De Novo*</ i>-Designed Antimicrobial Peptides. Antimicrobial Agents and Chemotherapy, 55(5): 2137–2145.

Bahar, A. A., and Ren, D. (2013). Antimicrobial peptides. Pharmaceuticals, 6(12): 1543–1575.

Bakare, O. O., Fadaka, A. O., Klein, A., and Pretorius, A. (2020). Dietary effects of antimicrobial peptides in therapeutics. All Life, 13(1): 78–91.

Barashkova, A. S., and Rogozhin, E. A. (2020). Isolation of antimicrobial peptides from different plant sources: Does a general extraction method exist? Plant Methods, 16(1): 143.

Barlow, P. G., Findlay, E. G., Currie, S. M., and Davidson, D. J. (2014). Antiviral potential of cathelicidins. Future Microbiology, 9(1): 55–73.

Bessin, Y., Saint, N., Marri, L., Marchini, D., and Molle, G. (2004). Antibacterial activity and pore-forming properties of ceratotoxins: a mechanism of action based on the barrel stave model. Biochimica et Biophysica Acta (BBA)-Biomembranes, 1667(2): 148–156.

Bond, P. J., Parton, D. L., Clark, J. F., and Sansom, M. S. P. (2008). Coarse-grained simulations of the membrane-active antimicrobial peptide maculatin 1.1. Biophysical Journal, 95(8): 3802–3815.

Breij, A.d., Riool, M., Cordfunke, R. A., Malanovic, N., Boer, L.d., Koning, R. I., Ravensbergen, E., Franken, M., Heijde, T.v.d., Boekema, B. K., Kwakman, P. H. S., Kamp, N., Ghalbzouri, A. E., Lohner, K., Zaat, S. A. J., Drijfhout, J. W., and Nibbering, P. H. (2018). The antimicrobial peptide SAAP-148 combats drug-resistant bacteria and biofilms. Science Translational Medicine, 10(423): eaan4044.

Breukink, E., Wiedemann, I., Kraaij, C.v., Kuipers, O. P., Sahl, H.-G., and Kruijff, B.d. (1999). Use of the cell wall precursor lipid II by a pore-forming peptide antibiotic. Science, 286(5448): 2361–2364.

Buchmann, J. P., and Holmes, E. C. (2015). Cell walls and the convergent evolution of the viral envelope. Microbiology and Molecular Biology Reviews, 79(4): 403–418.

Caboche, S., Leclère, V., Pupin, M., Kucherov, G., and Jacques, P. (2010). Diversity of monomers in nonribosomal peptides: towards the prediction of origin and biological activity. Journal of Bacteriology, 192(19): 5143–5150.

Cabrele, C., Martinek, T. A., Reiser, O., and Berlicki, Ł. (2014). Peptides containing β-amino acid patterns: challenges and successes in medicinal chemistry. Journal of Medicinal Chemistry, 57(23): 9718–9739.

Castle, M., Nazarian, A., and Tempst, P. (1999). Lethal effects of apidaecin on escherichia coliinvolve sequential molecular interactions with diverse targets. Journal of Biological Chemistry, 274(46): 32555–32564.

Chaitanya, K. (2019a). Genome and Genomics: Springer.

Chaitanya, K. (2019b). Structure and organization of virus genomes Genome and Genomics (pp. 1–30): Springer.

Chang, R., Ng, T. B., and Sun, W.-Z. (2020). Lactoferrin as potential preventative and adjunct treatment for COVID-19. International Journal of Antimicrobial Agents, 56(3): 106118.

Chang, T. L., Vargas, J., Jr., DelPortillo, A., and Klotman, M. E. (2005). Dual role of α-defensin-1 in anti–HIV-1 innate immunity. The Journal of Clinical Investigation, 115(3): 765–773.

Chen, C., Fan, H., Huang, Y., Peng, F., Fan, H., Yuan, S., and Tong, Y. (2014). Recombinant lysostaphin protects mice from methicillin-resistant *Staphylococcus aureus* pneumonia. BioMed Research International 2014.

Chen, F.-Y., Lee, M.-T., and Huang, H. W. (2002). Sigmoidal concentration dependence of antimicrobial peptide activities: a case study on alamethicin. Biophysical Journal, 82(2): 908–914.

Chen, J.-M., Fan, Y.-C., Lin, J.-W., Chen, Y.-Y., Hsu, W.-L., and Chiou, S.-S. (2017). Bovine lactoferrin inhibits dengue virus infectivity by interacting with heparan sulfate, low-density lipoprotein receptor, and DC-SIGN. International Journal of Molecular Sciences, 18(9): 1957.

Chen, Y., Wang, J., Li, G., Yang, Y., and Ding, W. (2021). Current advancements in sactipeptide natural products. Front Chem., 9: 595991–595991.

Cheng, J. T. J., Hale, J. D., Elliott, M., Hancock, R. E. W., and Straus, S. K. (2011). The importance of bacterial membrane composition in the structure and function of aurein 2.2 and selected variants. Biochimica et Biophysica Acta (BBA) - Biomembranes, 1808(3): 622–633.

Chernysh, S., Kim, S. I., Bekker, G., Pleskach, V. A., Filatova, N. A., Anikin, V. B., Platonov, V. G., and Bulet, P. (2002). Antiviral and antitumor peptides from insects. Proceedings of the National Academy of Sciences, 99(20): 12628–12632.

Chew, M.-F., Poh, K.-S., and Poh, C.-L. (2017). Peptides as therapeutic agents for dengue virus. Int. J. Med. Sci., 14(13): 1342–1359.

Chileveru, H. R., Lim, S. A., Chairatana, P., Wommack, A. J., Chiang, I. L., and Nolan, E. M. (2015). Visualizing attack of *Escherichia coli* by the antimicrobial peptide human defensin 5. Biochemistry, 54(9): 1767–1777.

Chowdhury, A. S., Reehl, S. M., Kehn-Hall, K., Bishop, B., and Webb-Robertson, B.-J. M. (2020). Better understanding and prediction of antiviral peptides through primary and secondary structure feature importance. Scientific Reports, 10(1): 19260.

Clements, J. D., Gusman, H., Travis, J., Helmerhorst, E. J., Potempa, J., Troxler, R. F., and Oppenheim, F. G. (2001). Salivary histatin 5 is an inhibitor of both host and bacterial enzymes implicated in periodontal disease. Infection and Immunity, 69(3): 1402–1408.

Couto, M. A., Harwigm, S. S., and Lehrer, R. I. (1993). Selective inhibition of microbial serine proteases by eNAP-2, an antimicrobial peptide from equine neutrophils. Infect. Immun., 61(7): 2991–2994.

Das, S., Ben Haj Salah, K., Djibo, M., and Inguimbert, N. (2018). Peptaibols as a model for the insertions of chemical modifications. Archives of Biochemistry and Biophysics, 658: 16–30.

Daussy, C. F., and Wodrich, H. (2020). "Repair Me if You Can": membrane damage, response, and control from the viral perspective. Cells, 9(9): 2042.

De Mandal, S., Panda, A. K., Murugan, C., Xu, X., Kumar, N.S., and Jin, F. (2021). Antimicrobial peptides: novel source and biological function with a special focus on entomopathogenic nematode/bacterium symbiotic complex. Frontiers in Microbiology, 12.

Dimitrov, D. S. (2004). Virus entry: molecular mechanisms and biomedical applications. Nature Reviews Microbiology, 2(2): 109–122.

Divyashree, M., Mani, M. K., Reddy, D., Kumavath, R., Ghosh, P., Azevedo, V., and Barh, D. (2020). Clinical applications of antimicrobial peptides (AMPs): where do we stand now? Protein and Peptide Letters, 27(2): 120–134.

Domingo, E. (2020). Introduction to virus origins and their role in biological evolution. Virus as Populations: 1–33.

Drayton, M., Deisinger, J. P., Ludwig, K. C., Raheem, N., Müller, A., Schneider, T., and Straus, S. K. (2021). Host defense peptides: dual antimicrobial and immunomodulatory action. International Journal of Molecular Sciences, 22(20): 11172.

Dugan, A. S., Maginnis, M. S., Jordan, J. A., Gasparovic, M. L., Manley, K., Page, R., Williams, G., Porter, E., O'Hara, B. A., and Atwood, W. J. (2008). Human α-defensins inhibit BK virus infection by aggregating virions and blocking binding to host cells. Journal of Biological Chemistry, 283(45): 31125–31132.

El-Gendy, A. O., Brede, D. A., Essam, T. M., Amin, M. A., Ahmed, S. H., Holo, H., Nes, I. F., and Shamikh, Y. I. (2021). Purification and characterization of bacteriocins-like inhibitory substances from food isolated *Enterococcus faecalis* OS13 with activity against nosocomial enterococci. Scientific Reports, 11(1): 3795.

Engelberg, Y., and Landau, M. (2020). The Human LL-37 (17-29) antimicrobial peptide reveals a functional supramolecular structure. Nature Communications, 11(1): 1–10.

Erdem Büyükkiraz, M., and Kesmen, Z. (2021). Antimicrobial peptides (AMPs): A promising class of antimicrobial compounds. Journal of Applied Microbiology n/a(n/a).

Felgueiras, H. P. (2021). An insight into biomolecules for the treatment of skin infectious diseases. Pharmaceutics, 13(7): 1012.

Felgueiras, H. P., and Amorim, M. T. P. (2017). Functionalization of electrospun polymeric wound dressings with antimicrobial peptides. Colloids and Surfaces B: Biointerfaces, 156: 133–148.

Felgueiras, H. P., Teixeira, M. A., Tavares, T. D., Homem, N. C., Zille, A., and Amorim, M. T. P. (2020). Antimicrobial action and clotting time of thin, hydrated poly(vinyl alcohol)/cellulose acetate films functionalized with LL37 for prospective wound-healing applications. Journal of Applied Polymer Science, 137(18): 48626.

Feng, M., Fei, S., Xia, J., Labropoulou, V., Swevers, L., and Sun, J. (2020). Antimicrobial Peptides as Potential Antiviral Factors in Insect Antiviral Immune Response. Frontiers in Immunology, 11(2030).

Fernandez, D. I., Le Brun, A. P., Whitwell, T. C., Sani, M.-A., James, M., and Separovic, F. (2012). The antimicrobial peptide aurein 1.2 disrupts model membranes via the carpet mechanism. Physical Chemistry Chemical Physics, 14(45): 15739–15751.

Fuente-Núñez, C.d.l., Korolik, V., Bains, M., Nguyen, U., Breidenstein, E.B.M., Horsman, S., Lewenza, S., Burrows, L., and Hancock, R. E. W. (2012). Inhibition of bacterial biofilm formation and swarming motility by a small synthetic cationic peptide. Antimicrobial Agents and Chemotherapy, 56(5): 2696–2704.

Funderburg, N., Lederman, M. M., Feng, Z., Drage, M. G., Jadlowsky, J., Harding, C. V., Weinberg, A., and Sieg, S. F. (2007). Human β-defensin-3 activates professional antigen-presenting cells via Toll-like receptors 1 and 2. Proceedings of the National Academy of Sciences, 104(47): 18631–18635.

Furci, L., Tolazzi, M., Sironi, F., Vassena, L., and Lusso, P. (2012). Inhibition of HIV-1 infection by human α-defensin-5, a natural antimicrobial peptide expressed in the genital and intestinal mucosae.

Furuse, Y., and Oshitani, H. (2020). Viruses that can and cannot coexist with humans and the future of SARS-CoV-2. Frontiers in Microbiology, 11(2211).

Gan, B. H., Gaynord, J., Rowe, S. M., Deingruber, T., and Spring, D. R. (2021). The multifaceted nature of antimicrobial peptides: current synthetic chemistry approaches and future directions. Chemical Society Reviews.

Gao, X., Ding, J., Liao, C., Xu, J., Liu, X., and Lu, W. (2021). Defensins: The natural peptide antibiotic. Advanced Drug Delivery Reviews, 179: 114008.

Gazit, E., Miller, I. R., Biggin, P. C., Sansom, M. S. P., and Shai, Y. (1996). Structure and Orientation of the Mammalian Antibacterial Peptide Cecropin P1 within Phospholipid Membranes. Journal of Molecular Biology, 258(5): 860–870.

Gerdol, M., Schmitt, P., Venier, P., Rocha, G., Rosa, R. D., and Destoumieux-Garzón, D. (2020). Functional insights from the evolutionary diversification of big defensins. Frontiers in Immunology, 11(758).

Ghosh, J. K., Shaool, D., Guillaud, P., Cicéron, L., Mazier, D., Kustanovich, I., Shai, Y., and Mor, A. (1997). Selective cytotoxicity of dermaseptin S3 toward intraerythrocyticplasmodium falciparum and the underlying molecular basis. Journal of Biological Chemistry, 272(50): 31609–31616.

Gong, T., Fu, J., Shi, L., Chen, X., and Zong, X. (2021). Antimicrobial peptides in gut health: a review. Front Nutr, 8: 751010–751010.

Gordon, Y. J., Romanowski, E. G., Shanks, R. M. Q., Yates, K. A., Hinsley, H., and Pereira, H. A. (2009). CAP37-derived antimicrobial peptides have *in vitro* antiviral activity against adenovirus and herpes simplex virus Type 1. Current Eye Research, 34(3): 241–249.

Gounder, A. P., Wiens, M. E., Wilson, S. S., Lu, W., and Smith, J. G. (2012). Critical determinants of human α-defensin 5 activity against non-enveloped viruses. Journal of Biological Chemistry, 287(29): 24554–24562.

Groot, F., Geijtenbeek, T. B. H., Sanders, R. W., Baldwin, C. E., Sanchez-Hernandez, M., Floris, R., Kooyk, Y.v., Jong, E. C. d., andBerkhout, B. (2005). Lactoferrin prevents dendritic cell-mediated human immunodeficiency virus type 1 transmission by blocking the DC-SIGN—gp120 Interaction. Journal of Virology, 79(5): 3009–3015.

Grove, J., and Marsh, M. (2011). The cell biology of receptor-mediated virus entry. Journal of Cell Biology, 195(7): 1071–1082.

Gulati, N. M., Miyagi, M., Wiens, M. E., Smith, J. G., and Stewart, P. L. (2019). α-Defensin HD5 stabilizes human papillomavirus 16 capsid/core interactions. Pathog. Immun., 4(2): 196–234.

Guyot, N., Meudal, H., Trapp, S., Iochmann, S., Silvestre, A., Jousset, G., Labas, V., Reverdiau, P., Loth, K., Hervé, V., Aucagne, V., Delmas, A. F., Rehault-Godbert, S., and Landon, C. (2020). Structure, function, and evolution of <em>Gga</em>-AvBD11, the archetype of the structural avian-double-β-defensin family. Proceedings of the National Academy of Sciences, 117(1): 337–345.

Habets, M. G. J. L., and Brockhurst, M. A. (2012). Therapeutic antimicrobial peptides may compromise natural immunity. Biology Letters, 8(3): 416–418.

Hagan, M. F. (2014). Modeling Viral Capsid Assembly. Adv. Chem. Phys., 155: 1–68.

Hall, K., Lee, T.-H., and Aguilar, M.-I. (2011). The role of electrostatic interactions in the membrane binding of melittin. Journal of Molecular Recognition, 24(1): 108–118.

Hao, G., Shi, Y.-H., Tang, Y.-L., and Le, G.-W. (2013). The intracellular mechanism of action on Escherichia coli of BF2-A/C, two analogues of the antimicrobial peptide Buforin 2. Journal of Microbiology, 51(2): 200–206.

Harant, H., Höfinger, S., Kricek, F., Ruf, C., Ruzsics, Z., Hengel, H., and Lindley, I. J. (2021). The peptide TAT-I24 with antiviral activity against DNA viruses binds double-stranded DNA with high affinity. Biologics, 1(1): 41–60.

Hassan, M., Kjos, M., Nes, I. F., Diep, D. B., and Lotfipour, F. (2012). Natural antimicrobial peptides from bacteria: characteristics and potential applications to fight against antibiotic resistance. Journal of Applied Microbiology, 113(4): 723–736.

Hassan, M. F., Qutb, A. M., and Dong, W. (2021). Prediction and activity of a cationic α-helix antimicrobial peptide ZM-804 from maize. International Journal of Molecular Sciences, 22(5): 2643.

Hazrati, E., Galen, B., Lu, W., Wang, W., Ouyang, Y., Keller, M. J., Lehrer, R. I., and Herold, B. C. (2006). Human α- and β-defensins block multiple steps in herpes simplex virus infection. The Journal of Immunology, 177(12): 8658–8666.

He, S., Yang, Z., Yu, W., Li, J., Li, Z., Wang, J., and Shan, A. (2020). Systematically studying the optimal amino acid distribution patterns of the amphiphilic structure by using the ultrashort amphiphiles. Frontiers in Microbiology, 2819.

Heidary, F., and Gharebaghi, R. (2020). Ivermectin: a systematic review from antiviral effects to COVID-19 complementary regimen. The Journal of Antibiotics, 73(9): 593–602.

Heilbronner, S., Krismer, B., Brötz-Oesterhelt, H., and Peschel, A. (2021). The microbiome-shaping roles of bacteriocins. Nature Reviews Microbiology, 19(11): 726–739.

Henriques, S. T., Huang, Y.-H., Rosengren, K. J., Franquelim, H. G., Carvalho, F. A., Johnson, A., Sonza, S., Tachedjian, G., Castanho, M. A., and Daly, N. L. 2011. Decoding the membrane activity of the cyclotide kalata B1: the importance of phosphatidylethanolamine phospholipids and lipid organization on hemolytic and anti-HIV activities. Journal of Biological Chemistry, 286(27): 24231–24241.

Henzler Wildman, K. A., Lee, D.-K., and Ramamoorthy, A. (2003). Mechanism of lipid bilayer disruption by the human antimicrobial peptide, LL-37. Biochemistry, 42(21): 6545–6558.

Hilpert, K., McLeod, B., Yu, J., Elliott, M. R., Rautenbach, M., Ruden, S., Bürck, J., Muhle-Goll, C., Ulrich, A. S., Keller, S., and Hancock, R. E. W. (2010). Short cationic antimicrobial peptides interact with ATP. Antimicrobial Agents and Chemotherapy, 54(10): 4480–4483.

Hoffmann, M., Kleine-Weber, H., Schroeder, S., Krüger, N., Herrler, T., Erichsen, S., Schiergens, T. S., Herrler, G., Wu, N.-H., Nitsche, A., Muller, M. A., Drosten, C., and Pöhlmann, S. (2020). SARS-CoV-2 cell entry depends on ACE2 and TMPRSS2 and is blocked by a clinically proven protease inhibitor. Cell, 181: 271–280.

Holthausen, D. J., Lee, S. H., Kumar, V. T. V., Bouvier, N. M., Krammer, F., Ellebedy, A. H., Wrammert, J., Lowen, A. C., George, S., Pillai, M. R., and Jacob, J. (2017). An Amphibian Host Defense Peptide Is Virucidal for Human H1 Hemagglutinin-Bearing Influenza Viruses. Immunity, 46(4): 587–595.

Höng, K., Austerlitz, T., Bohlmann, T., and Bohlmann, H. (2021). The thionin family of antimicrobial peptides. PLOS ONE, 16(7): e0254549.

Huan, Y., Kong, Q., Mou, H., and Yi, H. (2020). Antimicrobial peptides: classification, design, application and research progress in multiple fields. Frontiers in Microbiology, 11: 2559.

Järvå, M., Lay, F. T., Phan, T. K., Humble, C., Poon, I. K. H., Bleackley, M. R., Anderson, M. A., Hulett, M. D., and Kvansakul, M. (2018). X-ray structure of a carpet-like antimicrobial defensin–phospholipid membrane disruption complex. Nature Communications, 9(1): 1962.

Jean-François, F., Elezgaray, J., Berson, P., Vacher, P., and Dufourc, E. J. (2008). Pore formation induced by an antimicrobial peptide: electrostatic effects. Biophysical Journal, 95(12): 5748–5756.

Jenssen, H., Hamill, P., and Hancock, R. E. W. (2006). Peptide antimicrobial agents. Clinical Microbiology Reviews, 19(3): 491–511.

Jia, X., Guo, J., Yuan, W., Sun, L., Liu, Y., Zhou, M., Xiao, G., Lu, W., Garzino-Demo, A., Wang, W., and Heise, M. T. (2021). Mechanism through which retrocyclin targets flavivirus multiplication. Journal of Virology, 95(15): e00560–00521.

Jung, Y., Kong, B., Moon, S., Yu, S.-H., Chung, J., Ban, C., Chung, W.-J., Kim, S.-G., and Kweon, D.-H. (2019). Envelope-deforming antiviral peptide derived from influenza virus M2 protein. Biochemical and Biophysical Research Communications, 517(3): 507–512.

Kaur, G., Singh, T., and Malik, R. (2013). Antibacterial efficacy of Nisin, Pediocin 34 and Enterocin FH99 against Listeria monocytogenes and cross resistance of its bacteriocin resistant variants to common food preservatives. Brazilian Journal of Microbiology, 44: 63–71.

Kell, D. B., Heyden, E. L., and Pretorius, E. (2020). The biology of lactoferrin, an iron-binding protein that can help defend against viruses and bacteria. Frontiers in Immunology, 11(1221).

Kim, C., Spano, J., Park, E.-K., and Wi, S. (2009). Evidence of pores and thinned lipid bilayers induced in oriented lipid membranes interacting with the antimicrobial peptides, magainin-2 and aurein-3.3. Biochimica et Biophysica Acta (BBA) - Biomembranes, 1788(7): 1482–1496.

Koehbach, J., and Craik, D. J. (2019). The vast structural diversity of antimicrobial peptides. Trends in Pharmacological Sciences, 40(7): 517–528.

Koga, N., Koga, R., Liu, G., Castellanos, J., Montelione, G. T., and Baker, D. (2021). Role of backbone strain in *de novo* design of complex α/β protein structures. Nature Communications, 12(1): 3921.

Kozhikhova, K. V., Shilovskiy, I. P., Shatilov, A. A., Timofeeva, A. V., Turetskiy, E. A., Vishniakova, L. I., Nikolskii, A. A., Barvinskaya, E. D., Karthikeyan, S., and Smirnov, V. V. (2020). Linear and dendrimeric antiviral peptides: Design, chemical synthesis and activity against human respiratory syncytial virus. Journal of Materials Chemistry, B 8(13): 2607–2617.

Kragol, G., Lovas, S., Varadi, G., Condie, B. A., Hoffmann, R., and Otvos, L. (2001). The Antibacterial Peptide Pyrrhocoricin Inhibits the ATPase Actions of DnaK and Prevents Chaperone-Assisted Protein Folding. Biochemistry, 40(10): 3016–3026.

Kranjec, C., Kristensen, S. S., Bartkiewicz, K. T., Brønner, M., Cavanagh, J. P., Srikantam, A., Mathiesen, G., and Diep, D. B. (2021). A bacteriocin-based treatment option for Staphylococcus haemolyticus biofilms. Scientific Reports, 11(1): 13909.

Kranjec, C., Morales Angeles, D., Torrissen Mårli, M., Fernández, L., García, P., Kjos, M., and Diep, D. B. (2021). Staphylococcal biofilms: Challenges and novel therapeutic perspectives. Antibiotics, 10(2): 131.

Kulkarni, N. N., O'Neill, A. M., Dokoshi, T., Luo, E. W. C., Wong, G. C. L., and Gallo, R. L. (2021). Sequence determinants in the cathelicidin LL-37 that promote inflammation via presentation of RNA to scavenger receptors. Journal of Biological Chemistry, 297(1): 100828.

Kumar, A., Hossain, R. A., Yost, S. A., Bu, W., Wang, Y., Dearborn, A. D., Grakoui, A., Cohen, J. I., and Marcotrigiano, J. (2021). Structural insights into hepatitis C virus receptor binding and entry. Nature, 598(7881): 521–525.

Kumar, P., Kizhakkedathu, J. N., and Straus, S. K. (2018). Antimicrobial peptides: diversity, mechanism of action and strategies to improve the activity and biocompatibility *in vivo*. Biomolecules, 8(1): 4.

Kumariya, R., Garsa, A. K., Rajput, Y. S., Sood, S. K., Akhtar, N., and Patel, S. (2019). Bacteriocins: Classification, synthesis, mechanism of action and resistance development in food spoilage causing bacteria. Microbial Pathogenesis, 128: 171–177.

Kurpe, S. R., Grishin, S. Y., Surin, A. K., Panfilov, A. V., Slizen, M. V., Chowdhury, S. D., and Galzitskaya, O. V. (2020). Antimicrobial and amyloidogenic activity of peptides. Can antimicrobial peptides be used against sars-cov-2? International Journal of Molecular Sciences, 21(24): 9552.

Lee, M.-T., Chen, F.-Y., and Huang, H. W. (2004). Energetics of Pore Formation Induced by Membrane Active Peptides. Biochemistry, 43(12): 3590–3599.

Leeuw, E.d., Li, C., Zeng, P., Li, C., Buin, M. D.-d., Lu, W.-Y., Breukink, E., and Lu, W. (2010). Functional interaction of human neutrophil peptide-1 with the cell wall precursor lipid II. FEBS Letters, 584(8): 1543–1548.

Lei, J., Sun, L., Huang, S., Zhu, C., Li, P., He, J., Mackey, V., Coy, D. H., and He, Q. (2019). The antimicrobial peptides and their potential clinical applications. American Journal of Translational Research, 11(7): 3919.

Leikina, E., Delanoe-Ayari, H., Melikov, K., Cho, M.-S., Chen, A., Waring, A. J., Wang, W., Xie, Y., Loo, J. A., Lehrer, R. I., and Chernomordik, L. V. (2005). Carbohydrate-binding molecules inhibit viral fusion and entry by crosslinking membrane glycoproteins. Nature Immunology, 6(10): 995–1001.

Levinson, W., and Jawetz, E. (1996). Medical microbiology and immunology: examination and board review: Appleton & Lange.

Lewies, A., Du Plessis, L., and Wentzel, J. (2019). Probiotics Antimicrob. Proteins, 11: 370–381.

Li, F., Lang, Y., Ji, Z., Xia, Z., Han, Y., Cheng, Y., Liu, G., Sun, F., Zhao, Y., and Gao, M. (2019). A scorpion venom peptide Ev37 restricts viral late entry by alkalizing acidic organelles. Journal of Biological Chemistry, 294(1): 182–194.

Li, J., Hu, S., Jian, W., Xie, C., and Yang, X. (2021). Plant antimicrobial peptides: structures, functions, and applications. Botanical Studies, 62(1): 5.

Li, Q., Zhao, Z., Zhou, D., Chen, Y., Hong, W., Cao, L., Yang, J., Zhang, Y., Shi, W., Cao, Z., Wu, Y., Yan, H., and Li, W. (2011). Virucidal activity of a scorpion venom peptide variant mucroporin-M1 against measles, SARS-CoV and influenza H5N1 viruses. Peptides, 32(7): 1518–1525.

Li, Y., Liu, D., Wang, Y., Su, W., Liu, G., and Dong, W. (2021). The importance of glycans of viral and host proteins in enveloped virus infection. Frontiers in Immunology, 12(1544).

Lima, P. G., Oliveira, J. T., Amaral, J. L., Freitas, C. D., and Souza, P. F. (2021). Synthetic antimicrobial peptides: Characteristics, design, and potential as alternative molecules to overcome microbial resistance. Life Sciences, 119647.

Lin, F.-c., and Young, H. A. (2014). Interferons: Success in anti-viral immunotherapy. Cytokine & Growth Factor Reviews, 25(4): 369–376.

Liu, H., Diao, H., Hou, J., Yu, H., and Wen, H. (2021). Soluble expression and purification of human β-defensin DEFB136 in *Escherichia coli* and identification of its bioactivity. Protein Expression and Purification, 188: 105968.

Liu, R., Khan, R. A. A., Yue, Q., Jiao, Y., Yang, Y., Li, Y., and Xie, B. (2020). Discovery of a new antifungal lipopeptaibol from Purpureocillium lilacinum using MALDI-TOF-IMS. Biochemical and Biophysical Research Communications, 527(3): 689–695.

Liu, S., Bao, J., Lao, X., and Zheng, H. (2018). Novel 3D structure based model for activity prediction and design of antimicrobial peptides. Scientific Reports, 8(1): 11189.

Liu, X., Guo, C., Huang, Y., Zhang, X., and Chen, Y. (2015). Inhibition of porcine reproductive and respiratory syndrome virus by Cecropin D *in vitro*. Infection, Genetics and Evolution, 34: 7–16.

Liu, Y., Shi, J., Tong, Z., Jia, Y., Yang, B., and Wang, Z. (2021). The revitalization of antimicrobial peptides in the resistance era. Pharmacological Research, 163: 105276.

Lokhande, K. B., Banerjee, T., Swamy, K. V., Ghosh, P., and Deshpande, M. (2021). An *in silico* scientific basis for LL-37 as a therapeutic for Covid-19. Proteins: Structure, Function, and Bioinformatics n/a(n/a).

Lorin, C., Saidi, H., Belaid, A., Zairi, A., Baleux, F., Hocini, H., Bélec, L., Hani, K., and Tangy, F. (2005). The antimicrobial peptide dermaseptin S4 inhibits HIV-1 infectivity *in vitro*. Virology, 334(2): 264–275.

Luo, X., Ouyang, J., Wang, Y., Zhang, M., Fu, L., Xiao, N., Gao, L., Zhang, P., Zhou, J., and Wang, Y. (2021). A novel anionic cathelicidin lacking direct antimicrobial activity but with potent anti-inflammatory and wound healing activities from the salamander Tylototriton kweichowensis. Biochimie, 191: 37–50.

Luteijn, R. D., Praest, P., Thiele, F., Sadasivam, S. M., Singethan, K., Drijfhout, J. W., Bach, C., de Boer, S. M., Lebbink, R. J., and Tao, S. (2020). A broad-spectrum antiviral peptide blocks infection of viruses by binding to phosphatidylserine in the viral envelope. Cells, 9(9): 1989.

Madavaraju, K., Koganti, R., Volety, I., Yadavalli, T., and Shukla, D. (2021). Herpes simplex virus cell entry mechanisms: an update. Frontiers in Cellular and Infection Microbiology, 10(852).

Magana, M., Pushpanathan, M., Santos, A. L., Leanse, L., Fernandez, M., Ioannidis, A., Giulianotti, M. A., Apidianakis, Y., Bradfute, S., and Ferguson, A. L. (2020). The value of antimicrobial peptides in the age of resistance. The Lancet Infectious Diseases.

Maginnis, M. S. (2018). Virus–receptor interactions: the key to cellular invasion. Journal of Molecular Biology, 430(17): 2590–2611.

Mahlapuu, M., Björn, C., and Ekblom, J. (2020). Antimicrobial peptides as therapeutic agents: Opportunities and challenges. Critical Reviews in Biotechnology, 40(7): 978–992.

Mahlapuu, M., Hakansson, J., Ringstad, L., and Bjorn, C. (2016). Antimicrobial peptides: an emerging category of therapeutic agents. Front Cell Infect. Microbiol., 6: 194.

Maiti, B. K. (2020). Potential Role of Peptide-Based Antiviral Therapy Against SARS-CoV-2 Infection. ACS Pharmacology & Translational Science, 3(4): 783–785.

Mak, P., Szewczyk, A., Mickowska, B., Kicinska, A., and Dubin, A. (2001). Effect of antimicrobial apomyoglobin 56–131 peptide on liposomes and planar lipid bilayer membrane. International Journal of Antimicrobial Agents, 17(2): 137–142.

Mammari, N., Krier, Y., Albert, Q., Devocelle, M., and Varbanov, M. (2021). Plant-derived antimicrobial peptides as potential antiviral agents in systemic viral infections. Pharmaceuticals, 14(8): 774.

Manabe, T., and Kawasaki, K. 2017. D-form KLKLLLLLKLK-NH2 peptide exerts higher antimicrobial properties than its L-form counterpart via an association with bacterial cell wall components. Scientific Reports, 7(1): 43384.

Mandel, S., Michaeli, J., Nur, N., Erbetti, I., Zazoun, J., Ferrari, L., Felici, A., Cohen-Kutner, M., and Bachnoff, N. (2021). OMN6 a novel bioengineered peptide for the treatment of multidrug resistant Gram negative bacteria. Scientific Reports, 11(1): 6603.

Manzini, M. C., Perez, K. R., Riske, K. A., Bozelli, J. C., Santos, T. L., da Silva, M. A., Saraiva, G. K. V., Politi, M. J., Valente, A. P., Almeida, F. C. L., Chaimovich, H., Rodrigues, M. A., Bemquerer, M. P., Schreier, S., and Cuccovia, I. M. (2014). Peptide:lipid ratio and membrane surface charge determine the mechanism of action of the antimicrobial peptide BP100. Conformational and functional studies. Biochimica et Biophysica Acta (BBA) - Biomembranes, 1838(7): 1985–1999.

Marcocci, M. E., Amatore, D., Villa, S., Casciaro, B., Aimola, P., Franci, G., Grieco, P., Galdiero, M., Palamara, A. T., Mangoni, M. L., and Nencioni, L. (2018). The amphibian antimicrobial peptide temporin B Inhibits <i>*In Vitro*</i> Herpes Simplex Virus 1 Infection. Antimicrobial Agents and Chemotherapy, 62(5): e02367–02317.

Mardirossian, M., Barrière, Q., Timchenko, T., Müller, C., Pacor, S., Mergaert, P., Scocchi, M., and Wilson, D. N. (2018). Fragments of the nonlytic proline-rich antimicrobial peptide Bac5 Kill *Escherichia coli* cells by inhibiting protein synthesis. Antimicrobial Agents and Chemotherapy, 62(8): e00534–00518.

Mardirossian, M., Pérébaskine, N., Benincasa, M., Gambato, S., Hofmann, S., Huter, P., Müller, C., Hilpert, K., Innis, C. A., Tossi, A., and Wilson, D. N. (2018). The dolphin proline-rich antimicrobial peptide tur1a inhibits protein synthesis by targeting the bacterial ribosome. Cell Chemical Biology, 25(5): 530–539.e537.

Martínez-Núñez, M. (2016). y López VEL. Nonribosomal peptides synthetases and their applications in industry. Sustain Chem Process, 4(1): 13.

Matanic, V. C. A., and Castilla, V. (2006). Tracking Infectious Diseases since, 2006.

Matthews, T., Salgo, M., Greenberg, M., Chung, J., DeMasi, R., and Bolognesi, D. (2004). Enfuvirtide: the first therapy to inhibit the entry of HIV-1 into host CD4 lymphocytes. Nature Reviews Drug Discovery, 3(3): 215–225.

McMillan, K. A., and Coombs, M. R. (2020). Examining the natural role of amphibian antimicrobial peptide magainin. Molecules, 25(22): 5436.

Memariani, H., Memariani, M., Moravvej, H., and Shahidi-Dadras, M. (2020). Melittin: a venom-derived peptide with promising anti-viral properties. European Journal of Clinical Microbiology & Infectious Diseases, 39(1): 5–17.

Mertinková, P., Mochnáčová, E., Bhide, K., Kulkarni, A., Tkáčová, Z., Hruškovicová, J., and Bhide, M. (2021). Development of peptides targeting receptor binding site of the envelope glycoprotein to contain the West Nile virus infection. Scientific Reports, 11(1): 20131.

Mirabelli, C., Wotring, J. W., Zhang, C. J., McCarty, S. M., Fursmidt, R., Pretto, C. D., Qiao, Y., Zhang, Y., Frum, T., Kadambi, N. S., Amin, A. T., O'Meara, T. R., Spence, J. R., Huang, J., Alysandratos, K. D., Kotton, D. N., Handelman, S.,K., Wobus, C. E., Weatherwax, K. J., Mashour, G. A., O'Meara, M. J., Chinnaiyan, A. M., and Sexton, J. Z. (2021). Morphological cell profiling of SARS-CoV-2 infection identifies drug repurposing candidates for COVID-19. Proceedings of the National Academy of Sciences, 118(36): e2105815118.

Montero-Alejo, V., Corzo, G., Porro-Suardíaz, J., and Pardo-Ruiz, Z. (2016). Erick Perera 3, Leandro Rodríguez-Viera 4, Gabriela Sánchez-Díaz 5, Erix Wiliam 4 Hernández-Rodríguez 5, Carlos Álvarez 6, Steve Peigneur 7, Jan Tytgat 7, Rolando Perdomo-5 Morales 6.

Mookherjee, N., Anderson, M. A., Haagsman, H. P., and Davidson, D. J. (2020). Antimicrobial host defence peptides: functions and clinical potential. Nature Reviews Drug Discovery, 19(5): 311–332.

Moravej, H., Moravej, Z., Yazdanparast, M., Heiat, M., Mirhosseini, A., Moosazadeh, Moghaddam, M., and Mirnejad, R. (2018). Antimicrobial peptides: features, action, and their resistance mechanisms in bacteria. Microbial Drug Resistance, 24(6): 747–767.

Moreno-Habel, D. A., Biglang-awa, I. M., Dulce, A., Luu, D. D., Garcia, P., Weers, P. M. M., and Haas-Stapleton, E. J. (2012). Inactivation of the budded virus of Autographa californica M nucleopolyhedrovirus by gloverin. Journal of Invertebrate Pathology, 110(1): 92–101.

Moretta, A., Scieuzo, C., Petrone, A. M., Salvia, R., Manniello, M. D., Franco, A., Lucchetti, D., Vassallo, A., Vogel, H., and Sgambato, A. (2021). Antimicrobial peptides: A new hope in biomedical and pharmaceutical fields. Frontiers in Cellular and Infection Microbiology, 11: 453.

Mousavi Maleki, M. S., Rostamian, M., and Madanchi, H. (2021). Antimicrobial peptides and other peptide-like therapeutics as promising candidates to combat SARS-CoV-2. Expert Review of Anti-infective Therapy, 19(10): 1205–1217.

Mukherjee, P. K., Wiest, A., Ruiz, N., Keightley, A., Moran-Diez, M. E., McCluskey, K., Pouchus, Y. F., and Kenerley, C. M. (2011). Two classes of new peptaibols are synthesized by a single non-ribosomal peptide synthetase of *Trichoderma virens*. Journal of Biological Chemistry, 286(6): 4544–4554.

Müller, D. M., Carrasco, M. S., Simonetta, A. C., Beltramini, L. M., and Tonarelli, G. G. (2007). A synthetic analog of plantaricin 149 inhibiting food-borne pathogenic bacteria:evidence for α-helical conformation involved in bacteria–membrane interaction. Journal of Peptide Science, 13(3): 171–178.

Mwangi, J., Hao, X., Lai, R., and Zhang, Z.-Y. (2019). Antimicrobial peptides: new hope in the war against multidrug resistance. Zool. Res., 40(6): 488–505.

Naito, A., Nagao, T., Norisada, K., Mizuno, T., Tuzi, S., and Saitô, H. (2000). Conformation and Dynamics of Melittin Bound to Magnetically Oriented Lipid Bilayers by Solid-State 31P and 13C NMR Spectroscopy. Biophysical Journal, 78(5): 2405–2417.

Nath, A. (2021). Prediction for understanding the effectiveness of antiviral peptides. Computational Biology and Chemistry, 95: 107588.

Nguyen, E. K., Nemerow, G. R., and Smith, J. G. (2010). Direct evidence from single-cell analysis that human α-defensins block adenovirus uncoating to neutralize infection. Journal of Virology, 84(8): 4041–4049.

Niyonsaba, F., Iwabuchi, K., Someya, A., Hirata, M., Matsuda, H., Ogawa, H., and Nagaoka, I. (2002). A cathelicidin family of human antibacterial peptide LL-37 induces mast cell chemotaxis. Immunology, 106(1): 20–26.

O'Donnell, C. D., and Shukla, D. (2008). The importance of heparan sulfate in herpesvirus infection. Virologica Sinica, 23(6): 383–393.

Oliveira, N. G. J., Cardoso, M. H., Velikova, N., Giesbers, M., Wells, J. M., Rezende, T. M. B., de Vries, R., and Franco, O. L. (2020). Physicochemical-guided design of cathelicidin-derived peptides generates membrane active variants with therapeutic potential. Scientific Reports, 10(1): 9127.

Omidvar, R., Vosseler, N., Abbas, A., Gutmann, B., Grünwald-Gruber, C., Altmann, F., Siddique, S., and Bohlmann, H. (2021). Analysis of a gene family for PDF-like peptides from Arabidopsis. Scientific reports, 11(1): 1–16.

Ongey, E. L., Pflugmacher, S., and Neubauer, P. (2018). Bioinspired designs, molecular premise and tools for evaluating the ecological importance of antimicrobial peptides. Pharmaceuticals, 11(3): 68.

Otvos, L., O. I., Rogers, M. E., Consolvo, P. J., Condie, B. A., Lovas, S., Bulet, P., and Blaszczyk-Thurin, M. (2000). Interaction between heat shock proteins and antimicrobial peptides. Biochemistry, 39(46): 14150–14159.

Pahar, B., Madonna, S., Das, A., Albanesi, C., and Girolomoni, G. (2020). Immunomodulatory role of the antimicrobial LL-37 peptide in autoimmune diseases and viral infections. Vaccines, 8(3): 517.

Pan, Y.-L., Cheng, J. T .J., Hale, J., Pan, J., Hancock, R. E. W., and Straus, S. K. (2007). Characterization of the Structure and Membrane Interaction of the Antimicrobial Peptides Aurein 2.2 and 2.3 from Australian Southern Bell Frogs. Biophysical Journal, 92(8): 2854–2864.

Pandidan, S., and Mechler, A. (2019). Nano-viscosimetry analysis of the membrane disrupting action of the bee venom peptide melittin. Scientific Reports, 9(1): 10841.

Pandit, G., Chowdhury, N., Mohid, S. A., Bidkar, A. P., Bhunia, A., and Chatterjee, S. (2021). Effect of secondary structure and side chain length of hydrophobic amino acid residues on the antimicrobial activity and toxicity of 14-residue-long *de novo* AMPs. Chem. Med. Chem., 16(2): 355–367.

Parthasarathy, A., Borrego, E. J., Savka, M. A., Dobson, R. C. J., and Hudson, A. O. (2021). Amino acid-derived defense metabolites from plants: A potential source to facilitate novel antimicrobial development. Journal of Biological Chemistry, 296: 100438.

Pen, G., Yang, N., Teng, D., Mao, R., Hao, Y., and Wang, J. (2020). A review on the use of antimicrobial peptides to combat porcine viruses. Antibiotics, 9(11): 801.

Peng, J., Wu, Z., Liu, W., Long, H., Zhu, G., Guo, G., and Wu, J. (2019). Antimicrobial functional divergence of the cecropin antibacterial peptide gene family in Musca domestica. Parasites & Vectors, 12(1): 537.

Porcelli, F., Verardi, R., Shi, L., Henzler-Wildman, K. A., Ramamoorthy, A., and Veglia, G. (2008). NMR structure of the cathelicidin-derived human antimicrobial peptide LL-37 in dodecylphosphocholine micelles. Biochemistry, 47(20): 5565–5572.

Pouny, Y., Rapaport, D., Mor, A., Nicolas, P., and Shai, Y. (1992). Interaction of antimicrobial dermaseptin and its fluorescently labeled analogs with phospholipid membranes. Biochemistry, 31(49): 12416–12423.

Quintero-Gil, C., Parra-Suescún, J., Lopez-Herrera, A., and Orduz, S. 2017. *In-silico* design and molecular docking evaluation of peptides derivatives from bacteriocins and porcine beta defensin-2 as inhibitors of Hepatitis E virus capsid protein. VirusDisease, 28(3): 281–288.

Rahbar Saadat, Y., Hosseiniyan Khatibi, S. M., Zununi Vahed, S., and Ardalan, M. (2021). Host Serine Proteases: A Potential Targeted Therapy for COVID-19 and Influenza. Front Mol. Biosci., 8: 725528–725528.

Rapaport, D., and Shai, Y. (1991). Interaction of fluorescently labeled pardaxin and its analogues with lipid bilayers. The Journal of Biological Chemistry, 266(35): 23769–23775.

Rasmusson, A. G., and Moller, I. M. (1991). Effect of calcium ions and inhibitors on internal NAD(P)H dehydrogenases in plant mitochondria. European Journal of Biochemistry, 202(2): 617–623.

Ray, G., Noori, M. T., and Ghangrekar, M. M. (2017). Novel application of peptaibiotics derived from *Trichoderma* sp. for methanogenic suppression and enhanced power generation in microbial fuel cells. RSC Advances, 7(18): 10707–10717.

Reffuveille, F., Haney, E., Straus, S., and Hancock, R. 2014. Broad-spectrum anti-biofilm peptide that targets a cellular stress responses. PLoS Pathog, 10: e1004152.

Rima, M., Rima, M., Fajloun, Z., Sabatier, J.-M., Bechinger, B., and Naas, T. (2021). Antimicrobial peptides: a potent alternative to antibiotics. Antibiotics, 10(9): 1095.

Röhrl, J., Yang, D., Oppenheim, J. J., and Hehlgans, T. (2010). Human β-Defensin 2 and 3 and Their Mouse Orthologs Induce Chemotaxis through Interaction with CCR2. The Journal of Immunology, 184(12): 6688–6694.

Rončević, T., Puizina, J., and Tossi, A. (2019). Antimicrobial peptides as anti-infective agents in pre-post-antibiotic era? International Journal of Molecular Sciences, 20(22): 5713.

Ruzsics, Z., Hoffmann, K., Riedl, A., Krawczyk, A., Widera, M., Sertznig, H., Schipper, L., Kapper-Falcone, V., Debreczeny, M., Ernst, W., Grabherr, R., Hengel, H., and Harant, H. (2020). A novel, broad-acting peptide inhibitor of double-stranded DNA virus gene expression and replication. Frontiers in Microbiology, 11: 601555–601555.

Ryu, M., Park, J., Yeom, J.-H., Joo, M., and Lee, K. (2021). Rediscovery of antimicrobial peptides as therapeutic agents. Journal of Microbiology, 59(2): 113–123.

Sala, A., Ardizzoni, A., Ciociola, T., Magliani, W., Conti, S., Blasi, E., and Cermelli, C. (2018). Antiviral activity of synthetic peptides derived from physiological proteins. Intervirology, 61(4): 166–173.

Salomón, R. A., and Farías, R. N. (1992). Microcin 25, a novel antimicrobial peptide produced by *Escherichia coli*. Journal of Bacteriology, 174(22): 7428–7435.

Sani, M.-A., Whitwell, T., Gehman, J., Robins-Browne, R., Pantarat, N., Attard, T., Reynolds, E., O'Brien-Simpson, N., and Separovic, F. (2013). Maculatin 1.1 disrupts Staphylococcus aureus lipid membranes via a pore mechanism. Antimicrobial Agents and Chemotherapy, 57(8): 3593–3600.

Santana, F. L., Estrada, K., Ortiz, E., and Corzo, G. (2021). Reptilian β-defensins: Expanding the repertoire of known crocodylian peptides. Peptides, 136: 170473.

Scheenstra, M. R., van den Belt, M., Tjeerdsma-van Bokhoven, J. L. M., Schneider, V. A. F., Ordonez, S. R., van Dijk, A., Veldhuizen, E. J. A., and Haagsman, H. P. (2019). Cathelicidins PMAP-36, LL-37 and CATH-2 are similar peptides with different modes of action. Scientific Reports, 9(1): 4780.

Schibli, D. J., Epand, R. F., Vogel, H. J., and Epand, R. M. (2002). Tryptophan-rich antimicrobial peptides: comparative properties and membrane interactions. Biochemistry and Cell Biology, 80(5): 667–677.

Schito, A. M., Schito, G. C., and Alfei, S. (2021). Synthesis and antibacterial activity of cationic amino acid-conjugated dendrimers loaded with a mixture of two triterpenoid acids. Polymers, 13(4): 521.

Schmidt, R., Knappe, D., Wende, E., Ostorházi, E., and Hoffmann, R. (2017). *In vivo* efficacy and pharmacokinetics of optimized apidaecin analogs. Front Chem., 5(15).

Sengupta, D., Leontiadou, H., Mark, A. E., and Marrink, S.-J. (2008). Toroidal pores formed by antimicrobial peptides show significant disorder. Biochimica et Biophysica Acta (BBA) - Biomembranes, 1778(10): 2308–2317.

Shafee, T. M., Lay, F. T., Phan, T. K., Anderson, M. A., and Hulett, M. D. (2017). Convergent evolution of defensin sequence, structure and function. Cellular and Molecular Life Sciences, 74(4): 663–682.

Shafee, T. M. A., Lay, F. T., Hulett, M. D., and Anderson, M. A. (2016). The defensins consist of two independent, convergent protein superfamilies. Molecular Biology and Evolution, 33(9): 2345–2356.

Shenkarev, Z. O., Balandin, S. V., Trunov, K.,I., Paramonov, A. S., Sukhanov, S. V., Barsukov, L. I., Arseniev, A. S., and Ovchinnikova, T. V. (2011). Molecular mechanism of action of β-hairpin antimicrobial peptide arenicin: oligomeric structure in dodecylphosphocholine micelles and pore formation in planar lipid bilayers. Biochemistry, 50(28): 6255–6265.

Sherman, M. B., Smith, H. Q., and Smith, T. J. (2020). The dynamic life of virus capsids. Viruses, 12(6): 618.

Shi, G., Kang, X., Dong, F., Liu, Y., Zhu, N., Hu, Y., Xu, H., Lao, X., and Zheng, H. (2022). DRAMP 3.0: an enhanced comprehensive data repository of antimicrobial peptides. Nucleic Acids Res. 2022 Jan 7; 50(D1): D488–D496.

Shi, J., Ross, C. R., Leto, T. L., and Blecha, F. (1996). PR-39, a proline-rich antibacterial peptide that inhibits phagocyte NADPH oxidase activity by binding to Src homology 3 domains of p47 phox. Proceedings of the National Academy of Sciences of the United States of America, 93(12): 6014–6018.

Shinnar, A. E., Butler, K. L., and Park, H. J. (2003). Cathelicidin family of antimicrobial peptides: proteolytic processing and protease resistance. Bioorganic Chemistry, 31(6): 425–436.

Smith, J. G., Silvestry, M., Lindert, S., Lu, W., Nemerow, G. R., and Stewart, P. L. (2010). Insight into the mechanisms of adenovirus capsid disassembly from studies of defensin neutralization. PLoS pathogens, 6(6): e1000959.

Smyth, D. S. (2020). COVID-19, Ebola, and Measles: achieving sustainability in the era of emerging and reemerging infectious diseases. Environment: Science and Policy for Sustainable Development, 62(6): 31–40.

Solanki, S. S., Singh, P., Kashyap, P., Sansi, M. S., and Ali, S. A. (2021). Promising role of defensins peptides as therapeutics to combat against viral infection. Microbial. Pathogenesis, 104930.

Soltani, S., Hammami, R., Cotter, P. D., Rebuffat, S., Said, L. B., Gaudreau, H., Bédard, F., Biron, E., Drider, D., and Fliss, I. (2020). Bacteriocins as a new generation of antimicrobials: toxicity aspects and regulations. FEMS Microbiology Reviews, 45(1).

Song, C., Weichbrodt, C., Salnikov, E. S., Dynowski, M., Forsberg, B. O., Bechinger, B., Steinem, C., de Groot, B. L., Zachariae, U., and Zeth, K. (2013). Crystal structure and functional mechanism of a human antimicrobial membrane channel. Proceedings of the National Academy of Sciences, 110(12): 4586.

Soper, A., Kimura, I., Nagaoka, S., Konno, Y., Yamamoto, K., Koyanagi, Y., and Sato, K. (2018). Type I Interferon Responses by HIV-1 Infection: Association with Disease Progression and Control. Frontiers in Immunology, 8(1823).

Sousa, F. H., Casanova, V., Findlay, F., Stevens, C., Svoboda, P., Pohl, J., Proudfoot, L., and Barlow, P.G. (2017). Cathelicidins display conserved direct antiviral activity towards rhinovirus. Peptides, 95: 76–83.

Stambuk, F., Ojeda, C., Matos, G. M., Rosa, R. D., Mercado, L., and Schmitt, P. (2021). Big defensin from the scallop Argopecten purpuratus ApBD1 is an antimicrobial peptide which entraps bacteria through nanonets formation. Fish & Shellfish Immunology.

Su, X., Wang, Q., Wen, Y., Jiang, S., and Lu, L. (2020). Protein- and Peptide-Based Virus Inactivators: Inactivating Viruses Before Their Entry Into Cells. Frontiers in Microbiology, 11(1063).

Subbalakshmi, C., and Sitaram, N. (1998). Mechanism of antimicrobial action of indolicidin. FEMS Microbiology Letters, 160(1): 91–96.

Sutter, G. (2020). A vital gene for modified vaccinia virus Ankara replication in human cells. Proceedings of the National Academy of Sciences, 117(12): 6289–6291.

Szekeres, A., Leitgeb, B., Kredics, L., Antal, Z., Hatvani, L., Manczinger, L., and Vágvölgyi, C. (2005). Peptaibols and related peptaibiotics of Trichoderma. Acta microbiologica et immunologica Hungarica, 52(2): 137–168.

Takahashi, T., Kulkarni, N. N., Lee, E. Y., Zhang, L.-j., Wong, G. C., and Gallo, R. L. (2018). Cathelicidin promotes inflammation by enabling binding of self-RNA to cell surface scavenger receptors. Scientific Reports, 8(1): 1–13.

Takiguchi, T., Morizane, S., Yamamoto, T., Kajita, A., Ikeda, K., and Iwatsuki, K. (2014). Cathelicidin antimicrobial peptide LL-37 augments interferon-β expression and antiviral activity induced by double-stranded RNA in keratinocytes. British Journal of Dermatology, 171(3): 492–498.

Tang, M., and Hong, M. (2009). Structure and mechanism of β-hairpin antimicrobial peptides in lipid bilayers from solid-state NMR spectroscopy. Molecular BioSystems, 5(4): 317–322.

Tassanakajon, A., Somboonwiwat, K., and Amparyup, P. (2015). Sequence diversity and evolution of antimicrobial peptides in invertebrates. Developmental & Comparative Immunology, 48(2): 324–341.

Tavares, T. D., Antunes, J. C., Padrão, J., Ribeiro, A. I., Zille, A., Amorim, M. T. P., Ferreira, F., and Felgueiras, H. P. (2020). Activity of specialized biomolecules against gram-positive and gram-negative bacteria. Antibiotics, 9(6): 314.

Teixeira, C., Gomes, J. R. B., Gomes, P., and Maurel, F. (2011). Viral surface glycoproteins, gp120 and gp41, as potential drug targets against HIV-1: Brief overview one quarter of a century past the approval of zidovudine, the first anti-retroviral drug. European Journal of Medicinal Chemistry, 46(4): 979–992.

Timur, S. S., and Gürsoy, R. N. (2021). The role of peptide-based therapeutics in oncotherapy. Journal of Drug Targeting, 29(10): 1048–1062.

Tornesello, A. L., Borrelli, A., Buonaguro, L., Buonaguro, F. M., and Tornesello, M. L. (2020). Antimicrobial peptides as anticancer agents: Functional properties and biological activities. Molecules, 25(12): 2850.

Tosteson, M. T., and Tosteson, D. C. (1981). The sting. Melittin forms channels in lipid bilayers. Biophys. J., 36(1): 109–116.

Uyterhoeven, E. T., Butler, C. H., Ko, D., and Elmore, D. E. (2008). Investigating the nucleic acid interactions and antimicrobial mechanism of buforin II. FEBS Letters, 582(12): 1715–1718.

Van Wetering, S. (1997). Mannesse-Lazeroms SP, Van Sterkenburg MA, Daha MR, Dijkman JH, Hiemstra PS. Effect of defensins on interleukin-8 synthesis in airway epithelial cells. Am. J. Physiol. Lung Cell Mol. Physiol., 272: L888–L896.

Vedadghavami, A., Zhang, C., and Bajpayee, A. G. (2020). Overcoming negatively charged tissue barriers: Drug delivery using cationic peptides and proteins. Nano Today, 34: 100898.

Vilas Boas, L. C. P., Campos, M. L., Berlanda, R. L. A., de Carvalho Neves, N., and Franco, O. L. (2019). Antiviral peptides as promising therapeutic drugs. Cellular and Molecular Life Sciences, 76(18): 3525–3542.

Wang, C., Hong, T., Cui, P., Wang, J., and Xia, J. (2021). Antimicrobial peptides towards clinical application: delivery and formulation. Advanced Drug Delivery Reviews.

Wang, C., Wang, S., Li, D., Chen, P., Han, S., Zhao, G., Chen, Y., Zhao, J., Xiong, J., Qiu, J., Wei, D.-Q., Zhao, J., and Wang, J. (2021). Human Cathelicidin Inhibits SARS-CoV-2 Infection: Killing Two Birds with One Stone. ACS Infectious Diseases, 7(6): 1545–1554.

Wang, J., Dou, X., Song, J., Lyu, Y., Zhu, X., Xu, L., Li, W., and Shan, A. (2019). Antimicrobial peptides: Promising alternatives in the post feeding antibiotic era. Medicinal Research Reviews, 39(3): 831–859.

Wang, Y., Wang, M., Shan, A., and Feng, X. (2020). Avian host defense cathelicidins: structure, expression, biological functions, and potential therapeutic applications. Poultry Science, 99(12): 6434–6445.

Wielgat, P., Rogowski, K., Godlewska, K., and Car, H. (2020). Coronaviruses: is sialic acid a gate to the eye of cytokine storm? From the Entry to the Effects. Cells, 9(9): 1963.

Wilson, S. S., Wiens, M. E., and Smith, J. G. (2013). Antiviral mechanisms of human defensins. Journal of Molecular Biology, 425(24): 4965–4980.

Wimley, W. C. (2010). Describing the mechanism of antimicrobial peptide action with the interfacial activity model. ACS Chemical Biology, 5(10): 905–917.

Wong, H., Bowie, J. H., and Carver, J. A. (1997). The solution structure and activity of caerin 1.1, an antimicrobial peptide from the australian green tree frog, litoria splendida. European Journal of Biochemistry, 247(2): 545–557.

Wong, J. H., Legowska, A., Rolka, K., Ng, T. B., Hui, M., Cho, C. H., Lam, W. W. L., Au, S. W. N., Gu, O. W., and Wan, D. C. C. (2011). Effects of cathelicidin and its fragments on three key enzymes of HIV-1. Peptides, 32(6): 1117–1122.

Wong, J. H., Liu, Z., Law, K. W. K., Liu, F., Xia, L., Wan, D. C. C., and Ng, T. B. (2014). A study of effects of peptide fragments of bovine and human lactoferrins on activities of three key HIV-1 enzymes. Peptides, 62: 183–188.

Xiong, Y.-Q., Yeaman, M. R., and Bayer, A. S. (1999). *In vitro* antibacterial activities of platelet microbicidal protein and neutrophil defensin against *Staphylococcus aureus* are influenced by antibiotics differing in mechanism of action. Antimicrobial Agents and Chemotherapy, 43(5): 1111–1117.

Xu, D., and Lu, W. (2020). Defensins: a double-edged sword in host immunity. Frontiers in Immunology, 11(764).

Yamaguchi, S., Huster, D., Waring, A., Lehrer, R. I., Kearney, W., Tack, B. F., and Hong, M. (2001). Orientation and dynamics of an antimicrobial peptide in the lipid bilayer by solid-state NMR spectroscopy. Biophysical Journal, 81(4): 2203–2214.

Yang, D., Chen, Q., Schmidt, A. P., Anderson, G. M., Wang, J. M., Wooters, J., Oppenheim, J. J., and Chertov, O. (2000). Ll-37, the neutrophil granule–and epithelial cell–derived cathelicidin, utilizes formyl peptide receptor–like 1 (Fprl1) as a receptor to chemoattract human peripheral blood neutrophils, monocytes, and T Cells. Journal of Experimental Medicine, 192(7): 1069–1074.

Yang, L., Weiss, T. M., Lehrer, R. I., and Huang, H. W. (2000). Crystallization of Antimicrobial Pores in Membranes: Magainin and Protegrin. Biophysical Journal, 79(4): 2002–2009.

Yanmei, L., Xiang, Q., Zhang, Q., Huang, Y., and Su, Z. (2012). Overview on the recent study of antimicrobial peptides: Origins, functions, relative mechanisms and application. 2012. Oct., 37(2): 207–215.

Yazici, A., Ortucu, S., Taskin, M., and Marinelli, L. (2018). Natural-based antibiofilm and antimicrobial peptides from Micro-organisms. Current Topics in Medicinal Chemistry, 18(24): 2102–2107.

Yoneyama, F., Imura, Y., Ohno, K., Zendo, T., Nakayama, J., Matsuzaki, K., and Sonomoto, K. (2009). Peptide-lipid huge toroidal pore, a new antimicrobial mechanism mediated by a lactococcal bacteriocin, lacticin Q. Antimicrobial Agents and Chemotherapy, 53(8): 3211–3217.

Yu, Y., Cooper, C., Wang, G., Morwitzer, M., Kota, K., Tran, J., Bradfute, S., Liu, Y., Shao, J., and Zhang, A. (2020). Engineered human cathelicidin antimicrobial peptides inhibit Ebola virus infection. iScience, 23, 100999: DOI.

Zhang, L., Ghosh, S. K., Basavarajappa, S. C., Muller-Greven, J., Penfield, J., Brewer, A., Ramakrishnan, P., Buck, M., and Weinberg, A. (2021). Molecular dynamics simulations and functional studies reveal that hBD-2 binds SARS-CoV-2 spike RBD and blocks viral entry into ACE2 expressing cells. bioRxiv: 2021.2001.2007.425621.

Zhang, Q.-Y., Yan, Z.-B., Meng, Y.-M., Hong, X.-Y., Shao, G., Ma, J.-J., Cheng, X.-R., Liu, J., Kang, J., and Fu, C.-Y. (2021). Antimicrobial peptides: mechanism of action, activity and clinical potential. Military Medical Research, 8(1): 48.

Zhao, H., To, K. K., Sze, K.-H., Yung, T. T.-M., Bian, M., Lam, H., Li, C., Chu, H., and Yuen, K.-Y. (2020). A broad-spectrum virus-and host-targeting antiviral peptide against SARS-CoV-2 and other respiratory viruses.

Ziaja, M., Dziedzic, A., Szafraniec, K., and Piastowska-Ciesielska, A. (2020). Cecropins in cancer therapies-where we have been? European Journal of Pharmacology, 882: 173317.

Chapter 8

Therapeutic Applications of Cyclodextrins Against Trypanosomiases and Leishmaniasis

Zeinab Dirany,[1] *Rima El-Dirany*,[2,3] *Gustavo González-Gaitano*[1] and *Paul A. Nguewa*[2,*]

Cyclodextrins

Since their discovery at the end of the 19th century, cyclodextrins (CDs) have become a subject of particular interest in chemical and medical research, food science, the cosmetics industry, and the environment (Kurkov and Loftsson 2013). Initially described as crystalline substances produced by the bacterial fermentation of starch, cyclodextrins were then identified as a family of cyclic oligosaccharides composed of a number of glucopyranose residues joined by α-1,4 glycosidic bonds (Morin-Crini et al. 2021). The glucose subunits are assembled in a hollow truncated cone shape with the hydroxyl groups located at the rims, while the cavity is lined with hydrogen and ether oxygen atoms. This peculiar distribution bestows cyclodextrins water-solubility due to the hydrophilic exterior

[1] University of Navarra, Department of Chemistry, Faculty of Sciences, 31080 Pamplona, Navarra, Spain. Emails: zdiranyahma@alumni.unav.es; gaitano@unav.es

[2] University of Navarra, ISTUN Instituto de Salud Tropical, Department of Microbiology and Parasitology. IdiSNA (Navarra Institute for Health Research), c/Irunlarrea 1, 31008 Pamplona, Navarra, Spain. Email: reldirany@alumni.unav.es

[3] Lebanese University, Faculty of Sciences I, 1003 Hadath, Lebanon.

* Corresponding author: panguewa@unav.es

surface and the ability to host a variety of hydrophobic guests in the relatively hydrophobic cavity (Crini et al. 2021).

Native cyclodextrins, α-CD, β-CD, and γ-CD, are the most used, containing six, seven, and eight glucose residues, respectively. Although larger cyclodextrins exist, as δ-CD with nine glucose units, they are more difficult to purify and less studied (Larsen 2002). One major difference lies in their aqueous solubility, which is considerably lower for β- than for α- and γ-CDs. The solubility and complex formation with specific guests can be tuned by substituting the hydroxyl at the rims of the macrocycle with other functional groups, yielding a wide variety of derivatives. Such substitutions include hydroxypropylation, methylation, sulfonation, and monotosylation (Szente and Szejtli 1999, Tripodo et al. 2013, Muderawan et al. 2005), β-CD being the most frequently used macrocycle to this effect.

The formation of an inclusion complex may improve some properties of the guest molecule, which makes them useful as pharmaceutical excipients, in order to increase the solubility, bioavailability, and stability of the drug (Challa et al. 2005). Numerous marketed products include cyclodextrins in the formulation in various administration routes, depending on the cyclodextrin type (Jambhekar and Breen 2016). For example, while native cyclodextrins are not suitable for parenteral use, 2-hydroxypropyl-β-CD (HP-β-CD) and sulfobutylether-β-CD (SBE-β-CD) are considered safe for intravenous application and are approved by the U.S Food and Drug Administration (FDA) and the European Medicines Agency (EMA). On the other hand, α-, β-, and γ-CDs are suitable for oral intake due to the lack of absorption from the gastrointestinal tract and are 'generally recognized as safe' compounds according to the FDA. Nasal, ocular, topical, and rectal applications have also been described (Braga 2019, Loftsson 2021, Loftsson and Masson 2001, Medicines Agency 2017).

The ability of cyclodextrins to encapsulate lipophilic molecules has been exploited for the development of supramolecular systems based on polymers and nanoparticles, affording multifunctional medical devices with site-specific delivery and controlled drug release (Petitjean et al. 2021, Xu et al. 2021). These structures, as well as inclusion complexes of cyclodextrins, have shown promising results in anti-cancer therapies (Santos et al. 2021, Gandhi and Shende 2021) and several formulations are undergoing clinical trials for intravenous use (Anselmo and Mitragotri 2019).

Cyclodextrins also exhibit interesting applications in neurodegenerative diseases by increasing the brain delivery of the therapeutics agent and facilitating the cross of the blood-brain barrier to reach the central nervous system (Gosselet et al. 2021, Papakyriakopoulou et al. 2021).

Several publications have reported the use of cyclodextrins in antiviral therapies. The interaction with anti-HIV drugs improved their solubility and absorption (Braga et al. 2021), and sulfonated cyclodextrins showed a broad-spectrum virucidal activity capable of overcoming the cytotoxicity of the existing molecules (Jones et al. 2020). Cyclodextrin derivatives are also used as adjuvants in antiviral vaccines. HP-β-CD has been shown to enhance antibody production and induce the proliferation of T-helper type 2 lymphocytes, which play a prime role in immunization. Likewise, in the ongoing SARS-CoV-2 pandemic, HP-β-CD has been used in the development of the vaccine ad26.cov2.s, acting as a cryopreservative agent to prevent the damage of the viral particles caused during the freeze-drying method used for the preparation process (Braga et al. 2021).

The properties of cyclodextrins make them attractive in gene therapy strategies for the delivery of DNA and RNA fragments. The use of cyclodextrins protects the oligonucleotides from degradation by endonucleases, improving their cellular penetration and reducing their immunogenicity (Haley et al. 2020, Challa et al. 2005). Another interesting application of cyclodextrins is their association with proteins and peptides, via different conjugation reactions, in order to improve their pharmaceutical properties (Łagiewka et al. 2021). The conjugation of β-CD with insulin through N-hydroxysuccinimide (NHS) esters resulted in increased enzymatic and thermal stability of the protein without altering its secondary structure (Hirotsu et al. 2014). These are some selected examples of current and practical applications of cyclodextrins as pharmaceutical excipients and in the design of drug delivery systems, which show how they not only improve the properties of the drugs but also play a substantial part in their safe and targeted delivery. Scientific data on cyclodextrins and their applications in medicine, pharmacy, biology, green chemistry, analytical chemistry, food packaging, and in the research for coronavirus treatment, among other topics, are abundant and we refer the reader to some recent ones (Chodankar et al. 2021, Fatmi et al. 2021, Jeandet et al. 2021, Jicsinszky and Cravotto 2021, Liu et al. 2022, Suresh and London 2022, Umar et al. 2021, Yadav et al. 2021). In this work, we specifically review the therapeutic applications of cyclodextrins against leishmaniasis and trypanosomiasis.

Leishmaniasis and Trypanosomiases

Trypanosomatids are a group of flagellated parasites harboring a kinetoplast (Correa et al. 2020, Boucinha et al. 2020). Some of them are able to cause pathogenicity to animals, humans, and plants. The invertebrate hosts, normally insects, are the vectors transmitting the parasite to vertebrates or plants (Lukeš et al. 2018, Vizcaíno-Castillo et al. 2020). Trypanosoma and Leishmania are examples of dixenous Trypanosomatids causing relevant human infectious diseases, including American Trypanosomiasis (or "Chagas disease") caused by *Trypanosoma cruzi* (Sánchez-Valdéz et al. 2018), African Trypanosomiasis (also called "sleeping sickness") caused by *Trypanosoma brucei,* and Leishmaniasis caused by *Leishmania* species (Correa et al. 2020) (Boucinha et al. 2020). All the aforementioned illnesses are considered neglected tropical diseases (NTDs) (Correa et al. 2020).

Leishmaniasis

Leishmaniasis is a vector-borne disease that is caused by the protozoan parasite from the genus *Leishmania* and transmitted by the bite of infected female sand flies (Karunaweera and Ferreira 2018). Leishmaniasis is recognized by the World Health Organization (WHO) as one of the 20 neglected tropical diseases. The parasite exists in two main lifestyles: promastigote in the vector host (sand fly) and amastigote in mammalian hosts. Due to human activities such as migration, deforestation and climate change, leishmaniasis has spread worldwide (Kevric et al. 2015) (Rodriguez-Morales et al. 2009) and is currently endemic in 98 countries where more than 1 billion people live at risk of developing the disease. The pathology presents three clinical forms: cutaneous leishmaniasis (CL), mucocutaneous leishmaniasis (MCL), and visceral leishmaniasis (VL). CL is the most common form, affecting 0.7 to 1.2 million people per year (Centers for Disease Control and Prevention -CDC- 2020a), while VL is the most severe, exhibiting the second-highest rate of mortality among parasitic diseases after Malaria (Lozano et al. 2012). Table 1 summarizes the current drugs used in the clinic against leishmaniasis. Generic pentavalent antimonials have been so far the essential chemotherapy agents against this pathology, but they present several limitations such as drug resistance and severe side effects. Alternative therapeutic strategies, based on amphotericin B, miltefosine, and paromomycin have also been approved. Amphotericin B (a polyene antibiotic) and miltefosine (a recognized oral agent used for the treatment of breast cancer) offer a higher efficacy but a costlier option than pentavalent antimonials. In addition, some broad-spectrum antimicrobials such as paromomycin have also shown activity in the treatment of this disease, but they are limited by several side effects (hepatic and renal toxicity).

Chagas Disease

Chagas disease is a tropical pathology predominant in Latin America. It is spreading globally due to human migration and affecting 8–10 million people around the world (Mills 2020, Rios et al. 2019, Santé n.d.,

Table 1. Current drugs used in the clinic against Leishmaniasis.

Disease	Current Drugs	Mechanism of Action	Limitations	Observations	References
Leishmaniasis	Pentamidine	Inhibition of DNA synthesis	Resistance issues, diabetes mellitus, pancreatitis, hypoglycemia, hypotension, hyperkalemia and cardiac alterations	Intramuscular administration complementary for HIV co-infected patients	Nwaka and Hudson 2006, de Brito et al. 2020, Nguewa et al. 2005, Yang et al. 2016, Sundar and Chakravarty 2015
	Pentavalent antimonials	Interaction with sulfhydryl-containing molecules, DNA damage, Immunomodulatory function	Local irritation, anorexia, cardiac and hepatic alterations, nausea and vomiting	Intralesional (CL) Systemic (MCL)	Nwaka and Hudson 2006, de Brito et al. 2020
	Amphotericin B	Parasite membrane disruption, immunomodulatory function	Fever, myocarditis, nephrotoxicity and hypokalemia, high cost	Parenteral administration to VL and HIV co-infected, pregnant and transplanted patients	Nwaka and Hudson 2006, Sundar and Chakravarty 2013, Gray et al. 2012, Sundar et al. 2007a, Sundar et al. 2007b
	Miltefosine	Immune response modulation, mitochondrial alteration, apoptosis	Contraindicated in pregnancy, gastrointestinal issues, hepatotoxicity and teratogenic effects	Oral (combinatorial therapies)	Nwaka and Hudson 2006, Croft et al. 1987, Hilgard et al. 1997, Zeisig et al. 1995, Wadhone et al. 2009, Hochhuth et al. 1992, Eue 2002, Verma and Dey 2004, Srivastava et al. 2017, Sundar and Chakravarty 2015
	Paromomycin	Inhibition of protein synthesis and cellular respiration	mild pain on application site, hepatic and renal toxicity	Intramuscular (VL, India)	Petitjean et al. 2021, Anselmo and Mitragotri 2019, Kim et al. 2009

Basile et al. 2011). It is normally transmitted to humans through the bug of hematophagous insects belonging to Triatominae subfamily or by the consumption of food contaminated by triatomine feces (Meymandi et al. 2018). Other transmission ways such as organ transplantation, blood transfusion, and transmission from the infected mother to the fetus have also been reported (Rios et al. 2019, Coura 2015). American trypanosomiasis presents two phases: the acute phase, manifesting 1–2 weeks after infection, and the chronic phase. Once in the chronic phase, trypomastigotes transform into amastigotes and infiltrate through the organ tissues, specifically the heart and digestive tract (Meymandi et al. 2018). Consequently, 30–40% of patients affected with the chronic phase of Chagas disease will suffer cardiomyopathies, less often gastroenterological, and neurologic disorders. Chagas disease can ultimately cause death due to heart failure (Rios et al. 2019).

Benznidazole (BZN), a nitroimidazole derivative (N-Benzil 2 Nitro 1-Imidazolacetamide), and nifurtimox (NFX), a nitrofuran compound, are currently the only drugs available for treating Chagas disease (Table 2). Benznidazole is often considered the first-line therapy because of its better tolerability, but both drugs produce significant side effects. These nitroheterocyclic compounds inhibit the parasite's ability to replicate DNA and are effective against the trypomastigote and amastigote forms (Cançado 2002). In part due to the limited evidence of efficacy, and to the increasing frequency and severity of side effects in relation to patient age, treatment decisions have been based on age categories (Meymandi et al. 2018). In addition, the reason for the high variability in the efficacy of BZN and NFX has been primarily attributed to the broad genetic diversity of *T. cruzi* isolates. Recent studies demonstrated the occurrence of dormancy phenomenon in *T. cruzi* that may explain the drug resistance of this parasite (Sánchez-Valdéz et al. 2018).

Human African trypanosomiasis (HAT)

Sleeping sickness (or HAT) is another form of trypanosomiasis. It is endemic in sub-Saharan African countries and 65 million people are at risk of infection. Two subspecies of *Trypanosoma brucei* (*T. b.*), both transmitted by Tsetse fly, are responsible for two different patterns of the disease: chronic and acute African trypanosomiasis, caused by *T. b. gambiense,* and *T. b. rhodesiense*, respectively (World Health Organization -WHO- 2021). The illness manifests as a haemolymphatic first stage characterized by non-specific symptoms, such as headache and fever, and then the parasites migrate to the central nervous system, which leads to the encephalic stage. Symptoms in this stage include neuropsychological signs with disruption of the sleep cycle, hence the name 'sleeping sickness'. The progression

Table 2. Therapeutic options against Chagas disease and HAT.

Disease	Current Drugs	Mechanism of Action	Limitations	Observations	References
Chagas Disease (American Trypanosomiasis)	Benznidazole (BZN)	Inhibition of parasite's ability to replicate DNA	Dermatitis from hypersensitivity, digestive intolerance, polyneuritis depression of bone marrow, toxic hepatitis, lymphomas	Only active against blood form of the disease (in the acute stage), long-term treatment compliance	Maya et al. 2007, CANÇADO 2002, Viotti et al. 2009, Sosa-Estani et al. n.d., Nwaka and Hudson 2006, Coura and Castro 2002
	Nifurtimox (NFX)	Inhibition of parasite's ability to replicate DNA	Anorexia, loss of weight, psychic alterations, digestive manifestations (nausea, vomiting), intestinal colic and diarrhea	Only active against blood form of the disease (in the acute stage)	Maya et al. 2007, Nwaka and Hudson 2006, Coura and Castro 2002, Castro deMecca et al. 2006, Ferreira HO 1961, Coura et al. (1961, 1962)

Human African Trypanosomiasis (HAT)	Suramin	Alteration of cholesterol homeostasis of *T. Brucei*	Lack of efficacy against the late-stage HAT, safety concerns, injectable	Only active against *T.b. rhodesiense*	Franco et al. 2018, Nwaka and Hudson 2006, Kourbeli et al. 2021
	Pentamidine	Interference with DNA biosynthesis (inhibition of thymidylate synthetase in Trypanosomes)	No efficacy against the late-stage HAT, resistance issues, safety concerns, injectable	Only active against *T.b. gambiense*	Nwaka and Hudson 2006, Kourbeli et al. 2021, Sands et al. n.d.
	Melarsoprol	Interaction with thiol-containing enzymes of the parasite	Resistance issues, injectable, reactive encephalopathy	Used for the second stage of both *T.b. gambiense* and *rhodesiense*	Nwaka and Hudson 2006, Cullen and Mocerino 2017, Fairlamb and Horn 2018, Nok 2003
	Eflornithine	Specific and irreversible binding toornithine decarboxylase (ODC)	Resistance issues, intravenous administration the high cost (in countries with a lack of medical facilities)	Only active against *T.b. gambiense*, less toxic than melarsoprol	Nwaka and Hudson 2006, Cullen and Mocerino 2017, McCann and Pegg 1992
	Fexinidazole	Inhibition of DNA synthesis	Activity only against *T.b. gambiense*	Used against *T.b. gambiense* infection at both stages	Deeks 2019, Asher Mullard 2021

into the second stage depends on the parasite's subspecies and takes months or years for *T. b. gambiense* and a few weeks in the case of *T. b. rhodesiense* (Centers for Disease Control and Prevention -CDC- 2020b). The treatment of HAT relies on five drugs and depends on the type of sickness and stage of infection. Pentamidine and Suramin are considered the first-line drugs for the treatment of early-stage HAT caused by *T. b. gambiense* and *T. b. rhodesiense*, respectively. Both drugs are parenterally administrated and totally ionize after administration, rendering them ineffective against the late-stage of HAT since they are not able to traverse the blood-brain barrier (BBB) (Kourbeli et al. 2021). Melarsoprol is an arsenic derivative used for the second stage of both *gambiense* and *rhodesiense* infections. However, this drug presents life-threatening effects, the reactive encephalopathy being the most dramatic (Cullen and Mocerino 2017). In addition, there has been a spread of resistance to melarsoprol in trypanosomes found in central Africa. Another second-stage drug is eflornithine, only active against *T. b. gambiense*. Less toxic than melarsoprol, the main issue of eflornithine is its demanding way of administration that requires four daily intravenous infusions for 14 days, which in turn poses a serious problem in endemic regions suffering from a lack of medical care facilities. To reduce the duration of the treatment, eflornithine is usually combined with the oral drug nifurtimox, originally used for the treatment of American trypanosomiasis. Nevertheless, this therapy is still limited by intravenous administration and is only effective against *gambiense* subspecies (Cullen and Mocerino 2017). Recently, a new treatment has been approved by WHO and FDA as the first all-oral treatment for sleeping sickness, which is fexinidazole. This drug acts as a DNA synthesis inhibitor and it is used against *T. b. gambiense* infection at both stages (Deeks 2019, Asher Mullard 2021) (Table 2).

Applications of CDs Against Kinetoplastid-Caused Diseases

In this section, we will focus on the applications of cyclodextrins for the treatment of three neglected infectious diseases (Leishmaniasis, HAT, and American Trypanosomiasis). The ways of use and effects have been schematically summarized in Figure 1.

CDs and Leishmaniasis

The administration routes of the conventional treatments, accompanied by side effects and poor affordability, remain major issues in leishmaniasis therapy and have thus led to the further development of new candidates. In this context, cyclodextrins have been explored for their use in antileishmanial treatments. 2-hydroxypropyl-β-CD (HP-β-CD) grafted

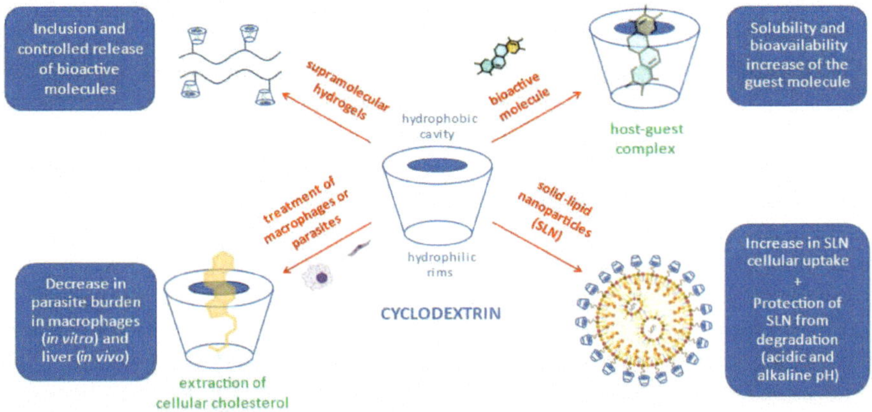

Figure 1. Current applications of cyclodextrins in the treatment of kinetoplastid protozoan infections: the ways of use and their effects.

solid-lipid nanoparticles (SLN) have been used as a combination treatment against leishmaniasis (Parvez et al. 2020). A formulation of Melatonin (Mel) and Amphotericin B (AmB) was assessed against visceral leishmaniasis via oral administration. *In vitro* tests showed a significant inhibition (98%) of amastigote growth at 1 μg/ml. At a concentration of 10 mg/kg for 5 days, HP-β-CD/Mel-AmB/SLN efficiently inhibited the liver parasite burden in *L. donovani*-infected BALB/c mice (99%) with no hepato- and nephrotoxicity. Besides, entrapment efficiency and loading capacity of both AmB and Mel were considerably high, and the developed nanoparticles revealed a sustained drug release of up to 66% of AmB and 73% of Mel after 72 hours. Consequently, HP-β-CD contributed to the enhancement of the oral bioavailability of this nano-formulation with less cytotoxicity in comparison with the drugs alone (Parvez et al. 2021). Similarly, AmB and Paromomycin (PM) were loaded into HP-β-CD modified nanoparticles and evaluated for antileishmanial activity. Stability studies in gastrointestinal simulated fluids proved that HP-β-CD was able to protect the lipid nanoparticles from degradation both in acidic (pH = 1.6) and in alkaline (pH = 6.5) media with a negligible drug leaching. Furthermore, HP-β-CD coating enhanced the cellular uptake of SLNs in J774A.1 macrophages. As a result, the effect on *L. donovani* amastigotes was significantly more efficient than either AmB-SLN, free AmB, or free PM. The oral administration for five days at a concentration of 20 mg/kg in *L. donovani* infected BALB/c mice strongly inhibited the liver parasite burden up to 91% (Parvez et al. 2020).

The development of liposomes incorporating a cyclodextrin host-guest complex has also proven advantageous to prevent the low solubility and toxicity of 7-allylamino-17-demethoxygeldanamycin (17-AAG), an HSP90-inhibitor widely studied as an anti-cancer agent, and also against several Leishmania species (Petersen et al. 2012, Santos et al. 2014). The interactions between 17-AAG and HP-β-CD were studied by FTIR and thermoanalytical methods, reporting a 33-fold enhancement of the solubility. The 17-AAG:HP-β-CD loaded liposomes showed long-term stability with high loading capacity and encapsulation efficiency of the inhibitor. Concerning their leishmanicidal activities, both 17-AAG:HP-β-CD complex and the liposomal formulation presented lower IC_{50}s and higher selectivity indexes when compared to the 17-AAG alone and led to a total amastigote clearance in infected macrophages after 48 hours of treatment at a concentration of 6 pM (Petersen et al. 2018).

Other investigations have focused on the use of cyclodextrins to improve the water solubility of leishmanial lipophilic drugs. This is the case of furazolidone (FZD), which was complexed with β-CD and HP-β-CD in the solid-state. The interactions of FZD with the macrocycles were verified by spectroscopic techniques. However, thermoanalytical assays demonstrated that only the solid compound composed of FZD:HP-β-CD in a 1:2 molar ratio formed an inclusion complex. All the tested formulations resulted non-toxic and active against *L. amazonensis* promastigotes with IC_{50} values lower than those of the free drug (Carvalho et al. 2020). Similarly, pentamidine isethionate (PNT) was nano encapsulated using β-CD. NMR spectroscopy proved the inclusion of both the hydrophobic and aliphatic moieties of PNT inside the host cavity. The oral administration of the solid complex resulted in a substantial decrease in the parasite load in the liver and spleen of *L. infantum Chagasi*-infected mice compared to free PNT (de Paula et al. 2012).

Topical formulations with cyclodextrins have been also described. For example, a gel preparation composed of 12.5% γ-CD, 0.125% AmB, and 3% methylcellulose, which showed a strong association between the macrocycle and the guest (stability constant 1,129 M^{-1}) and high stability over six months. The AmB-CD gel was effective against extra and intra cellular forms of several *Leishmania* species, with greater selectivity indexes compared to the reference AmB deoxycholate. Furthermore, the topical application of the gel to a murine model of CL significantly minimized the size of the skin lesions after 21 days of treatment (Ruiz et al. 2014).

Treatments based on natural compounds with the inherent benefit of low-cost synthesis compared to more expensive drugs or their synthetic analogs have been increasingly studied against leishmaniasis. Isopentyl caffeate (ICaf), a bioactive organic ester, was found active against cutaneous and visceral leishmaniasis but exhibited poor water solubility. Therefore, ICaf was combined with β-CD using different methods. The 1:1 ICaf-β-CD complex obtained by co-evaporation showed a more efficient complexation, favoring the incorporation of the drug in the cyclodextrin cavity. Dissolution assays revealed an improvement of the complex solubility compared to the pure ICaf along with high leishmanicidal activity against *L. amazonensis* (IC_{50}= 3.8 µg/ml) and *L. Chagasi* (IC_{50}= 2.7 µg/ml) (Marques et al. 2020). In another study, β-CD was used to form a host-guest complex with the synthetic curcumin derivative Dibenzalacetone (DBA). Complexes were prepared by kneading or lyophilization in 1:1 and 1:2 DBA/CD ratios. The solid-state characterization by near-infrared spectroscopy, differential scanning calorimetry and NMR proved the inclusion of DBA into the β-CD cavity in all cases, freeze-drying at a 1:1 ratio being the most stable formulation. The aqueous solubility of DBA increased around 40-fold in the presence of β-CD, with a complete release from the LIO 1:1 complex after 120 minutes. Interestingly, only 1:1 DBA/CD complexes inhibited the growth of *L. major* promastigotes, with an IC_{50} = 51.3 µg/mL, without harming the THP-1 macrophage cell line and showed a high selectivity index for *L. major* and *L. infantum* strains (Pinto et al. 2020). Similarly, Piperolactam A (PL), a compound extracted from *Piper betle* roots, was complexed with HPβ-CD and prepared as nanoparticles to improve its solubility, delivery rate, and biological activity. The resulting formulation exhibited a significant cellular uptake in peritoneal macrophages within one hour of incubation. Nano encapsulated PL displayed potent antileishmanial activity against *L. donovani* wild-type and drug-resistant strains with higher selectivity indexes when compared to PL alone (Bhattacharya et al. 2016). Other authors have reported a new formulation of 2-n-propylquinoline (2-n-PQ), a bioactive compound isolated from the medicinal tree *Galilea longiflora*, and traditionally applied to cure cutaneous leishmaniasis lesions. The intravenous administration of this compound was achieved by the complexation of the drug with HPβ-CD. Cytotoxicity assays confirmed the biocompatibility of the formulation (2-n-PQ-HPC) with a notable antileishmanial activity against axenic and intracellular amastigotes of wild-type and drug-resistant *L. donovani* parasites. Moreover, the intravenous injection of 2-n-PQ-HPC exhibited the same effect as the oral intake of the reference drug, miltefosine (Balaraman et al. 2015).

Apart from their use in combination with leishmanicidal drugs, cyclodextrins themselves have been tested for the development of novel leishmaniasis therapies. Given their feature of binding lipophilic molecules, cyclodextrins are capable of forming complexes with cholesterol (Castagne et al. 2009, Frijlink et al. 1991), which is a major component of the plasmatic membrane of eukaryotic organisms. In *Leishmania* sp., the components that contribute to the pathogenicity of the parasite and enable the infection of the host cell are proteins associated with the lipidic domains of the membrane. Cholesterol plays a crucial role in the interactions between the parasite and the mammalian host cell (Chattopadhyay Amitabha and Jafurulla 2012, Pucadyil et al. 2004). Some studies have shown that the treatment of either the metacyclic parasites or the mammalian macrophages with methyl-β-CD contributed to the extraction of membrane sterols and resulted in the reduction of parasite loads in macrophages and liver of BALB/c infected mice, which demonstrated the importance of cholesterol for the infectivity of the parasites and their attachment to the host membrane (Yao et al. 2013, Pucadyil et al. 2004).

CDs and American Trypanosomiasis

Despite their toxic side effects, Benznidazole and Nifurtimox are currently the only available drugs for the treatment of American trypanosomiasis. One of the most severe effects of Bz is the alteration of bone marrow functions induced and manifested by an abnormal decrease in the number of platelets (thrombocytopenia) and granulocytes (agranulocytosis). The efficacy of Bz is reduced due to its poor aqueous solubility and the emergence of drug resistance (Castro and de Mecca et al. 2006). Hence, some investigations have focused on the physicochemical characterization of the inclusion complexes formed to develop safer and more effective formulations of this drug (Ndayishimiye et al. 2021, Maximiano et al. 2011, Leonardi et al. 2013, Priotti et al. 2015, Vinuesa et al. 2017, Lyra et al. 2012). For instance, solid complexes of Bz and γ-CD were prepared by supercritical carbon dioxide extraction ($scCO_2$) or freeze-drying. X-ray diffraction and FTIR studies demonstrated that the complexation took place by both preparation methods with higher complexation efficiency and loading capacity in $scCO_2$-processed samples. The dissolution rate of complexed Bz was higher than for free Bz and increased with the γ-CD content (Ndayishimiye et al. 2021). Another formulation involved effervescent tablets of Bz using an inclusion complex with cyclodextrin. Among the tested CDs, HP-β-CD was selected as the best candidate for improving the aqueous solubility of Bz and nine formulations of effervescent tablets with varying proportions of HP-β-CD and effervescent mixture (EM) were then prepared. Notably, the dissolution rate of all the tablets was faster than the commercial Bz tablet (Rochagan®), the mixture of 48% CD, and 20% EM being the optimal formulation (Maximiano et al. 2011). Another study reported the impact of β-CD, HPβ-CD, and methylated β-CD on the bioavailability of Bz. The administration of 10 mg/kg of Bz: CD complexes to Wistar rats resulted in increased drug plasma levels compared to Bz alone. Differential scanning calorimetry profiles showed that methylated β-CD exhibited the strongest interaction with Bz, manifested in the complete disappearance of the endothermal peak of the drug, while the characteristic peaks of Bz were still observed in the complexes with β-CD and HPβ-CD (Leonardi et al. 2013). Similarly, solid-state NMR studies of Bz:β-CD, Bz:methyl β-CD, and Bz:HP-β-CD complexes proved that only methyl β-CD formed a stable complex with Bz by the inclusion of the benzene ring into the macrocycle cavity, while no significant spectral changes were detected with β-CD and HP-β-CD (Priotti et al. 2015). A further study proved that a formulation of HP-β-CD containing 12% of Bz (CD12) lowered the toxicity of Bz in the L929 fibroblast cell lines without reducing the biological activity against extra and intracellular forms of *T. cruzi*. Morphological observations by scanning electronic microscopy showed a homogenous nanostructure of the complex (Vinuesa et al. 2017). The complexation with randomly methylated-β-CD (RAMEB) did not hamper the trypanocidal activity of Bz but considerably reduced the cytotoxic effect in murine peritoneal macrophages. Additionally, the complexation with RAMEB increased the drug solubility and generated a photostable complex (Lyra et al. 2012).

Lastly, another modified CD, 6-amino-β-CD (6-NH2-β-CD) was used to complex four substituted amido coumarins (ACS), known for their trypanocidal activity but of limited activity due to their low solubility. The analysis of the thermodynamic parameters of binding indicated the formation of stable inclusion complexes of 1:1 stoichiometry in all cases. The complexation enhanced the biological activity of the non-hydroxylated coumarin derivatives in addition to the increase in permeability in artificial membranes. In this same study, a pH-responsive hydrogel formulation with 6-NH2-β-CD was prepared to evaluate the release of the ACS, and proved that pH = 8 were the conditions of maximum release of the drug (Moncada-Basualto et al. 2019).

CDs and African Trypanosomiasis

A new therapeutic approach based on the functionalization of cyclodextrin has been proposed to target the parasitic agent *Trypanosoma brucei*, responsible for African trypanosomiasis. In order to synthesize a flexible drug delivery system, able to form a stable complex and transport a wide variety of drugs, 6-O-monotosyl-6-β-CD was conjugated with a vinyl sulfone group (VS), known to interact with the amine groups present in the biomolecules. The resulting structure (VSCD) was then coupled to the poly-Histidine-tagged nanobody 'cAb-An33 Nb'. This nanobody is an antibody fragment consisting of the variable antigen domain and served as a targeted element to recognize the antigen present on the surface of *T. brucei*. Subsequently, the construct (Nb-VSCD) was used to load the drug selected for the treatment of HAT and nitrofurazone. Before the drug loading, the interaction of VS with the nanobody was confirmed by SDS PAGE and the binding of Nb-VS with *T. brucei* antigen by fluorescence spectroscopy. The biological evaluation of the formulations revealed the growth inhibition of *T. brucei* parasites after treatment with Nb-VSCD + NF (15 μM Nb-VSCD carrying 5 μM NF), while the incubation with PBS + NF and Nb + NF did not show any trypanocidal effect, which indicated that the complexation was essential to achieve the targeted delivery of NF and the desired biological activity (del Castillo et al. 2014).

Currently, Melarsoprol (Mel) is the only effective drug available on the market for the treatment of the late stage of HAT. Mel is poorly soluble in water and presents serious adverse effects. Consequently, its complexation with different substituted cyclodextrins (HP-β-CD and RAMEB) could prove advantageous. To perform the *in vivo* studies, the complexes were orally delivered to *T. b. brucei*-infected mice after 21 days of infection, when the parasites have already invaded the central nervous system (CNS). The observations of blood smears before and after treatment demonstrated the curative effect of the cyclodextrin complexes in contrast to the meagre effect of conventional Mel and its derivatives. More importantly, Mel-CDs complexes reduced the trypanosomes burden in the brain by more than 80% after 24 hours of therapy, decreased the neuroinflammatory response, and restored the blood-brain barrier integrity as visualized by magnetic resonance imaging (MRI). Mice were monitored during the treatment and no apparent signs of toxicity were observed (Rodgers et al. 2011).

Conclusions

The improvement of the properties of drugs, like their solubility, enhanced chemical and photochemical stability, or cytocompatibility via complexation with native and modified cyclodextrins has been studied for decades and found important applications in the Pharmaceutical industry. Over the last few years, the interest has shifted to the construction of more complex structures that incorporate surfactants, polymers, or nanoparticles to produce drug-delivery systems capable of site-specific delivery and stimuli-responsiveness, like temperature and pH, in a 'beyond the cyclodextrin' approach. However, the use of cyclodextrins for the treatment of kinetoplastid-caused infections has been little explored by any of these approaches. These neglected diseases are facing serious problems in their treatments, including secondary effects, long-term treatment, lack of oral availability, and high costs. This work provides evidence that cyclodextrins-based therapies represent a promising tool in the research of anti-parasitic treatments through their ability to establish host-guest interactions with the bioactive molecules. Apart from the solubility enhancement of the lipophilic agents that help in their administration, the explored formulations have proven to exert efficient antiparasitic effects with reduced cytotoxicity. Based on the multifunctional properties of cyclodextrins, like the ability to form peptide/protein-CD conjugates for gene delivery, cyclodextrins can contribute to the development of antiparasitic vaccines and may serve as vehicles for the natural peptides that exhibit antiparasitic potentials as antimicrobial peptides.

Funding and Acknowledgments

This research was funded by the Spanish ministry MINECO (project PID2020-112713RB-C21). PN thanks *Fundación* La Caixa (LCF/PR/PR13/51080005), Fundación Roviralta, Ubesol, SPD Foundation, and COST Actions CA18217 (ENOVAT) and CA18218, CA21105 and CA21111, and EU Project unCoVer (DLV-101016216) for their support. A PhD scholarship was granted by the *Asociación Amigos de la Universidad de Navarra* to Zeinab Dirany.

References

Anselmo, A. C., and Mitragotri, S. (2019). Nanoparticles in the clinic: An update. Bioengineering and Translational Medicine, 4(3). https://doi.org/10.1002/btm2.10143.

Asher Mullard. (2021). FDA approves first all-oral sleeping sickness drug. Nature Reviews Drug Discovery 20: 658. https://doi.org/https://doi.org/10.1038/d41573.021-00140-5.

Balaraman, K., Vieira, N. C., Moussa, F., Vacus, J., Cojean, S., Pomel, S., Bories, C., Figadère, B., Kesavan, V., and Loiseau, P. M. (2015). *In vitro* and *in vivo* antileishmanial properties of a 2-n-propylquinoline hydroxypropyl β-cyclodextrin formulation and pharmacokinetics via intravenous route. Biomedicine and Pharmacotherapy, 76: 127–133. https://doi.org/10.1016/j.biopha.2015.10.028.

Basile, L., Jansà, J. M., Carlier, Y., Salamanca, D. D., Angheben, A., Bartoloni, A., Seixas, J., van Gool, T., Cañavate, C., Flores-Chávez, M., Jackson, Y., and Gool, V. T. (2011). Chagas disease in European countries: the challenge of a surveillance system. In Euro Surveill (Vol. 16, Issue 37). www.eurosurveillance.org:pii=19968.Availableonline:http://www.eurosurveillance.org/ViewArticle.aspx?ArticleId=19968.

Bhattacharya, P., Mondal, S., Basak, S., Das, P., Saha, A., and Bera, T. (2016). *In vitro* susceptibilities of wild and drug resistant Leishmania donovani amastigotes to piperolactam A loaded hydroxypropyl-β-cyclodextrin nanoparticles. Acta Tropica, 158, 97–106. https://doi.org/10.1016/j.actatropica.2016.02.017.

Boucinha, C., Caetano, A. R., Santos, H. L., Helaers, R., Vikkula, M., Branquinha, M. H., dos Santos, A. L. S., Grellier, P., Morelli, K. A., and d'Avila-Levy, C. M. (2020). Analysing ambiguities in trypanosomatids taxonomy by barcoding. Memorias Do Instituto Oswaldo Cruz, 115: e200504–e200504. https://doi.org/10.1590/0074-02760200504.

Braga, S. S. (2019). Cyclodextrins: Emerging medicines of the new millennium. In Biomolecules (Vol. 9, Issue 12). MDPI AG. https://doi.org/10.3390/biom9120801.

Braga, S. S., Barbosa, J. S., Santos, N. E., El-Saleh, F., and Paz, F. A. A. (2021). Cyclodextrins in antiviral therapeutics and vaccines. Pharmaceutics, 13(3). https://doi.org/10.3390/pharmaceutics13030409.

CANÇADO, J. R. (2002). Long term evaluation of etiological treatment of Chagas disease with benznidazole. Revista Do Instituto de Medicina Tropical de São Paulo, 44(1): 29–37. https://doi.org/10.1590/S0036-46652002000100006.

Carvalho, S. G., Cipriano, D. F., de Freitas, J. C. C., Junior, M. Â. S., Ocaris, E. R. Y., Teles, C. B. G., de Jesus Gouveia, A., Rodrigues, R. P., Zanini, M. S., and Villanova, J. C. O. (2020). Physicochemical characterization and *in vitro* biological evaluation of solid compounds from furazolidone-based cyclodextrins for use as leishmanicidal agents. Drug Delivery and Translational Research, 10(6): 1788–1809. https://doi.org/10.1007/s13346-020-00841-1.

Castagne, D., Fillet, M., Delattre, L., Evrard, B., Nusgens, B., and Piel, G. (2009). Study of the cholesterol extraction capacity of β-cyclodextrin and its derivatives, relationships with their effects on endothelial cell viability and on membrane models. Journal of Inclusion Phenomena and Macrocyclic Chemistry, 63(3-4): 225–231. https://doi.org/10.1007/s10847-008-9510-9.

Castro, J. A., de Mecca, M. M., and Bartel, L. C. (2006). Toxic side effects of drugs used to treat Chagas' disease (American trypanosomiasis). Human and Experimental Toxicology, 25(8): 471–479. https://doi.org/10.1191/0960327106het653oa.

Centers for Disease Control and Prevention (CDC). (2020a, February 18). Leishmaniasis. Epidemiology and Risk Factors. Internet. https://www.cdc.gov/parasites/leishmaniasis/epi.html.

Centers for Disease Control and Prevention (CDC). (2020b, April 28). Disease- African Trypanosomiasis.

Challa, R., Ahuja, A., Ali, J., and Khar, R. K. (2005). Cyclodextrins in Drug Delivery: An Updated Review. http://www.aapspharmscitech.org.

Chattopadhyay Amitabha, and Jafurulla, Md. (2012). Role of membrane cholesterol in leishmanial infection. pp. 201–213. *In*: Sudhakaran, A., Perumana, R., and Surolia, A. (Ed.). Biochemical Roles of Eukaryotic Cell Surface Macromolecules. Springer New York.

Chodankar, D., Vora, A., and Kanhed, A. (2021). β-cyclodextrin and its derivatives: application in wastewater treatment. Environmental Science and Pollution Research. https://doi.org/10.1007/s11356-021-17014-3.

Correa, J. P., Bacigalupo, A., Yefi-Quinteros, E., Rojo, G., Solari, A., Cattan, P. E., and Botto-Mahan, C. (2020). Trypanosomatid Infections among Vertebrates of Chile: A Systematic Review. Pathogens (Basel, Switzerland), 9(8): 661. https://doi.org/10.3390/pathogens9080661.

Coura, J. R., Ferreira, L. F., Saad, E. A., Mortel, R. E., and Silva, J. R. (1961). Tentativa terapêutica com a nitrofurazona (Furacin) na forma crônica da doença de Chagas. O Hospital 60: 425–429.

Coura, J. R., Ferreira, L. F., and Silva, J. R. (1962). Experiências com nitrofurazona na fase crônica da doença de Chagas. O Hospital 62: 957–964.

Coura, J. R., and Castro, S. L. de. (2002). A critical review on chagas disease chemotherapy. Memórias Do Instituto Oswaldo Cruz, 97(1): 3–24. https://doi.org/10.1590/S0074-02762002000100001.

Coura, J. R. (2015). The main sceneries of Chagas disease transmission. The vectors, blood and oral transmissions--a comprehensive review. Memorias Do Instituto Oswaldo Cruz, 110(3): 277–282. https://doi.org/10.1590/0074-0276140362.

Crini, G., French, A. D., Kainuma, K., Jane, J. lin, and Szente, L. (2021). Contributions of Dexter French (1918–1981) to cycloamylose/cyclodextrin and starch science. In Carbohydrate Polymers (Vol. 257). Elsevier Ltd. https://doi.org/10.1016/j.carbpol.2021.117620.

Croft, S. L., Neal, R. A., Pendergast, W., and Chan, J. H. (1987). The activity of alkyl phosphorylcholines and related derivatives against Leishmania donovani. Biochemical

Pharmacology, 36(16): 2633–2636. https://doi.org/https://doi.org/10.1016/0006-2952(87)90543-0.

Cullen, D. R., and Mocerino, M. (2017). Current medicinal chemistry the international journal for timely in-depth reviews in medicinal chemistry BENTHAM SCIENCE Send Orders for Reprints to reprints@benthamscience.ae A Brief Review of Drug Discovery Research for Human African Trypanosomiasis. Current Medicinal Chemistry, 24: 701–717. https://doi.org/10.2174/092986732466617012.

de Brito, R. C. F., Aguiar-Soares, R. D. de O., Cardoso, J. M. de O., Coura-Vital, W., Roatt, B. M., and Reis, A. B. (2020). Recent advances and new strategies in *Leishmaniasis diagnosis*. Applied Microbiology and Biotechnology, 104(19): 8105–8116. https://doi.org/10.1007/s00253-020-10846-y.

de Paula, E. E. B., de Sousa, F. B., da Silva, J. C. C., Fernandes, F. R., Melo, M. N., Frézard, F., Grazul, R. M., Sinisterra, R. D., and MacHado, F. C. (2012). Insights into the multi-equilibrium, superstructure system based on β-cyclodextrin and a highly water soluble guest. International Journal of Pharmaceutics, 439(1-2): 207–215. https://doi.org/10.1016/j.ijpharm.2012.09.039.

Deeks, E. D. (2019). Fexinidazole: First Global Approval. In Drugs (Vol. 79, Issue 2, pp. 215–220). Springer International Publishing. https://doi.org/10.1007/s40265-019-1051-6.

del Castillo, T., Marales-Sanfrutos, J., Santoyo-Gonzlez, F., Magez, S., Lopez-Jaramillo, F. J., and Garcia-Salcedo, J. A. (2014). Monovinyl sulfone B-cyclodextrin. a flexible drug carrier system. Chem. Med. Chem., 9(2): 383–389. https://doi.org/10.1002/cmdc.201300385.

Eue, I. (2002). Hexadecylphosphocholine selectively upregulates expression of intracellular adhesion molecule-1 and class I major histocompatibility complex antigen in human monocytes. Journal of Experimental Therapeutics and Oncology, 2(6): 333–336. https://doi.org/https://doi.org/10.1046/j.1359-4117.2002.01048.x.

Fairlamb, A. H., and Horn, D. (2018). Melarsoprol resistance in african trypanosomiasis. Trends in Parasitology, 34(6): 481–492. https://doi.org/https://doi.org/10.1016/j.pt.2018.04.002.

Fatmi, S., Taouzinet, L., Skiba, M., and Iguer-Ouada, M. (2021). The Use of Cyclodextrin or its Complexes as a Potential Treatment Against the 2019 Novel Coronavirus: A Mini-Review. Current Drug Delivery, 18(4). https://doi.org/10.2174/1567201817666200917124241.

Ferreira, H. O. (1961). Forma aguda da doenca de Chagas tratada pela nitrofurzona. Rev. Inst. Med. Trop Sao Paulo, 3: 287–289.

Franco, J., Scarone, L., and Comini, M. A. (2018). Chapter three—Drugs and drug resistance in african and american trypanosomiasis. pp. 97–133. *In*: Botta, M. (Ed.). Annual Reports in Medicinal Chemistry (Vol. 51). Academic Press. https://doi.org/https://doi.org/10.1016/bs.armc.2018.08.003.

Frijlink, H. W., Eissens, A. C., Hefting, N. R., Poelstra, K., Lerk, C. F., and Meijer, D. K. F. (1991). The Effect of Parenterally Administered Cyclodextrins on Cholesterol Levels in the Rat. Pharmaceutical Research, 8(1): 9–16. https://doi.org/10.1023/A:1015861719134.

Gandhi, S., and Shende, P. (2021). Cyclodextrins-modified metallic nanoparticles for effective cancer therapy. In Journal of Controlled Release (Vol. 339, pp. 41–50). Elsevier B.V. https://doi.org/10.1016/j.jconrel.2021.09.025.

Gosselet, F., Loiola, R. A., Roig, A., Rosell, A., and Culot, M. (2021). Central nervous system delivery of molecules across the blood-brain barrier. Neurochemistry International, 144. https://doi.org/10.1016/j.neuint.2020.104952.

Gray, K. C., Palacios, D. S., Dailey, I., Endo, M. M., Uno, B. E., Wilcock, B. C., and Burke, M. D. (2012). Amphotericin primarily kills yeast by simply binding ergosterol. Proceedings of the National Academy of Sciences, 109(7): 2234. https://doi.org/10.1073/pnas.1117280109.

Haley, R. M., Gottardi, R., Langer, R., and Mitchell, M. J. (2020). Cyclodextrins in drug delivery: applications in gene and combination therapy. Drug Delivery and Translational Research, 10(3): 661–677. https://doi.org/10.1007/s13346-020-00724-5.

Hilgard, P., Klenner, T., Stekar, J., Nössner, G., Kutscher, B., and Engel, J. (1997). D-21266, a new heterocyclic alkylphospholipid with antitumour activity. European Journal of Cancer, 33(3): 442–446. https://doi.org/https://doi.org/10.1016/S0959-8049(97)89020-X.

Hirotsu, T., Higashi, T., Motoyama, K., Hirayama, F., Uekama, K., and Arima, H. (2014). Improvement of pharmaceutical properties of insulin through conjugation with glucuronylglucosyl-β-cyclodextrin. Journal of Inclusion Phenomena and Macrocyclic Chemistry, 80(1-2): 107–112. https://doi.org/10.1007/s10847-014-0407-5.

Hochhuth, C. H., Vehmeyer, K., Eibl, H., and Unger, C. (1992). Hexadecylphosphocholine induces interferon-γ secretion and expression of GM-CSF mRNA in human mononuclear cells. Cellular Immunology, 141(1): 161–168. https://doi.org/https://doi.org/10.1016/0008-8749(92)90135-C.

Jambhekar, S. S., and Breen, P. (2016). Cyclodextrins in pharmaceutical formulations I: Structure and physicochemical properties, formation of complexes, and types of complex. In Drug Discovery Today (Vol. 21, Issue 2, pp. 356–362). Elsevier Ltd. https://doi.org/10.1016/j.drudis.2015.11.017.

Jeandet, P., Sobarzo-Sánchez, E., Uddin, M. S., Bru, R., Clément, C., Jacquard, C., Nabavi, S. F., Khayatkashani, M., Batiha, G. E. S., Khan, H., Morkunas, I., Trotta, F., Matencio, A., and Nabavi, S. M. (2021). Resveratrol and cyclodextrins, an easy alliance: Applications in nanomedicine, green chemistry and biotechnology. In Biotechnology Advances (Vol. 53). Elsevier Inc. https://doi.org/10.1016/j.biotechadv.2021.107844.

Jicsinszky, L., and Cravotto, G. (2021). Toward a greener world—cyclodextrin derivatization by mechanochemistry. In Molecules (Vol. 26, Issue 17). MDPI AG. https://doi.org/10.3390/molecules26175193.

Jones, S. T., Cagno, V., Janeček, M., Ortiz, D., Gasilova, N., Piret, J., Gasbarri, M., Constant, D. A., Han, Y., Vuković, L., Král, P., Kaiser, L., Huang, S., Constant, S., Kirkegaard, K., Boivin, G., Stellacci, F., and Tapparel, C. (2020). Modified Cyclodextrins as Broad-spectrum Antivirals.

Karunaweera, N. D., and Ferreira, M. U. (2018). Leishmaniasis: Current challenges and prospects for elimination with special focus on the South Asian region. In Parasitology. https://doi.org/10.1017/S0031182018000471.

Kevric, I., Cappel, M. A., and Keeling, J. H. (2015). New world and old world leishmania infections: a practical review. Dermatologic Clinics, 33(3): 579–593. https://doi.org/https://doi.org/10.1016/j.det.2015.03.018.

Kim, D. H., Chung, H. J., Bleys, J., and Ghohestani, R. F. (2009). Is paromomycin an effective and safe treatment against cutaneous leishmaniasis? A meta-analysis of 14 randomized controlled trials. PLoS Neglected Tropical Diseases, 3(2): e381–e381. https://doi.org/10.1371/journal.pntd.0000381.

Kourbeli, V., Chontzopoulou, E., Moschovou, K., Pavlos, D., Mavromoustakos, T., and Papanastasiou, I. P. (2021). An overview on target-based drug design against kinetoplastid protozoan infections: Human african trypanosomiasis, chagas disease and leishmaniases. In Molecules (Vol. 26, Issue 15). MDPI AG. https://doi.org/10.3390/molecules26154629.

Kurkov, S. v., and Loftsson, T. (2013). Cyclodextrins. In International Journal of Pharmaceutics (Vol. 453, Issue 1, pp. 167–180). Elsevier B.V. https://doi.org/10.1016/j.ijpharm.2012.06.055.

Łagiewka, J., Girek, T., and Ciesielski, W. (2021). Cyclodextrins-peptides/proteins conjugates: Synthesis, properties and applications. In Polymers (Vol. 13, Issue 11). MDPI AG. https://doi.org/10.3390/polym13111759.

Larsen, K. L. (2002). Large cyclodextrins. Journal of Inclusion Phenomena and Macrocyclic Chemistry, 43(1): 1–13. https://doi.org/10.1023/A:1020494503684.

Leonardi, D., Bombardiere, M. E., and Salomon, C. J. (2013). Effects of benznidazole: Cyclodextrin complexes on the drug bioavailability upon oral administration to rats. International Journal of Biological Macromolecules, 62: 543–548. https://doi.org/10.1016/j.ijbiomac.2013.10.007.

Liu, Y., Sameen, D. E., Ahmed, S., Wang, Y., Lu, R., Dai, J., Li, S., and Qin, W. (2022). Recent advances in cyclodextrin-based films for food packaging. In Food Chemistry (Vol. 370). Elsevier Ltd. https://doi.org/10.1016/j.foodchem.2021.131026.

Loftsson, T., and Masson, M. (2001). Cyclodextrins in topical drug formulations: theory and practice. In International Journal of Pharmaceutics (Vol. 225). www.elsevier.com/locate/ijpharm.

Loftsson, T. (2021). Cyclodextrins in parenteral formulations. In Journal of Pharmaceutical Sciences (Vol. 110, Issue 2, pp. 654–664). Elsevier B.V. https://doi.org/10.1016/j.xphs.2020.10.026.

Lozano, R., Naghavi, M., Foreman, K., Lim, S., Shibuya, K., Aboyans, V., Abraham, J., Adair, T., Aggarwal, R., Ahn, S. Y., Alvarado, M., Anderson, H. R., Anderson, L. M., Andrews, K. G., Atkinson, C., Baddour, L. M., Barker-Collo, S., Bartels, D. H., Bell, M. L., … Memish, Z. A. (2012). Global and regional mortality from 235 causes of death for 20 age groups in 1990 and 2010: a systematic analysis for the Global Burden of Disease Study 2010. Lancet (London, England), 380(9859): 2095–2128. https://doi.org/10.1016/S0140-6736(12)61728-0.

Luis De Oliveira, A., Petersen, A., Sampaio Guedes, C. E., Leite Versoza, C., Geraldo, J., Lima, B., Antô, L., Rodrigues De Freitas, N., Ria Matos Borges, V., Sampaio, P., and Veras, T. (n.d.). 7-AAG Kills Intracellular Leishmania amazonensis while Reducing Inflammatory Responses in Infected Macrophages. https://doi.org/10.1371/journal.pone.0049496.

Lukeš, J., Butenko, A., Hashimi, H., Maslov, D. A., Votýpka, J., and Yurchenko, V. (2018). Trypanosomatids are much more than just trypanosomes: clues from the expanded family tree. Trends in Parasitology, 34(6): 466–480. https://doi.org/https://doi.org/10.1016/j.pt.2018.03.002.

Lyra, M. A. M., Soares-Sobrinho, J. L., Figueiredo, R. C. B. Q., Sandes, J. M., Lima, Á. A. N., Tenório, R. P., Fontes, D. A. F., Santos, F. L. A., Rolim, L. A., and Rolim-Neto, P. J. (2012). Study of benznidazole-cyclodextrin inclusion complexes, cytotoxicity and trypanocidal activity. Journal of Inclusion Phenomena and Macrocyclic Chemistry, 73(1–4): 397–404. https://doi.org/10.1007/s10847-011-0077-5.

Marques, C. S. F., Barreto, N. S., de Oliveira, S. S. C., Santos, A. L. S., Branquinha, M. H., de Sousa, D. P., Castro, M., Andrade, L. N., Pereira, M. M., da Silva, C. F., Chaud, M. v., Jain, S., Fricks, A. T., Souto, E. B., and Severino, P. (2020). β-Cyclodextrin/Isopentyl Caffeate Inclusion Complex: Synthesis, Characterization and Antileishmanial Activity. Molecules, 25(18). https://doi.org/10.3390/molecules25184181.

Maximiano, F. P., Costa, G. H. Y., de Sá Barreto, L. C. L., Bahia, M. T., and Cunha-Filho, M. S. S. (2011). Development of effervescent tablets containing benznidazole complexed with cyclodextrin. Journal of Pharmacy and Pharmacology, 63(6): 786–793. https://doi.org/10.1111/j.2042-7158.2011.01284.x.

Maya, J. D., Cassels, B. K., Iturriaga-Vásquez, P., Ferreira, J., Faúndez, M., Galanti, N., Ferreira, A., and Morello, A. (2007). Mode of action of natural and synthetic drugs against Trypanosoma cruzi and their interaction with the mammalian host. Comparative Biochemistry and Physiology Part A: Molecular and Integrative Physiology, 146(4): 601–620. https://doi.org/https://doi.org/10.1016/j.cbpa.2006.03.004.

McCann, P. P., and Pegg, A. E. (1992). Ornithine decarboxylase as an enzyme target for therapy. Pharmacology and Therapeutics, 54(2): 195–215. https://doi.org/https://doi.org/10.1016/0163-7258(92)90032-U.

Medicines Agency, E. (2017). Committee for Human Medicinal Products (CHMP). www.ema.europa.eu/contact.

Meymandi, S., Hernandez, S., Park, S., Sanchez, D. R., and Forsyth, C. (2018). Treatment of Chagas Disease in the United States. Current Treatment Options in Infectious Diseases, 10(3): 373–388. https://doi.org/10.1007/s40506-018-0170-z.

Mills, R. M. (2020). Chagas Disease: Epidemiology and Barriers to Treatment. The American Journal of Medicine, 133(11): 1262–1265. https://doi.org/https://doi.org/10.1016/j.amjmed.2020.05.022.

Moncada-Basualto, M., Matsuhiro, B., Mansilla, A., Lapier, M., Maya, J. D., and Olea-Azar, C. (2019). Supramolecular hydrogels of β-cyclodextrin linked to calcium homopoly-L-guluronate for release of coumarins with trypanocidal activity. Carbohydrate Polymers, 204: 170–181. https://doi.org/10.1016/j.carbpol.2018.10.010.

Morin-Crini, N., Fourmentin, S., Fenyvesi, É., Lichtfouse, E., Torri, G., Fourmentin, M., and Crini, G. (2021). 130 years of cyclodextrin discovery for health, food, agriculture, and the industry: a review. In Environmental Chemistry Letters (Vol. 19, Issue 3, pp. 2581–2617). Springer Science and Business Media Deutschland GmbH. https://doi.org/10.1007/s10311-020-01156-w.

Muderawan, I. W., Ong, T. T., Teck, C. L., Young, D. J., Chi, B. C., and Ng, S. C. (2005). A reliable synthesis of 2- and 6-amino-β-cyclodextrin and permethylated-β-cyclodextrin. Tetrahedron Letters, 46(46),: 7905–7907. https://doi.org/10.1016/j.tetlet.2005.09.099.

Mura, P. (2014). Analytical techniques for characterization of cyclodextrin complexes in aqueous solution: A review. In Journal of Pharmaceutical and Biomedical Analysis (Vol. 101, pp. 238–250). Elsevier B.V. https://doi.org/10.1016/j.jpba.2014.02.022.

Ndayishimiye, J., Popat, A., Kumeria, T., Blaskovich, M. A. T., and Robert Falconer, J. (2021). Supercritical carbon dioxide assisted complexation of benznidazole: γ-cyclodextrin for improved dissolution. International Journal of Pharmaceutics, 596. https://doi.org/10.1016/j.ijpharm.2021.120240.

Nguewa, P. A., Fuertes, M. A., Cepeda, V., Iborra, S., Carrión, J., Valladares, B., Alonso, C., and Pérez, J. M. (2005). Pentamidine Is an Antiparasitic and Apoptotic Drug That Selectively Modifies Ubiquitin. Chemistry and Biodiversity, 2(10): 1387–1400. https://doi.org/https://doi.org/10.1002/cbdv.200590111.

Nok, A. J. (2003). Arsenicals (melarsoprol), pentamidine and suramin in the treatment of human African trypanosomiasis. Parasitology Research, 90(1): 71–79. https://doi.org/10.1007/s00436-002-0799-9.

Nwaka, S., and Hudson, A. (2006). Innovative lead discovery strategies for tropical diseases. Nature Reviews Drug Discovery, 5(11): 941–955. https://doi.org/10.1038/nrd2144.

Papakyriakopoulou, P., Manta, K., Kostantini, C., Kikionis, S., Banella, S., Ioannou, E., Christodoulou, E., Rekkas, D. M., Dallas, P., Vertzoni, M., Valsami, G., and Colombo, G. (2021). Nasal powders of quercetin-β-cyclodextrin derivatives complexes with mannitol/lecithin microparticles for Nose-to-Brain delivery: *In vitro* and *ex vivo* evaluation. International Journal of Pharmaceutics, 607. https://doi.org/10.1016/j.ijpharm.2021.121016.

Parvez, S., Yadagiri, G., Gedda, M. R., Singh, A., Singh, O. P., Verma, A., Sundar, S., and Mudavath, S. L. (2020). Modified solid lipid nanoparticles encapsulated with Amphotericin B and Paromomycin: an effective oral combination against experimental murine visceral leishmaniasis. Scientific Reports, 10(1). https://doi.org/10.1038/s41598-020-69276-5.

Parvez, S., Yadagiri, G., Arora, K., Javaid, A., Kushwaha, A. K., Singh, O. P., Sundar, S., and Mudavath, S. L. (2021). Coalition of Biological Agent (Melatonin) With

Chemotherapeutic Agent (Amphotericin B) for Combating Visceral Leishmaniasis via Oral Administration of Modified Solid Lipid Nanoparticles. ACS Biomaterials Science and Engineering. https://doi.org/10.1021/acsbiomaterials.1c00859.

Petersen, A. L., Guedes, C. E., Versoza, C. L., Lima, J. G., de Freitas, L. A., Borges, V. M., and Veras, P. S. (2012). 17-AAG kills intracellular Leishmania amazonensis while reducing inflammatory responses in infected macrophages. PLoS One, 7(11): e49496. doi: 10.1371/journal.pone.0049496. Epub 2012 Nov 13.

Petersen, A. L. de O. A., Campos, T. A., Santos Dantas, D. A. dos, Rebouças, J. de S., da Silva, J. C., de Menezes, J. P. B., Formiga, F. R., de Melo, J. v., Machado, G., and Veras, P. S. T. (2018). Encapsulation of the HSP-90 chaperone inhibitor 17-AAG in stable liposome allow increasing the therapeutic index as assessed, *in vitro*, on Leishmania (L) amazonensis amastigotes-hosted in mouse CBA macrophages. Frontiers in Cellular and Infection Microbiology, 8(AUG). https://doi.org/10.3389/fcimb.2018.00303.

Petitjean, M., García-Zubiri, I. X., and Isasi, J. R. (2021). History of cyclodextrin-based polymers in food and pharmacy: a review. In Environmental Chemistry Letters (Vol. 19, Issue 4, pp. 3465–3476). Springer Science and Business Media Deutschland GmbH. https://doi.org/10.1007/s10311-021-01244-5.

Pinto, L. M. A., Adeoye, O., Thomasi, S. S., Francisco, A. P., Carvalheiro, M. C., and Cabral-Marques, H. (2020). Preparation and characterization of a synthetic curcumin analog inclusion complex and preliminary evaluation of in vitro antileishmanial activity. International Journal of Pharmaceutics, 589. https://doi.org/10.1016/j.ijpharm.2020.119764.

Priotti, J., Ferreira, M. J. G., Lamas, M. C., Leonardi, D., Salomon, C. J., and Nunes, T. G. (2015). First solid-state NMR spectroscopy evaluation of complexes of benznidazole with cyclodextrin derivatives. Carbohydrate Polymers, 131: 90–97. https://doi.org/10.1016/j.carbpol.2015.05.045.

Pucadyil, T. J., Tewary, P., Madhubala, R., and Chattopadhyay, A. (2004). Cholesterol is required for Leishmania donovani infection: Implications in leishmaniasis. Molecular and Biochemical Parasitology, 133(2): 145–152. https://doi.org/10.1016/j.molbiopara.2003.10.002.

Rios, L. E., Vázquez-Chagoyán, J. C., Pacheco, A. O., Zago, M. P., and Garg, N. J. (2019). Immunity and vaccine development efforts against Trypanosoma cruzi. Acta Tropica, 200, 105168. https://doi.org/10.1016/j.actatropica.2019.105168.

Rodgers, J., Jones, A., Gibaud, S., Bradley, B., McCabe, C., Barrett, M. P., Gettinby, G., and Kennedy, P. G. E. (2011). Melarsoprol cyclodextrin inclusion complexes as promising oral candidates for the treatment of human African trypanosomiasis. PLoS Neglected Tropical Diseases, 5(9). https://doi.org/10.1371/journal.pntd.0001308.

Rodriguez-Morales, A. J., Silvestre, J., and Cazorla-Perfetti, D. J. (2009). Imported Leishmaniasis in Australia. Journal of Travel Medicine, 16(2): 144–145. https://doi.org/10.1111/j.1708-8305.2008.00296_1.x.

Ruiz, H. K., Serrano, D. R., Dea-Ayuela, M. A., Bilbao-Ramos, P. E., Bolás-Fernández, F., Torrado, J. J., and Molero, G. (2014). New amphotericin B-gamma cyclodextrin formulation for topical use with synergistic activity against diverse fungal species and *Leishmania* spp. International Journal of Pharmaceutics, 473(1-2): 148–157. https://doi.org/10.1016/j.ijpharm.2014.07.004.

Sánchez-Valdéz, F. J., Padilla, A., Wang, W., Orr, D., and Tarleton, R. L. (2018). Spontaneous dormancy protects Trypanosoma cruzi during extended drug exposure. ELife, 7: e34039. https://doi.org/10.7554/eLife.34039.

Sands, M., Kron, M. A., and Brown, R. B. (n.d.). Pentamidine: A Review. In REVIEWS OF INFECTIOUS DISEASES (Vol. 7). http://cid.oxfordjournals.org/.

Santé, W. H. O. = O. mondiale de la. (n.d.). Chagas disease in Latin America : an epidemiological update based on 2010 estimates = Maladie de Chagas en Amérique latine : le point épidémiologique basé sur les estimations de 2010. In Weekly Epidemiological Record = Relevé épidémiologique hebdomadaire (Vol. 90, Issue 06, pp. 33–44). World Health Organization = Organisation mondiale de la Santé. https://apps.who.int/iris/handle/10665/242316.

Santos, A. C., Costa, D., Ferreira, L., Guerra, C., Pereira-Silva, M., Pereira, I., Peixoto, D., Ferreira, N. R., and Veiga, F. (2021). Cyclodextrin-based delivery systems for *in vivo*-tested anticancer therapies. In Drug Delivery and Translational Research (Vol. 11, Issue 1, pp. 49–71). Springer. https://doi.org/10.1007/s13346-020-00778-5.

Santos, D. M., Petersen, A. L. O. A., Celes, F. S., Borges, V. M., Veras, P. S. T., and de Oliveira, C. I. (n.d.). Chemotherapeutic Potential of 17-AAG against Cutaneous Leishmaniasis Caused by Leishmania (Viannia) braziliensis. https://doi.org/10.1371/journal.pntd.0003275.

Sosa-Estani, S., Armenti, A., Araujo, G., Viotti, R., Lococo, B., Ruiz Vera, B., Vigliano, C., de Rissio, A. M., and Segura, E. L. (n.d.). Articulo Original Tratamiento De La Enfermedad De Chagas Con Benznidazol Y Acido Tioctico.

Srivastava, S., Mishra, J., Gupta, A. K., Singh, A., Shankar, P., and Singh, S. (2017). Laboratory confirmed miltefosine resistant cases of visceral leishmaniasis from India. Parasites and Vectors, 10(1): 49. https://doi.org/10.1186/s13071-017-1969-z.

Sundar, S., Chakravarty, J., Rai, V. K., Agrawal, N., Singh, S. P., Chauhan, V., and Murray, H. W. (2007a). Amphotericin B Treatment for Indian Visceral Leishmaniasis: Response to 15 Daily versus Alternate-Day Infusions. Clinical Infectious Diseases, 45(5): 556–561. https://doi.org/10.1086/520665.

Sundar, S., Jha, T. K., Thakur, C. P., Sinha, P. K., and Bhattacharya, S. K. (2007b). Injectable Paromomycin for Visceral Leishmaniasis in India. New England Journal of Medicine, 356(25): 2571–2581. https://doi.org/10.1056/NEJMoa066536.

Sundar, S., and Chakravarty, J. (2013). Leishmaniasis: an update of current pharmacotherapy. Expert Opinion on Pharmacotherapy, 14(1): 53–63. https://doi.org/10.1517/14656566.2013.755515.

Sundar, S., and Chakravarty, J. (2015). An update on pharmacotherapy for leishmaniasis. Expert Opinion on Pharmacotherapy, 16(2): 237–252. https://doi.org/10.1517/14656566.2015.973850.

Suresh, P., and London, E. (2022). Using cyclodextrin-induced lipid substitution to study membrane lipid and ordered membrane domain (raft) function in cells. In Biochimica et Biophysica Acta - Biomembranes (Vol. 1864, Issue 1). Elsevier B.V. https://doi.org/10.1016/j.bbamem.2021.183774.

Szente, L., and Szejtli, J. (1999). Highly soluble cyclodextrin derivatives: chemistry, properties, and trends in development. In Advanced Drug Delivery Reviews (Vol. 36).

Tripodo, G., Wischke, C., Neffe, A. T., and Lendlein, A. (2013). Efficient synthesis of pure monotosylated beta-cyclodextrin and its dimers. Carbohydrate Research, 381: 59–63. https://doi.org/10.1016/j.carres.2013.08.018.

Umar, Y., Al-Batty, S., Rahman, H., Ashwaq, O., Sarief, A., Sadique, Z., Sreekumar, P. A., and Haque, S. K. M. (2021). Polymeric materials as potential inhibitors against SARS-CoV-2. In Journal of Polymers and the Environment. Springer. https://doi.org/10.1007/s10924-021-02272-6.

Verma, N. K., and Dey, C. S. (2004). Possible mechanism of miltefosine-mediated death of Leishmania donovani. Antimicrobial Agents and Chemotherapy, 48(8): 3010–3015. https://doi.org/10.1128/AAC.48.8.3010-3015.2004.

Vinuesa, T., Herráez, R., Oliver, L., Elizondo, E., Acarregui, A., Esquisabel, A., Pedraz, J. L., Ventosa, N., Veciana, J., and Viñas, M. (2017). Benznidazole nanoformulates: A chance to improve therapeutics for Chagas disease. American Journal of Tropical Medicine and Hygiene, 97(5): 1469–1476. https://doi.org/10.4269/ajtmh.17-0044.

Viotti, R., Vigliano, C., Lococo, B., Alvarez, M. G., Petti, M., Bertocchi, G., and Armenti, A. (2009). Side effects of benznidazole as treatment in chronic Chagas disease: fears and realities. Expert Review of Anti-Infective Therapy, 7(2): 157–163. https://doi.org/10.1586/14787210.7.2.157.

Vizcaíno-Castillo, A., Osorio-Méndez, J. F., Ambrosio, J. R., Hernández, R., and Cevallos, A. M. (2020). The complexity and diversity of the actin cytoskeleton of trypanosomatids. Molecular and Biochemical Parasitology, 237: 111278. https://doi.org/https://doi.org/10.1016/j.molbiopara.2020.111278.

Wadhone, P., Maiti, M., Agarwal, R., Kamat, V., Martin, S., and Saha, B. (2009). Miltefosine Promotes IFN-γ-Dominated Anti-Leishmanial Immune Response. The Journal of Immunology, 182(11): 7146. https://doi.org/10.4049/jimmunol.0803859.

World Health Organization WHO. (2021, May 18). Trypanosomiasis, human African (sleeping sickness).

Xu, W., Li, X., Wang, L., Li, S., Chu, S., Wang, J., Li, Y., Hou, J., Luo, Q., and Liu, J. (2021). Design of cyclodextrin-based functional systems for biomedical applications. In Frontiers in Chemistry (Vol. 9). Frontiers Media S.A. https://doi.org/10.3389/fchem.2021.635507.

Yadav, M., Thakore, S., and Jadeja, R. (2021). A review on remediation technologies using functionalized Cyclodextrin. In Environmental Science and Pollution Research. Springer Science and Business Media Deutschland GmbH. https://doi.org/10.1007/s11356-021-15887-y.

Yang, G., Choi, G., and No, J. H. (2016). Antileishmanial mechanism of diamidines involves targeting kinetoplasts. Antimicrobial Agents and Chemotherapy, 60(11): 6828–6836. https://doi.org/10.1128/AAC.01129-16.

Yao, C., Dixit, U. G., Barker, J. H., Teesch, L. M., Love-Homan, L., Donelson, J. E., and Wilson, M. E. (2013). Attenuation of Leishmania infantum chagasi metacyclic promastigotes by sterol depletion. Infection and Immunity, 81(7): 2507–2517. https://doi.org/10.1128/IAI.00214-13.

Zeisig, R., Rudolf, M., Eue Dietrich Arndt, I., Zeisig, R., Eue-D Arndt, I., and Rudolf, M. (1995). Influence of hexadecylphosphocholine on the release of tumor necrosis factor and nitroxide from peritoneal macrophages *in vitro*. In. J. Cancer Res. Clin. Oncol. (Vol. 121). Springer-Verlag.

CHAPTER 9

Biosensors as a Diagnostic Tool for Analysis and Clinical Determination of Biomolecules

Asmaa Missoum

Introduction

Biomolecules, which are involved in various biochemical and biological processes, are currently applied as 'biomarkers' to diagnose health conditions, such as strokes, diabetes, and even cancers. These can be detected by various clinical laboratory tests (Pandey and Malhotra 2019). However, conventional protocols have time constraints and often require skilled technical staff as well as costly equipment. In recent decades, the biopharmaceutical sector has flourished thanks to an ever-increasing knowledge of disease mechanisms, biomedical engineering, and device technology. This has empowered the development of much cheaper and faster methods that can be utilized remotely to provide a cost-effective and rapid diagnosis (Ali et al. 2020). Among these methods, biosensors are the most explored as they offer simplified analyses without the requirement of reagents, making them more convenient for a variety of industrial and biomedical applications. Due to their portability, miniaturization, and ability to convert data from a biological process into a measurable electrical signal, their importance in developing point-of-care technologies has expanded. Thus, biosensors could contribute to efficient and continuous monitoring of physiologically important analytes, obtained from biological samples, in clinical routine practice (Srinivasan and Tung 2015).

Université Paris-Saclay, Faculté des Sciences d'Orsay, Orsay,France.
Email: amissoum93@gmail.com

In the early 1900s, the history of biosensors began with the introduction of the pH concept by M. Cremer and S. Sørensen. A decade later, W.S. Hughes developed an electrode for pH measurements. However, the first biosensor ever was an 'oxygen-detecting electrode' that was invented by L. Clark Jr. in 1956. This remarkable work was followed by the discovery of an amperometric electrode in 1962 for the detection of glucose levels. Ever since the first commercialization of this type of biosensor by Yellow Spring Instruments (YSI) in 1975, noteworthy progress has been accomplished in the multidisciplinary field of biosensors which played a vital role in clinical applications (Bhalla et al. 2016). Clearly, biosensors are an essential tool which has the potential to detect the outbreak of infectious diseases caused by bacteria, viruses, and other microscopic agents. The spread of these emerging diseases has caused significant interest in recent years since they are the second leading cause of death worldwide after heart diseases. A complex interaction of factors, such as demographic shifts, global movement of people and animals, ecological and climate changes, etc., have led to the emergence of an expanding number of new diseases. Therefore, there is an urgent need in monitoring and controlling the proliferation of emerging infectious diseases (EID) worldwide, facilitated by the advanced technology of biosensors (Pejcic et al. 2006).

Because there is not much scientific literature that covers an overview of biosensor types, particularly in relation to emerging infectious diseases, this chapter is dedicated to their key role and applications in this global concern. In this context, principles of different types of biosensors and their analytical performance for clinically important biomolecules are presented. Recent trends, emerging novel protocols as well as major future challenges in the field of biosensors technology are also discussed in this chapter.

Fundamentals of Biosensors

Definition, Components, and Characteristics

According to the IUPAC, a biosensor can be defined as a device that utilizes specific biological reactions mediated by whole cells, immunosystems, tissues, isolated enzymes, or organelles to detect biological compounds by thermal, optical, or electrical signals (Figure 1) (Altintas 2018). Therefore, this device consists of three principal components: a recognition element that will selectively bind with the target analyte, a transducing element that will convert the biological binding event into a measurable signal, and a method for detecting and quantifying the resulting signal into beneficial information (Ali et al. 2020). The ideal biosensor for clinical applications

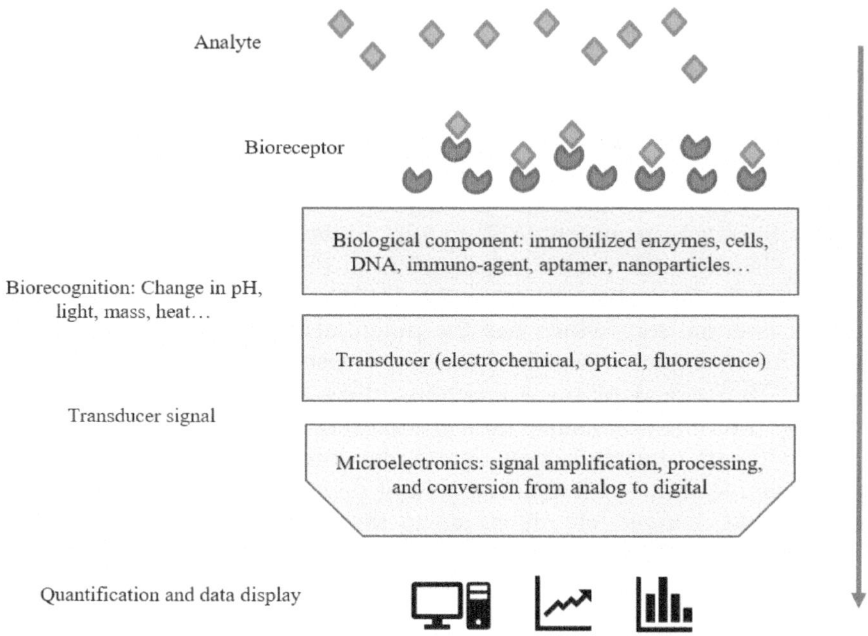

Figure 1. Schematic representation of a biosensor. First, the analyte is specifically recognized by the bioreceptor, which could be an enzyme, deoxyribonucleic acid (DNA), or aptamers. Then, signalization takes place as the transducer converts the energy of the bio-recognition event into a detectable, measurable signal. Most transducers produce either electrical or optical signals. Subsequently, the microelectronics process this signal and prepare for its display after quantification. The output signal on the display unit can be graphic or numeric, depending on the conditions required by the user (Altintas 2018).

should monitor and measure biomarkers in real-time directly from body fluids in a continuous manner. However, there are certain characteristics to be considered in order to validate the performance of the biosensor and ensure the quality of its results. Generally, the following merits are attributed (Justino et al. 2016):

Sensitivity: when a minor change in analyte concentration leads to a great change in response. This is an important property for detecting analyte concentrations as low as ng/ml to confirm its presence in a sample, which could eliminate the need for invasive biopsies.

Selectivity: when the response of the analyte of interest can be differentiated from all responses. In other words, a specific analyte is detected in a sample containing other contaminants and admixtures. This can be portrayed by the interaction of an antigen in a buffer solution with the antibody immobilized on the transducer's surface.

Stability: the degree of susceptibility to surrounding disturbances throughout the biosensing system, such as temperature or bioreceptor affinity and shelf-life. These can cause a shift in the output signals and lead to an error in the measured concentration and affect the accuracy and precision of the biosensor.

Reproducibility: the ability to generate identical responses for duplicated experimental trials, performed in different conditions related to apparatus, operators, and/or intervals of time analysis. It is also characterized by the accuracy and precision of the transducer and microelectronics in a biosensor.

Repeatability: the ability to generate identical measurements of the same parameter, which were performed in the same conditions in terms of apparatus, operators, and/or intervals of time analysis.

Linearity: it illustrates the accuracy of a measured response to a straight line that is represented as $y = mc$, where y is the output signal, m is the sensitivity of the biosensor, and c is the concentration of the analyte. Accordingly, the biosensor response changes linearly with the range of analyte concentrations.

The characteristics mentioned above should be assessed during the development phase and must be verified regularly during routine use. This would estimate capacity, variability, and reliability, which are important analytical performance features (Bhalla et al. 2016).

Biorecognition Elements and Types of Biosensors

In order for a biosensor to function efficiently, the biological analytes should be attached properly to the transducers. For this reason, the appropriate molecular environment must be provided to ensure adequate biocatalyst activities as well as to sustain the stability of the biomolecule using the right immobilisation method. This depends on the type of transducer used and on the physicochemical characteristics of the analyte. The four main methods for biomolecular immobilization include adsorption, covalent bonding, crosslinking, and entrapment (Saylan et al. 2019). Furthermore, the biological recognition elements, also acknowledged as target receptors, are a crucial part of a biosensor as they act as 'mediators' of specific responses after a binding event. Since more and more research is carried out in this field, these will have a leading position in the development of advanced biosensing devices (Figure 2). Regarding the recognition behaviour of the biomolecules, the recognition elements are categorized into three principal groups: natural (enzymes and antibodies), semi-synthetic (aptamers and nucleic acids), and synthetic

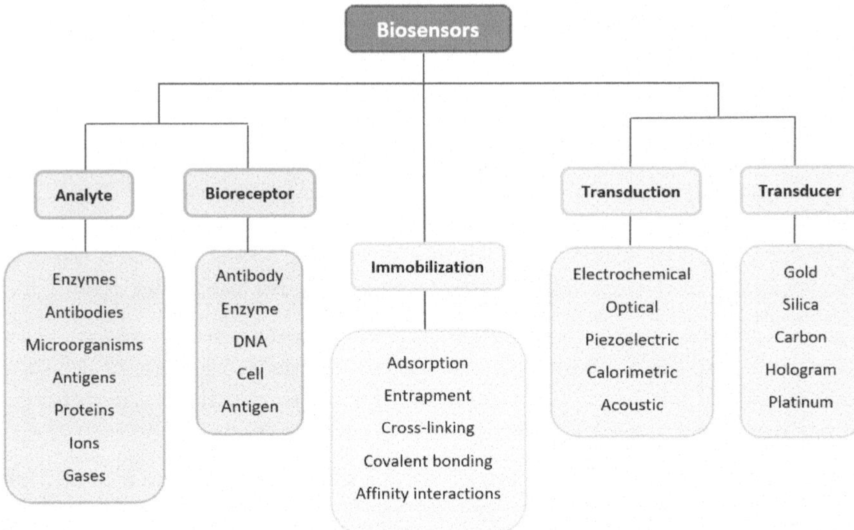

Figure 2. Summary of various biocomponents and transducers employed in the manufacture of biosensors. According to the mode of transduction, sensing devices are classified and characterized into numerous categories.

such as molecularly imprinted polymers. These 'MIPs' are manufactured by electrochemical polymerization, and they are also highly selective, stable, and thermostable (Justino et al. 2015). Nevertheless, biosensors can be classified into various types based on the principles of a transducing element:

Electrochemical (EC): Monitor any alterations in charge distribution and dielectric properties while the analyte-receptor complex is formed on the electrode surface, which plays a central role in the overall performance. Thus, the detection ability depends on the properties of the electrode such as material, dimension, and surface modification. EC biosensors are easy to use, portable, and robust analytical systems that would provide highly sensitive readouts. Since they can also operate in turbid media, they have been very useful in detecting viral infections and tropical diseases (Campuzano et al. 2017a). Furthermore, different electrochemical techniques are applied within this category of biosensors. Such techniques include amperometry, electrochemical impedance spectroscopy (EIS) as well as potentiometry. Recent differential pulse voltammetry (DPV), square-wave voltammetry, and cyclic voltammetry (CV) are also taken into consideration. These inexpensive biosensors have been employed to detect countless biological targets, including cancer biomarkers, proteins, as well as glycosylation patterns, which could indeed provide a warning signal about disease progression (Pihíková et al. 2015).

Optical: Measure changes in the optical features of the transducer surface when the biorecognition element and the analyte form a complex. These present a satisfactory assay versatility, and they can sense numerous types of biomolecules from physiological specimens. For example, *Streptococcus pneumoniae* and *Salmonella Typhimurium* have been successfully detected in clinical samples by incorporating a developed colorimetric strategy for standardized visual bioanalysis (Chen et al. 2015). Nevertheless, optical biosensors can be divided into two categories: direct and indirect. In the first category, the signal generation depends on the complex formation of the transducer surface. For example, the immobilization of biomolecules influences the orientation of optical cells containing liquid crystals, in a quantitative and label-free optical sensing diagnosis (Nguyen et al. 2015). The second category includes sensors that use florescent labels to identify the binding events and then amplify the signal. Although indirect optical biosensing can generate higher signal levels and is ideal for dynamic measurements, it has many drawbacks such as costly reagents for the labelling step and non-specific binding (Morris 2013).

Bioluminescent: This is a subtype of an optical biosensor where luminescence is measured instead of fluorescence emissions. The sensed interaction between microbes and the analytes is correlated to the concentration of the microbial population. Since adenosine triphosphate (ATP) is considered as the energy currency of living microbes, it can be utilized as a fast indicator of microbial viability (Eed et al. 2015). ATP bioluminescence is based on the conversion of luciferin to oxyluciferin, a biochemical reaction catalyzed by the luciferase enzyme. This takes place in the presence of magnesium cation (Mg^{++}) and oxygen (O_2), where light is emitted after the conversion of ATP into adenosine monophosphate (AMP). The overall reaction is described according to the following equation:

$$\text{Luciferin} + \text{ATP} + O_2 \xrightarrow{\textit{luciferase, Mg} + 2} \text{Oxyluciferin} + \text{AMP} + \textit{products} + \text{light}$$

This light output can be measured in a luminometer using sensitive photons of light meters. The intensity of emitted light in this reaction is expressed in RLU (relative light units). Thus, the highest the amount of ATP, the greater amount of light will be generated by the assay and consequently the highest the RLU value will be produced (Ali et al. 2020). Unlike conventional approaches, such as the plate-counting method, ATP-based bioluminescence sensor provides a rapid, sensitive, and real-time assay. Moreover, this method is non-destructive as it requires little to no sample pretreatment, making it less expensive and less laborious (Eed et al. 2016). ATP-based bioluminescence sensors have frequently been used for the investigation of food-borne pathogens and for assessing the efficiency of quality control procedures. Interestingly, this revolutionary method has also been optimized as a 'whole cell' biosensor to detect toxins, heavy metals, and chemicals that disrupt cellular respiration as well as ATP production processes (Rahimirad et al. 2019).

Piezoelectric: It is based on a physical phenomenon known as piezoelectricity, which states that when a material is mechanically stressed, it can eventually produce voltage. Therefore, this alternating voltage leads to mechanical oscillations of quartz crystals and then their frequency is measured. Any change in mass, density, and/or viscosity of the bound analyte to the crystal's surface results in change of oscillation frequency. Due to elevated sensitivity to environmental surroundings, the biosensing mechanism necessitates isolation apparatus that minimizes any interfering factors, such as vibration (Pohanka 2018). Here, quartz crystals are used as a transduction material and the platform is termed quartz crystals microbalance (QCM). Because of its simplicity, it has been embraced as a competitive tool for immuno-sensing applications. The benefits of using this transducer are label-free detection, real-time monitoring, and ease of use. However, lack of specificity and sensitivity, as well as calibration problems and excessive interference, are some of the drawbacks that must be overcome. Hence, these sensors are considered inferior and are not receiving much attention compared to electrochemical and optical biosensors (Marrazza 2014). The significant improvement of piezoelectric sensors is the use of thin films as the transduction material instead of quartz crystals. Film bulk acoustic resonators (FBAR) were recently developed and are employed in the detection of proteins, DNA, and small ligands. These possess better piezoelectric properties, such as low acoustic loss, high acoustic velocity, and high electromechanical coupling coefficient (*k*2). Zinc oxide and aluminium nitride are the most optimizing materials utilized for the manufacture of FBARs owing to their physical properties, low cost, and simple fabrication processes. Last but not least, QCMs can operate on a resonance frequency of 5 to 20 MHz, while FBARs have a range of 2–10 GHz (Zhang et al. 2018).

Calorimetric: Detects the heat (enthalpy change) of biochemical reaction between the analyte and the biorecognition element, which can be monitored in a continuous manner. Since most biological processes are naturally either endothermic or exothermic, the key advantage of using this sensing mechanism is universality. Gaddes et al. have developed a thermal sensor consisting of a Y-cut quartz crystal resonator (QCR) as a temperature sensor positioned near a fluidic chamber filled with an immobilized enzyme. Accordingly, the transducer element is physically isolated from the analyte solution and is therefore unaffected by other effects occurring on the sensor surface. This configuration yields a sensitive, robust, and fast biosensing system that is capable of continuous injection 'flow' analysis of molecules, such as urea, glucose, and creatinine (Gaddes et al. 2015).

Surface Plasmon Resonance (SPR): This type of biosensor uses electromagnetic surface plasmon waves, which detect changes when the target analyte interacts with the immobilized biomolecule on the sensor surface. During this interaction, a change in the refractive index takes place at the sensor surface. This is because the optical waves will excite plasmon waves and cause them to absorb their resonant energy (Figure 3). Consequently, changes in the propagation constant of plasmon waves will be detected by a spectrophotometer. Two principal advantages of SPR-based sensors are real-time signal interpretation by the sensorgram and

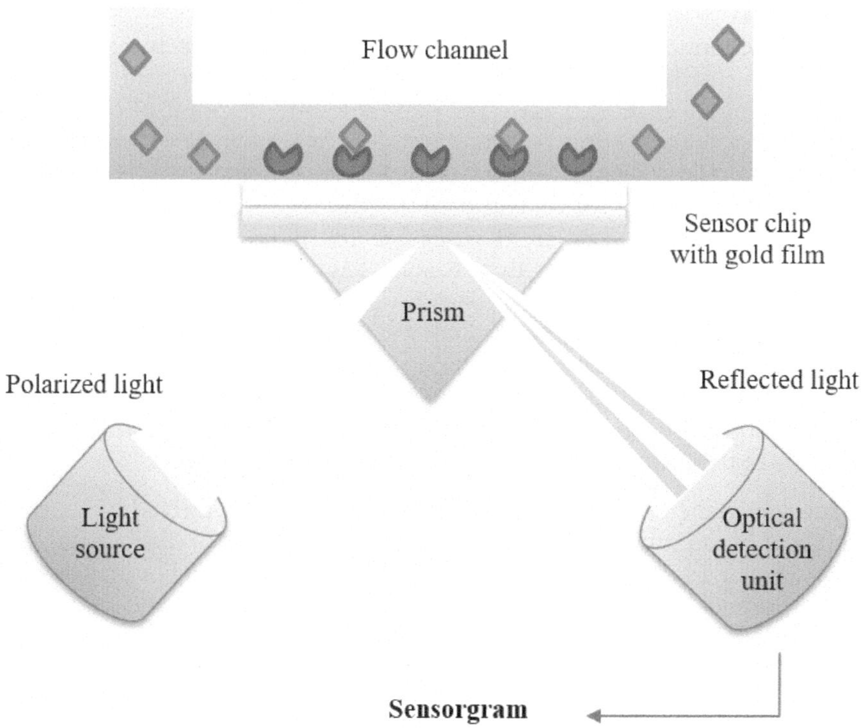

Figure 3. Schematic diagram of a Surface Plasmon Resonance (SPR) biosensor. Basic components include a glass slide and a thin gold coating that is mounted on a prism. When light passes through the prism and the slide, it reflects off the gold surface and passes back through the prism to the optical detector. When the analyte is passed through a flow cell over the immobilized receptor on the sensor chip, a change in mass will take place due to the interaction. This is detected as a variation in the angle of the incident light required to generate the SPR phenomenon at the gold interface (Prabowo et al. 2018).

label-free sensing without fluorescence or radioactive tagging. However, SPR sensors lack specificity and are not suitable for monitoring small analytes as they generate inadequate responses (Prabowo et al. 2018). Therefore, the use of a pure sample of the analyte is desirable and its pre-processing is essential. With recent improvements in microfluidic techniques, it has been possible to purify samples, e.g., separation of plasma or blood cells from whole blood. Such techniques are integrated into the miniaturized SPR biosensors and could include microcentrifuges, microporous filters, and CD-type platforms. As a result, this advancement has contributed to their remarkable role in point-of-care diagnostics (Wang and Fan 2016).

Detection of Biomolecules for Clinical Analysis

The detection of disease biomarkers is significant for the prognostic monitoring of health and will deliver a suitable method of treatment. As mentioned earlier, biosensors play a major in point-of-care (POC) diagnostics as the testing is carried out on-site allowing medical staff and patients to obtain results rapidly in a cost-effective manner. Biomolecules that act as indicators for certain pathological conditions such as diabetes, cancer, and cardiovascular diseases are found in body fluids in very minute concentrations. For this reason, it is crucial to sense their levels at an early stage to prevent disease progression. There is a wide variety of biomolecules, ranging from small molecules to whole cells, which are demonstrated with current detecting systems (Metkar and Girigoswami 2018). The following subsections will briefly report the configuration of the various biosensors used to detect physiologically significant analytes. Studies of their analytical performances are also considered.

Glucose: Diabetes is a chronic health condition characterized by elevated levels of blood glucose. It is one of the leading causes of disabilities, such as nerve degeneration, blindness, and kidney failure. For this reason, early diagnosis would help to avert the progression of the disease. In fact, one of the first developed biosensors was for detecting glucose and it was testified by Clark and Lyons in 1962. Since then, expensive and complicated technologies have been utilized such as Raman spectroscopy, liquid chromatography, and polarimetry. A promising tool for monitoring glucose is the capacitive biosensor, based on a polyvinylidene fluoride (PVDF) film. This type of low-cost sensor measures the changes in dielectric properties at the interface between the electrolyte and the electrode when glucose interacts with the immobilized oxidase enzyme on the isolated dielectric layer (Hartono et al. 2018). However, enzyme-based glucose biosensors present numerous challenges due to the low efficiency of enzyme immobilisation on the electrode. To overcome this problem, ZnO nanorod arrays were manufactured and incorporated as an amperometric electrode in which the oxidase enzyme was immobilized with high stability through physical adsorption. Due to its high surface-area-to-volume ratio and isoelectric point (IEP), loading capacity and sensitivity were also enhanced owing to the efficient electron transfer from oxidase to the electrode. According to the study, the modified sensor was tested using human blood samples and presented comparable results with excellent performance displaying a linear range of 0.05–1 mM (Ridhuan et al. 2018). Furthermore, it has been reported that an advanced self-powered biosensor has successfully quantified plasma glucose levels with a detection limit of 2.31 ± 0.3 mM. This device, which can be easily miniaturized, consists of a biofuel cell component that generated a power density correlated to plasma glucose at a cell voltage of 0.213 V. In other words, the glucose is detected by monitoring the charge and discharge frequency (Hz) of the transducer. Hence, these findings show that human plasma can be exploited as an energy source for biofuel cells and simultaneously sense plasma glucose levels (Slaughter and Kulkarni 2019).

Cardiac Biomarkers: Cholesterol is a crucial sterol that acts as a precursor for the biosynthesis of vitamin D, bile acids, and steroid hormones. When its levels reach above 240 mg/dL, it is considered a biomarker for heart attack and strokes, high blood pressure, and cardiovascular diseases (CVD). Accordingly, biosensors are ideal for monitoring cholesterol levels for clinical diagnosis as well as treatment of coronary diseases. Recently proposed sensors demonstrate a detection limit in the range of 0.000002–4 mM and work within a cholesterol concentration range of 0.000025–700 mM. They include electrochemical, conductometric, piezoelectric, as well as optical biosensors. Most of them mainly used nanoparticles, graphene materials, and microfluids to enhance the analytical response. Additionally, current devices are not portable, cannot be used by patients, and fail to monitor cholesterol in a real-time manner. For this reason, upcoming research should focus on developing miniaturized cholesterol sensors at a low cost. It could also consider fully automatic lab-on-chip devices to be easily used by patients for quick screening (Narwal et al. 2019). Nevertheless, there are other significant cardiac biomarkers that are detected by electrochemical-based biosensors and are utilized for the prediction of cardiovascular diseases, such as acute myocardial infarction. These can be ordered as C-reactive protein, interleukins 1 and 6, lipoprotein-associated phospholipase, low-density lipoprotein, myeloperoxidase, myoglobin, troponin I or T, and tumor necrosis factor-alpha (TNF-α). Besides, electrochemical devices present several advantages including low cost, portability, rapidity, reliability, robustness, the requirement of small samples, and multianalyte testing.

This would also minimize the cost of healthcare services. It has been reported that electrochemical impedance spectroscopy (EIS) was used for monitoring TNF-α, cyclic voltammetry (CV) for troponin I, and carbon-based screen-printed electrodes (SPCEs) for C-reactive protein. Other electrochemical methods, such as square wave voltammetry (SWV) and differential pulse voltammetry (DPV), are being enhanced for better sensitivity and selectivity (Bakirhan et al. 2018).

Cancer biomarkers: Cancer is a group of diseases in which cells with abnormal proliferation form tumors with the potential to invade nearby tissues and organs. Due to their high mortality rate and increasing prevalence, early diagnosis with non-invasive tools is crucial. However, currently used biopsy and clinical devices such as X-ray, computerized tomography (CT), and magnetic resonance imaging (MRI) are invasive, time-consuming, and limited to certain laboratories in large hospitals. In addition, molecular tools including polymerase chain reaction (PCR), flow cytometry, enzyme-linked immunosorbent assay (ELISA), and immunohistochemistry (IHC) also present several drawbacks. For this reason, biosensors represent a non-invasive and inexpensive alternative for cancer theranostics. Among these, DPV and SWV electrochemical biosensors are best-suited thanks to their simplicity, rapidity, and simultaneous multi-analyte analysis. Their sensitivity is equally considered as the reported limits of detection for molecular biomarkers and cancer cell detection are fg mL^{-1} and 1 cell mL^{-1} magnitude, respectively. The EIS also provides a dynamic long-term and real-time monitoring of cancerous cells and biomarkers (Cui et al. 2020). Throughout the process of tumorigenesis and metastasis, numerous and diverse cancer biomarkers are produced. These would compromise altered genomic circulating tumor DNA, cancer antigens, circulating tumor cells (CTCs), exosomes, microRNA, and even matrix metalloproteinases (MMPs). In response to cancer cells, the immune cells secrete special cytokines, which are different microproteins acting as immunomodulating agents. A dysregulation in their levels has been correlated with the onset of various types of cancer. Interestingly, electrochemical paper-based biosensors have been successfully applied in quantifying cytokines. Owing to their portability, accompanied by idiosyncratic properties, they offer fast analysis without the need to access a laboratory. The main challenge in manufacturing this type of biosensor is its ability to sense multiple cytokines simultaneously in clinical samples. Consequently, improving the performance of paper-based biosensors would contribute to optimistic societal implications in developing countries, where access to healthcare services and resources is limited. Apart from this, it is important to mention that biosensors can be also applied in phenotypic and metabolism analysis of cancer cells and in addition to chemotherapy drug sensitivity monitoring (Loo and Pui 2020).

Hormones: The medical management of endocrine disorders relies on timely and accurate measurements of hormones, such as cortisol and 17ß-estradiol for example. Cortisol is known to be a potential biological marker for estimating psychological stress. For instance, abnormal levels indicate chronic conditions, such as Addison's disease and Cushing's syndrome. The detection of cortisol is commonly limited to SPR and QCM. However, these techniques necessitate complex sample preparation and take time to present results. Due to the recent electrochemical immunosensor with nanomaterials, cortisol antibodies were immobilized and cortisol was selectively detected in saliva samples (Justino et al. 2016). On the other hand, 17ß-estradiol is known to be an endocrine disruptor with major impacts on reproductive functioning. Its quantification is crucial to clinical evaluations for postmenopausal status, fertility treatments, and even ovarian cancer. Voltammetry, chemiluminescence, and high-performance liquid chromatography (HPLC) are utilized for 17ß-estradiol detection. Since these expensive analytical techniques are insensitive and laborious, Wang et al. have developed an enzyme biosensor in which the oxidization of 17ß-estradiol is catalyzed via poly L-lysine films and the cross-linkage of glutaraldehyde with laccase. These modifications helped to achieve better conductivity and consequently resulted in more selectable, sensitive biosensing. Excellent performance was also demonstrated in the human urine sample analysis with a low detection limit of 1.3×10^{-13} M (Wang et al. 2019). In addition, the parathyroid hormone (PTH) is a secreted polypeptide that regulates calcium homeostasis in the human body. The presence of numerous molecular forms of the hormone complicates its measurement, which aids in the investigation of calcium levels as well as parathyroid gland disorders. Current immunoassay techniques lack quick response time and necessitate a large sample volume. Nevertheless, for rapid assessment of PTH, a novel biosensing device has been designed. This electrochemical nano-device has increased the immobilization of the capture probe allowing enhanced PTH interaction in a label-free manner. Besides, PTH levels were quantified with high sensitivity exhibiting a detection limit of 10 pg/ml in plasma and whole blood samples and 1 pg/ml in human serum samples. This new point-of-care sensor is devoid of sample pre-treatment and can be also suitable in surgical settings (Tanak et al. 2019).

Small molecules: To study cellular metabolisms, it is unquestionably crucial to detect the levels of small metabolites and other signaling molecules. Particularly, fluorescent sensors that are genetically encoded and may be expressed within specific intracellular locations and cell types. Nevertheless, low-abundance biomolecules are hard to detect since they are often incapable to bind to their recognition elements. Built on the principles of RNA-based biosensors, researchers from Cornell University

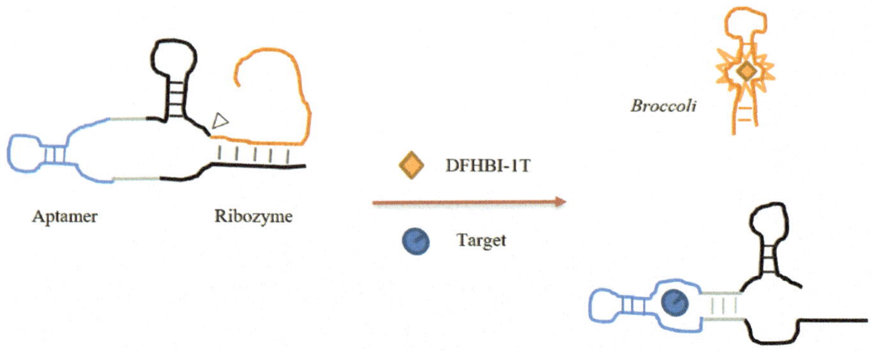

Figure 4. Illustration of a novel genetically encoded fluorescent biosensor. The RNA integrator is one of the main components as it enables the detection of low-abundance metabolites in living cells and *in vitro*. The RNA integrator consists of a ribozyme and a fluorogenic unfolded aptamer. When nonfluorescent cell-membrane permeable fluorophores such as DFHBI-1T are introduced, they bind to the fluorogenic aptamers, e.g., *Broccoli* making them fold and exhibit florescent (You et al. 2019).

have recently designed and integrated an RNA integrator which is an RNA sequence containing a ribozyme and a fluorogenic aptamer in an unfolded form. When it binds to its target, the ribozyme is cleaved and releases the aptamer, which then becomes fluorescent after it folds (Figure 4). Accordingly, RNA integrators offer better temporal resolution than most protein-based biosensors, and their expression in cells permits the imaging of metabolite levels with a large dynamic range and high sensitivity. Without a doubt, they provide an alternative and a better approach to sensing small metabolites in living bacterial cells (You et al. 2019).

Applications of Biosensors in Detecting Infectious Diseases

Apart from the molecular detection of particles, the majority of biosensors are extensively exploited for the medical diagnosis of whole cells and other infectious agents. These include viruses and bacterial cells in which some of them are summarized in Table 1 (Pashchenko et al. 2018). Each year, the spread of infectious diseases leads to millions of deaths, and hence needs to be prevented as early as possible using their specific pathogenic biomarkers (Lakshmipriya and Gopinath 2019). In the next subsections, various diagnostic modalities based on currently developed biosensors are discussed with a brief description of the most important pathogens and their biological markers.

Bacteria: Although the vast majority of bacteria are harmless or often beneficial, it is estimated that only fewer than a hundred species contribute

Table 1. Comparison between different types of biosensors in detecting infectious diseases caused by various pathogens (Pashchenko et al. 2018).

Technique	Target Pathogen	LOD	Time spent in detection	Additional Features
Faradaic electrochemical impedance spectroscopy (EIS)	*Escherichia coli*	1E2 to 1E3 CFU/mL	30 minutes	Label-free detection
Graphene-enabled biosensor using specific immobilized monoclonal antibody	Zika Virus	450 pM	5 minutes	Portable
Microfluidic immunosensor with Cholera toxin subunit B (CTB)-specific antibodies immobilized onto super-paramagnetic beads	Cholera toxin	9.0 ng/mL	1.5 hour	Detection in fecal samples
Electrochemical immunoassay using a layer-by-layer coating of carbon nanotubes and graphene oxide as a nanocarrier	*Clostridium difficile* (toxin B)	0.7 pg/mL	45 minutes	Detection in human stool samples
Microfluidic electrochemical quantitative loop-mediated isothermal amplification (MEQ-LAMP)	*Salmonella* spp. DNA	16 copies of DNA	< 50 minutes	Combines preparation, amplification, and detection in one multiplex platform
Thin-film transistor nanoribbon sensor	*Klebsiella pneumoniae*	10 copies of DNA	5 minutes	Real-time nucleic acid analysis
Lateral flow biosensor using loop-mediated isothermal amplification and gold nanoparticles	*Staphylococcus aureus*	680 CFU/mL	75 minutes	Detection in spiked blood samples
Multiplexed kit capable with target-specific fluorescently tagged strand displaceable probes with RT-LAMP	Dengue virus	~ 1.22 pfu equivalent viral RNAs	30 minutes	Detection in urine and plasma

to infectious diseases in humans. Among these are tuberculosis, meningitis, and leprosy. Different approaches have been established for detecting pathogenic bacteria. These include Matrix-Assisted Laser Desorption/Ionization Time-Of-Flight Mass Spectrometry (MALDI-TOF MS) in addition to Next Generation Sequencing (NGS). However, biosensors are promising diagnostic technologies since they are fast, efficient, and do not require specialized technicians for operation (Rodovalho et al. 2015).

Tuberculosis (TB) is caused by *Mycobacterium tuberculosis* and is currently the leading infectious cause of death. According to the World Health Organization, around 5.7 million cases were reported worldwide in 2013. Recently, SPR-based optical biosensors were employed in the detection of *M. tuberculosis* IS6110 DNA, as they decrease test turnover time and eliminate the need for expensive labels. They can also distinguish active TB infections through numerous biomarkers. For example, the specific secretory antigen CFP-10 was successfully identified in clinical urine samples by a prism coupled system with a sensitivity of 100 pg/mL. The signal of the SPR sensor was enhanced by magneto-plasmonic nanoparticles and an immunoassay was carried out to combine this high sensitivity with the specificity of antibody-antigen interactions (Li et al. 2019). From the same genus of *Mycobacterium, M. leprae* is another pathogenic bacterium that causes leprosy, a chronic disease leading to serious nerve damage. Early diagnosis can be achieved by electrochemical genosensor, which is based on the immobilization of specific single-stranded DNA oligonucleotide or two antigens to detect IgM and IgG antibodies (Campuzano et al. 2017b).

Other bacterial pathogens include *Neisseria meningitidis, Listeria monocytogenes,* and *Streptococcus pneumoniae.* These different species can cause meningitis, an acute inflammation affecting the protective membranes of brain and spinal cord. It has been shown that Omp85, a virulence protein of *N. meningitidis,* can be utilized as a biological recognition element in the QCM immunosensor (Rodovalho et al. 2015). On the other hand, *S. pneumoniae* can be determined by several types of biosensors, such as amperometric magneto-immunosensor with a LOD of 3.0×10^3 cfus, SWV for detecting pneumococcal surface protein A (PspA) peptide with a LOD of 0.218 ng mL^{-1}, and a disposable DNA sensor involving asymmetric polymerase chain reaction (aPCR) to target a precise region in the pneumococcal *lytA* gene. The latter has permitted discrimination between *S. pneumoniae* and closely related *streptococci* (Campuzano et al. 2017a).

Bacteriophages, simply known as phages, are viruses that infect bacteria. Since they are used in molecular biology experiments such as gene cloning, there has been a recent interest in developing phage-based biosensors for bacterial detection in clinical samples. This type of sensor uses phages as bio-probes with high selectivity and specificity. Moreover, different analytical approaches including optical, micromechanical, and electrochemical can be employed. For example, the P9b phage was successfully utilized to detect *Pseudomonas aeruginosa* in urine samples using SERS (surface-enhanced Raman spectroscopy) as a transducer with a detection limit of 10^3 CFU/mL. The whole assay took less than an hour and did not require selective media or supplementary biochemical tests (Farooq et al. 2019).

Viruses: These are infectious agents that infect all kinds of life forms and are responsible for numerous diseases in humans such as human papillomavirus (HPV), dengue virus, influenza A H1N1, hepatitis E, and human immunodeficiency virus (HIV). Currently, employed methods for the detection of viruses suffer from low specificity. To overcome this drawback, Chowdhury and his co-workers developed a highly sensitive electrochemical biosensor for hepatitis E virus (HEV), using gold-embedded polyaniline nanowires and graphene quantum dots. The proposed sensor is induced by electrical pulses and shows high selectivity owing to the usage of antibody-antigen interaction. It has detected HEV-like particles with a low detection limit of 0.8 fg mL^{-1} (Chowdhury et al. 2019). Hai and collaborators engineered a label-free biosensor for the detection of influenza A H1N1 virus. It was based on the notion of hemagglutinin binding specificity for galactose residue of human sialic acid, using PEDOT (poly-3,4-ethylenedioxythiophene) as a conducting polymer. Consequently, QCM and potentiometry were applied in the analysis of A (H1N1) pdm09 virus binding with the galactose residue on the conducting polymer. The developed system is a good candidate for point-of-care testing due to its mass productivity and processability. This assay has resulted in doubled sensitivity when compared with conventional kits (Hai et al. 2017). Furthermore, it has been reported that HIV was successfully diagnosed in a clinical study involving 133 patient samples via a surface acoustic wave (SAW) biosensor. This portable, low cost, and miniaturized sensor required only 6 μL plasma and used dual-channel biochips with an *in-situ* reference channel to detect either anti-gp41 or anti-p24 antibodies. It also showed excellent specificity and sensitivity in addition to rapid time to result (Gray et al. 2018).

Parasites: Protozoa are single-celled eukaryotes that cause several diseases in humans including malaria, leishmaniasis, trypanosomiasis, and toxoplasmosis. According to the WHO, about 3.2 billion people in 97 countries are still at risk of being infected with malaria. This tropical disease is caused by a protozoan of the genus *Plasmodium,* which is transmitted by different species of female anopheline mosquitoes. While conventional methods such as microscopic identification and flow cytometry fail to detect mixed infections and quartz crystal microbalance (QCM) was employed to differentiate between malaria species. Secreted antigens including histidine-rich protein-II and lactate dehydrogenase are also used as biomarkers in colorimetric, electrochemical, and optical immunosensors (Campuzano et al. 2017b).

American trypanosomiasis is another neglected tropical disease that is vector-borne. It is also known as 'Chagas disease' and is caused by *Trypanosoma cruzi* that is spread by *Triatominae*, blood-sucking buds. Recently, an amperometric immunosensor was used for serological diagnosis by immobilising *T. cruzi* proteins from epimastigote membranes on gold electrodes modified with thiol. Hence, the interaction between antibodies from sera of patients and the immobilized antigens is monitored, resulting in 12.4 ng mL^{-1} of IgG as the detection limit. When this assay was optimized by integrating a microfluidic system and a modified electrode with gold nanoparticles, the detection limit was improved to 3.065 ng mL^{-1} (Rodovalho et al. 2015).

Leishmaniasis is a zoonotic disease spread by 'sand flies' of the genus *Phlebotomus*, carrying an intracellular parasite of the genus *Leishmania*. It has three clinical manifestations: cutaneous, mucocutaneous, and visceral Leishmaniasis (VL). The latter is caused by *L. infantum* and its diagnosis has been performed by means of a developed piezoelectric immunosensor, which sensed 1.8×10^4 amastigotes/g of infected tissue. VL is usually confirmed by microscopic identification of the presence of amastigotes, the reproductive form of the parasite, in spleen aspirates or bone marrow. Therefore, biosensors could provide a faster and more specific diagnosis in humans as well as in domestic dogs as they are the main host of *L. infantum* (Cabral-Miranda et al. 2014).

Toxoplasma gondii is another obligate intracellular protozoan that infects humans causing toxoplasmosis. This disease presents flu-like symptoms and mimics other infectious diseases. Since clinical signs are non-specific and enough for a definite diagnosis, immunoassay-based biosensors are developed for detecting anti-*T. gondii* antibodies. For instance, He and his research team have managed to detect *T. gondii* specific DNA oligonucleotides via magnetic fluorescent nanoparticles (Fe_3O_4/CdTe), having an average size of 10 nm. The genosensor, which was fabricated based on the FRET mechanism, enabled *in vivo* and real-time monitoring with a limit of detection of 8.339 nM (He et al. 2015).

Recent Trends in Biosensors

Current advances in biosensor technologies for infectious diseases have the potential to convey point-of-care diagnostics, which surpass conventional methods regarding cost, time, and accuracy. Recent biosensors exploit nanofabrication and magnetic technologies as well as diverse sensing strategies including electrical, mass-based, and optical transducers (Fu et al. 2019). Detection strategies are broadly classified as either label-free or labeled assays. The first category quantifies an analyte directly through biological reactions on a transducer surface. For the second one, the analyte is packed in between detector and capture agents with a specific label on the detector agent for signal output. This label could be an enzyme, a fluorophore, or even a radioisotope (Sin et al. 2014). In order to develop a biosensor, a few important stages must be considered. First, it must be used in detecting only one certain marker and characterized in terms of quantitative determination. Next, the biosensor and its analytical procedure are validated by comparing results with another technique, such as ELISA methodology. Afterwards, the established biosensor and the related analytical procedure are employed for the investigation of the target marker in significant series of clinical samples, including control experiments of healthy donors (Gorodkiewicz and Lukaszewski 2018). Nevertheless, this categorization proves to be valuable in relation to the recent research summarized in the following subsections. It represents high innovative potential in terms of biosensor construction and could result in fully developed procedures in near future.

Nanotechnology: Nanomaterials compromise any materials whose single units, particles, or constitutes range from 1 to 100 nm in size. They have drawn enormous attention for ameliorating diagnostic devices owing to their electrical, chemical, physical, optical, and size-dependent properties. However, systems based on these materials are more likely to suffer from the unfavorable effects of nonspecific adsorption, making their application an important topic to consider. Since 1981, nanotechnology has gone through a prosperous development that resulted in its incorporation into the field of biosensors. Advantages such as high stability and surface energy, superior biocompatibility, and strong amplification effect on signals have significantly enhanced sensitivity as well as other analytical characteristics of biosensors. Gold nanoparticles (AuNPs) are one of the most frequently used as they intensify the detection signal of low-concentration analytes. Other nanomaterials that stood out include photonic crystals, graphene, and carbon nanotubes (Su et al. 2017).

Quantum Dots: These are semiconductor nanocrystals that act as optical imaging agents owing to their ultra-stability, sensitive fluorescence intensity, broad absorption spectrum, and outstanding quantum confinement effects. These properties also facilitate fluorescence transduction based on physical or biochemical interaction, as it would occur on the surface either through quenching or direct photoluminescent activation. Despite the toxic effects of some quantum dots, their applications in tissue engineering to detect DNAs, microRNAs, enzymes, and proteins are noteworthy achievements of biosensing research (Hasan et al. 2014).

Graphene-based biosensors: Graphene is a one-atom two-dimensional (2D) sheet of sp2-hybridized carbon atoms organized in a honeycomb network. Owing to its unique characteristics such as high electrical conductivity due to fast electron transportation, outstanding electrochemical stability, large specific surface area, and low production cost, this material has attracted noteworthy technological interests in the field of biosensors. Moreover, the 2D structure of graphene promotes a long-range π-electron conjugation, which results in good biocompatibility, excellent thermal conductivity, and strong mechanical strength. These properties contribute to the high selectivity, sensitivity, and low detection limit of graphene-based sensors. For example, a high specific surface area allows graphene to easily conjugate with biological molecules (enzymes, RNA, single-strand DNA, antibodies, receptors, etc.) with high efficiency. Besides, graphite can be oxidized via exfoliation to generate stable aqueous graphene oxide possessing carboxyl, epoxide, and hydroxyl functional groups. Biosensors can use them to enhance specific recognition sites (Zheng et al. 2014).

Carbon Nanotubes: CNTs, which are also considered allotropes of carbon, are a network of carbon atoms assembled into continuous cylinders of one or more layers in a hexagonal tiling pattern with either closed or open ends. The unique chemical and physical properties of carbon nanotubes have introduced many novels and enhanced sensing devices. These include high surface area, very good electrical conductivity, electrochemically stability, excellent mechanical strength, and high thermal conductivity. CNT-based electrochemical transduction proves significant improvements in the activity of immunosensors, nucleic-acid sensing biosensors, and amperometric enzyme electrodes. For example, their hollow structure is convenient for the adsorption of enzymes and their large surface area enables the immobilization of numerous receptor moieties for biosensing (Sireesha et al. 2018).

Magnetic particles: The integration of magnetic particles, whether of nano- or micro-sized dimensions, within biosensors improves their performance in terms of assay time, selectivity, and sensitivity. This is mainly due to the reactivity, increased surface area as well as high effectiveness of blocking reagents of the magnetic carriers, which can be employed as a solid support or as labels for detection. Other advantages include efficient and selective capture of the target analyte, removal of the interfering matrix, and suitability for automation due to magnetic actuation. Furthermore, the use of magnetic particles simplifies the analytical procedure as purification can be facilitated by applying an external magnet, unlike multi-step classical centrifugation and chromatography separation strategies (Kim et al. 2020).

Microfluidic biosensors: Microfluidics is a science of systems processing small amounts of fluids using micrometer-sized channels. It is characterized by tiny dimensions of structures and the minute consumption of liquid samples. Thus, integrating it into the biosensor technology offers a closed and stable environment so to enhance sensitivity, assimilation of multiple functions within a single device, separation and mutual processing of numerous assays with one or numerous samples simultaneously, high throughput automation, and precise control over experimental conditions. In addition, the field of point-of-care biosensors has been recently expanded due to the exploitation of microfluidic technologies. However, the miniaturization of this system can be very challenging from a manufacturing and operational point of view, given that numerous experimental attempts are reported (Wang et al. 2020).

Lab-on-a-Chip (LoC): This is a miniaturized device that automates and integrates several laboratory techniques into one system that fits on a few square centimeters sized chip. Thus, the combination of biosensor technology with LoC offers significant advancements in the overall performance of the sensing system, such as low sample and reagent consumption, speed of analysis, ease of use, and high reproducibility. LoC-based sensors have been applied as POC diagnostic platforms for the detection of DNA, proteins, and hormones. Biosite's blood test device represents a good example where only a drop of a patient's blood is required for the assay, which takes place on a single platform yielding diagnostic results about the patient's cardiac profile in just a few minutes (Azizipour et al. 2020).

Despite clinical demand, an adaptation of biosensing devices from research laboratories into clinical applications has been restricted to a few prominent examples, e.g., the glucose sensor. Nevertheless, novel research outcomes are expected to be continuously announced.

Challenges and Future Prospects

Despite the growing interest, scientific literature reports, and market availability, the field of biosensor technology and its implementation in point-of-care diagnostics still faces numerous challenges. Some of these include matrix effects, system integration, and sample preparation. For example, the exact size of graphene's dimension and its depth are hard to control. Thus, researchers should focus on developing a new regulation system for its synthesis, as well as on the surface chemistry throughout the conversion of graphene to graphene oxide (Zheng et al. 2014). Moreover, the integration of microfluidic systems and miniaturization are still at an early stage. The creation of automated microfluidic platforms with real-time monitoring capabilities will benefit the translation of such systems to clinical applications. This will require a standardization of the systems and a full assessment of their parameters (Hasan et al. 2014). As mentioned earlier, sample preparation is also a significant issue. Although magnetic beads have been utilized to pre-purify the analyte of interest before sensing, the recent temperature-responsive binary reagent system was proven to be more efficient and faster. Hence, such approaches will likely require to be assimilated into the sensing process to enable specific and rapid enrichment of target markers in complex clinical samples (Fu et al. 2019).

Other critical challenges are posed by non-specific binding, low concentration of target molecule, and bioreceptor alignment. The recognition of an analyte should be specific and not affected by other molecules, chemicals, or cells. This is burdensome as sample components are often complex and mixed with varieties of molecules. In addition, biological fouling introduces high signal noise due to non-specific binding, which can be eliminated by washing the binding surface with a buffer. A low concentration of target analyte within a low-volume sample can also be a significant issue. For this reason, many biomedical engineers are working on improving current biosensors to achieve lower detection limits (Sin et al. 2014). Concerning bioreceptor alignment, the repeatability of operation may vary even when following the same protocol. One reason behind this is the immobilization of numerous layers of molecules on the sensor surface, which induces destructive interference and results in a less sensitive detection. Thus, surface treatment protocols (e.g., activation, functionalization, and modification) should be experimented with and tested. Eventually, the design of the biosensor assay matrix is another crucial topic to be considered. This is important for all binding events, especially when the tested sample components are complex. For example, non-specific binding from unrelated components can be eliminated by integrating the appropriate reference binding sites. Such a proposition requires both imaging and data processing capabilities of the biosensor. In the meantime, surface treatment will be more difficult as complications in sensor surface properties necessitate better system responsivity and compatibility (Wang et al. 2020).

Regarding the commercialization of biosensors, academic research plays a mandatory role. This means that it is driven by propositions of scientific peer review, as well as funding agencies that determine priorities and are often characterized by numerous conflicts of interest. In this aspect, the field of biosensors has a certain distinction as they are pointed as practical devices to be used, even though they can barely be justified as 'curiosity-driven' research. On the other hand, production barriers such as performance testing, stability, costs, ease of manufacturing, and use-associated hazards are rapidly being broken. Huge investment has been also directed to translational research, predominantly, for healthcare applications. This has bought academic research closer to the industry with the purpose of providing commercially viable products and has improved the way scientists of different disciplines work across boundaries. Such an alliance is a very attractive proposal that will certainly contribute to biosensing development projects bringing advanced products to the market (Bhalla et al. 2016).

All the issues mentioned above remain rate-limiting factors for clinical translation and point-of-care diagnostics. Both researchers and industry are seeking for overcoming them. With the advantage of advances in microchip technology, microfluids and nanotechnology, innovative commercial biosensors will soon take their place in the medical devices market for medical applications. Soon, biosensor technology will surely demonstrate better sensitivity, measurement time, and reliability over conventional testing devices. Therefore, the engineering of automated, cost-effective, and easy-to-use sensors has become more crucial now than ever.

Conclusion

This chapter describes the fundamentals of several biosensors, which have been categorized based on their characteristics. The general classes include but are not limited to electrochemical, piezoelectric, optical, and calorimetric. These kinds of sensors are being developed into portable and automated systems in order to benefit point-of-care purposes. Nevertheless, a successful biosensor must have high-performance features such as affordability, sensitivity, selectivity, and user-friendliness. To validate the high performance of a sensing device, the choice of the target analyte is mandatory. There is a variety of biomolecules, ranging from small to whole cells, that have been demonstrated with these detecting systems. Pathological biomarkers are very crucial, as they are relevant to clinical diagnoses, particularly for detecting infectious diseases. To date, biosensor technology has been in progress for around 50 years, and there is no doubt that it has recently undergone noteworthy improvements in terms of its reliability and detection limits. Novel trends involve the introduction of nanotechnology, magnetic particles, graphene materials, microfluidics, and even miniaturization into lab-on-chip platforms. In spite of these advances, the application of biosensor technology in the diagnosis of infectious diseases is still in its early stages. Many scientific, technical, and commercial challenges still exist regardless of the type of sensing system. Yet, these are carefully being addressed by experts from different scientific domains and owing to their cooperation there will be soon more effective biosensing devices commercialized for clinical purposes.

References

Ali, A. A., Altemimi, A. B., Alhelfi, N., and Ibrahim, S. A. (2020). Application of Biosensors for Detection of Pathogenic Food Bacteria: A Review. Biosensors, 10(6): 58. https://doi.org/10.3390/bios10060058.

Altintas, Z. (2018). Biosensors and nanotechnology: Applications in health care diagnostics. Hoboken, NJ: Wiley.

Azizipour, N., Avazpour, R., Rosenzweig, D. H., Sawan, M., and Ajji, A. (2020). Evolution of biochip technology: a review from lab-on-a-chip to organ-on-a-chip. Micromachines, 11(6): 599. https://doi.org/10.3390/mi11060599.

Bakirhan, N. K., Ozcelikay, G., and Ozkan, S. A. (2018). Recent progress on the sensitive detection of cardiovascular disease markers by electrochemical-based biosensors. Journal of Pharmaceutical and Biomedical Analysis, 159: 406–424. https://doi.org/10.1016/j.jpba.2018.07.021.

Bhalla, N., Jolly, P., Formisano, N., and Estrela, P. (2016). Introduction to biosensors. Essays in Biochemistry, 60(1): 1–8. https://doi.org/10.1042/EBC20150001.

Cabral-Miranda, G., de Jesus, J. R., Oliveira, P. R., Britto, G. S., Pontes-de-Carvalho, L. C., Dutra, R. F., and Alcântara-Neves, N. M. (2014). Detection of parasite antigens in Leishmania infantum-infected spleen tissue by monoclonal antibody-, piezoelectric-based immunosensors. The Journal of Parasitology, 100(1): 73–78. https://doi.org/10.1645/GE-3052.1.

Campuzano, S., Yáñez-Sedeño, P., and Pingarrón, J. M. (2017a). Molecular biosensors for electrochemical detection of infectious pathogens in liquid biopsies: current trends and challenges. Sensors (Basel, Switzerland), 17(11): 2533. https://doi.org/10.3390/s17112533.

Campuzano, S., Yáñez-Sedeño, P., and Pingarrón, J. M. (2017b). Electrochemical biosensing for the diagnosis of viral infections and tropical diseases. ChemElectroChem, 4(4): 753–777. https://doi.org/10.1002/celc.201600805.

Chen, W., Yan, Y., Zhang, Y., Zhang, X., Yin, Y., and Ding, S. (2015). DNA transducer-triggered signal switch for visual colorimetric bioanalysis. Scientific reports, 5: 11190. https://doi.org/10.1038/srep11190.

Chowdhury, A. D., Takemura, K., Li, T. C., Suzuki, T., and Park, E. Y. (2019). Electrical pulse-induced electrochemical biosensor for hepatitis E virus detection. Nature communications, 10(1): 3737. https://doi.org/10.1038/s41467-019-11644-5.

Cui, F., Zhou, Z., and Zhou, H. S. (2020). Review—measurement and analysis of cancer biomarkers based on electrochemical biosensors. Journal of The Electrochemical Society, 167(3): 037525. https://doi.org/10.1149/2.0252003jes.

Eed, H. R., Abdel-Kader, N. S., El Tahan, M. H., Dai, T., and Amin, R. (2016). Bioluminescence-sensing assay for microbial growth recognition. Journal of Sensors, 2016, 1–5. https://doi.org/10.1155/2016/1492467.

Farooq, A., Ullah, M. W., Yang, Q., and Wang, S. (2019). Applications of phage-based biosensors in the diagnosis of infectious diseases, food safety, and environmental monitoring. *In*: Rinken, T., and Kivirand, K. (Ed.). Biosensors for Environmental Monitoring. IntechOpen. https://doi.org/10.5772/intechopen.88644.

Fu, Z., Lu, Y. C., and Lai, J. J. (2019). Recent advances in biosensors for nucleic acid and exosome detection. Chonnam Medical Journal, 55(2): 86–98. https://doi.org/10.4068/cmj.2019.55.2.86.

Gaddes, D. E., Demirel, M. C., Reeves, W. B., and Tadigadapa, S. (2015). Remote calorimetric detection of urea via flow injection analysis. Analyst. 140: 8033–40. https://doi.org/10.1039/c5an01306b.

Gorodkiewicz, E., and Lukaszewski, Z. (2018). Recent progress in surface plasmon resonance biosensors (2016 to Mid-2018). Biosensors, 8(4): 132. https://doi.org/10.3390/bios8040132.

Gray, E. R., Turbé, V., Lawson, V. E., Page, R. H., Cook, Z. C., Ferns, R. B., Nastouli, E., Pillay, D., Yatsuda, H., Athey, D., and McKendry, R. A. (2018). Ultra-rapid, sensitive and specific digital diagnosis of HIV with a dual-channel SAW biosensor in a pilot clinical study. NPJ digital medicine, 1: 35. https://doi.org/10.1038/s41746-018-0041-5.

Hai, W., Goda, T., Takeuchi, H., Yamaoka, S., Horiguchi, Y., Matsumoto, A., and Miyahara, Y. (2017). Specific recognition of human influenza virus with PEDOT bearing sialic acid-terminated trisaccharides. ACS Applied Materials & Interfaces, 9(16): 14162–14170. https://doi.org/10.1021/acsami.7b02523.

Hartono, A., Sanjaya, E., and Ramli, R. (2018). Glucose sensing using capacitive biosensor based on polyvinylidene fluoride thin film. Biosensors, 8(1): 12. https://doi.org/10.3390/bios8010012.

Hasan, A., Nurunnabi, M., Morshed, M., Paul, A., Polini, A., Kuila, T., Al Hariri, M., Lee, Y. K., and Jaffa, A. A. (2014). Recent advances in application of biosensors in tissue engineering. BioMed Research International, 2014, 307519. https://doi.org/10.1155/2014/307519.

He, L., Ni, L., Zhang, X., Zhang, C., and Xu, S. (2015). Fluorescent detection of specific DNA sequences related to toxoplasma gondii based on magnetic fluorescent nanoparticles Fe3O4/CdTe biosensor. International Journal of Biochemistry Research & Review, 6(3): 130–139. https://doi.org/10.9734/IJBCRR/2015/15254.

https://doi.org/10.1016/b978-0-12-813900-4.00001-4.

Justino, C. I. L., Freitas, A. C., Pereira, R., Duarte, A. C., and Rocha Santos, T. A. P. (2015). Recent developments in recognition elements for chemical sensors and biosensors. Trends in Analytical Chemistry, 68: 2–17. https://doi.org/10.1016/j.trac.2015.03.006.

Justino, C. I. L., Duarte, A. C., and Rocha-Santos, T. A. P. (2016). Critical overview on the application of sensors and biosensors for clinical analysis. Trends in Analytical Chemistry, 85: 36–60. https://doi.org/10.1016/j.trac.2016.04.004.

Kim, S. E., Tieu, M. V., Hwang, S. Y., and Lee, M. H. (2020). Magnetic particles: their applications from sample preparations to biosensing platforms. Micromachines, 11(3): 302. https://doi.org/10.3390/mi11030302.

Lakshmipriya, T., and Gopinath, S. C. B. (2019). An introduction to biosensors and biomolecules. Nanobiosensors for Biomolecular Targeting, 1–21.

Li, Z., Leustean, L., Inci, F., Zheng, M., Demirci, U., and Wang, S. (2019). Plasmonic-based platforms for diagnosis of infectious diseases at the point-of-care. Biotechnology Advances, 37(8): 107440. https://doi.org/10.1016/j.biotechadv.2019.107440.

Loo, S. W., and Pui, T. S. (2020). Cytokine and cancer biomarkers detection: the dawn of electrochemical paper-based biosensor. Sensors (Basel, Switzerland), 20(7): 1854. https://doi.org/10.3390/s20071854.

Marrazza, G. (2014). Piezoelectric biosensors for organophosphate and carbamate pesticides: a review. Biosensors, 4(3): 301–317. https://doi.org/10.3390/bios4030301.

Metkar, S. K., and Girigoswami, K. (2018). Diagnostic biosensors in medicine - a review. Biocatalysis and Agricultural Biotechnology, 17: 271–283. https://doi.org/10.1016/j.bcab.2018.11.029.

Morris, M. C. (2013). Spotlight on fluorescent biosensors-tools for diagnostics and drug discovery. ACS Medicinal Chemistry Letters, 5(2): 99–101. https://doi.org/10.1021/ml400472e.

Narwal, V., Deswal, R., Batra, B., Kalra, V., Hooda, R., Sharma, M., and Rana, J. S. (2019). Cholesterol biosensors: A review. Steroids, 143: 6–17. https://doi.org/10.1016/j.steroids.2018.12.003.

Nguyen, T. T., Han, G. R., Jang, C. H., and Ju, H. (2015). Optical birefringence of liquid crystals for label-free optical biosensing diagnosis. International Journal of Nanomedicine, 10 Spec Iss(Spec Iss), 25–32. https://doi.org/10.2147/IJN.S88286.

Pandey, C. M., and Malhotra, B. D. (2019). Biosensors: Fundamentals and applications (2nd ed.). Delhi, India: De Gruyter.

Pashchenko, O., Shelby, T., Banerjee, T., and Santra, S. (2018). A comparison of optical, electrochemical, magnetic, and colorimetric point-of-care biosensors for infectious disease diagnosis. ACS Infectious Diseases, 4(8): 1162–1178. https://doi.org/10.1021/acsinfecdis.8b00023.

Pejcic, B., De Marco, R., and Parkinson, G. (2006). The role of biosensors in the detection of emerging infectious diseases. The Analyst, 131(10): 1079–1090. https://doi.org/10.1039/b603402k.

Pihíková, D., Kasák, P., and Tkac, J. (2015). Glycoprofiling of cancer biomarkers: Label-free electrochemical lectin-based biosensors. Open Chemistry, 13(1): 636–655. https://doi.org/10.1515/chem-2015-0082.

Pohanka, M. (2018). Overview of piezoelectric biosensors, immunosensors and DNA sensors and their applications. Materials (Basel, Switzerland), 11(3): 448. https://doi.org/10.3390/ma11030448.

Prabowo, B. A., Purwidyantri, A., and Liu, K. C. (2018). Surface plasmon resonance optical sensor: a review on light source technology. Biosensors, 8(3): 80. https://doi.org/10.3390/bios8030080.

Rahimirad, N., Kavoosi, S., Shirzad, H., and Sadeghizadeh, M. (2019). Design and application of a bioluminescent biosensor for detection of toxicity using Huh7-CMV-Luc cell line. Iranian journal of pharmaceutical research: IJPR, 18(2): 686–695. https://doi.org/10.22037/ijpr.2019.1100687.

Ridhuan, N. S., Abdul Razak, K., and Lockman, Z. (2018). Fabrication and characterization of glucose biosensors by using hydrothermally grown ZnO Nanorods. Scientific Reports, 8(1): 13722. https://doi.org/10.1038/s41598-018-32127-5.

Rodovalho, V. R., Alves, L. M., Castro, A. C. H., Madurro, J. M., Brito-Madurro, A. G., and Santos, A. R. (2015). Biosensors applied to diagnosis of infectious diseases – an update. Journal of Biosensors & Bioelectronics, 1(3): 1015.

Saylan, Y., Erdem, Ö., Ünal, S., and Denizli, A. (2019). An alternative medical diagnosis method: biosensors for virus detection. Biosensors, 9(2): 65. https://doi.org/10.3390/bios9020065.

Sin, M. L., Mach, K. E., Wong, P. K., and Liao, J. C. (2014). Advances and challenges in biosensor-based diagnosis of infectious diseases. Expert Review of Molecular Diagnostics, 14(2): 225–244. https://doi.org/10.1586/14737159.2014.888313.

Sireesha, M., Babu, V. J., Kiran, A. S., and Ramakrishna, S. (2018). A review on carbon nanotubes in biosensor devices and their applications in medicine. Nanocomposites, 4(2): 36–57. doi:10.1080/20550324.2018.1478765.

Slaughter, G., and Kulkarni, T. (2019). Detection of human plasma glucose using a self-powered glucose biosensor. Energies, 12(5): 825. doi:10.3390/en12050825.

Srinivasan, B., and Tung, S. (2015). Development and applications of portable biosensors. Journal of Laboratory Sutomation, 20(4): 365–389. https://doi.org/10.1177/2211068215581349.

Su, H., Li, S., Jin, Y., Xian, Z.Y., Yang, D., Zhou, W., Mangaran, F., Leung, F., Sithamparanathan, G., and Kerman, K. (2017). Nanomaterial-based biosensors for biological detections. Advanced Health Care Technologies, 3: 19–29. https://doi.org/10.2147/AHCT.S94025.

Tanak, A. S., Muthukumar, S., Hashim, I. A., and Prasad, S. (2019). Rapid electrochemical device for single-drop point-of-use screening of parathyroid hormone. Bioelectronics in Medicine, 2(1): 13–27. https://doi.org/10.2217/bem-2019-0011.

Wang, A., Ding, Y., Li, L., Duan, D., Mei, Q., Zhuang, Q., Cui, S., and He, X. (2019). A novel electrochemical enzyme biosensor for detection of 17β-estradiol by mediated electron-transfer system. Talanta, 192: 478–485. https://doi.org/10.1016/j.talanta.2018.09.018.

Wang, D. S., and Fan, S. K. (2016). Microfluidic surface plasmon resonance sensors: from principles to point-of-care applications. Sensors (Basel, Switzerland), 16(8): 1175. https://doi.org/10.3390/s16081175.

Wang, J., Ren, Y., and Zhang, B. (2020). Application of microfluidics in biosensors. *In*: Ren, Y. (Ed.). Advances in Microfluidic Technologies for Energy and Environmental Applications. London, UK: IntechOpen. http://dx.doi.org/10.5772/intechopen.91929.

You, M., Litke, J. L., Wu, R., and Jaffrey, S. R. (2019). Detection of low-abundance metabolites in live cells using an rna integrator. Cell Chemical Biology, 26(4): 471–481.e3. https://doi.org/10.1016/j.chembiol.2019.01.005.

Zhang, Y., Luo, J., Flewitt, A. J., Cai, Z., and Zhao, X. (2018). Film bulk acoustic resonators (FBARs) as biosensors: A review. Biosensors & Bioelectronics, 116: 1–15. https://doi.org/10.1016/j.bios.2018.05.028.

Zheng, Q., Wu, H., Wang, N., Yan, R., Ma, Y., Guang, W., Wang, J., and Ding, K. (2014). Graphene-based biosensors for biomolecules detection. Current Nanoscience, 10(5): 627–637. doi:10.2174/1573413710666140422231701.

CHAPTER 10

Future Challenges in the Use of Biomolecules

Microbial and Natural Origin

Bidita Khandelwal, Arundhati Bag, Chamma Gupta* and *Abhishek Byahut*

Introduction

Biomolecules, the compounds produced by living organisms, may be grouped into small and large molecules. Small biomolecules, which are the intermediates and products of cell metabolism and have an important role in cell functioning, are often called metabolites. They can range from primary metabolites (amino acids, sugars, nucleic acids, and lipids) to secondary metabolites (alkaloids, essential oils, toxins, terpenoids, pigments, etc.). Primary metabolites are involved in basic life functions in all living organisms, which are hundreds of other small molecules called secondary metabolites that are found in the plant kingdom. Plants produce secondary metabolites through metabolic pathways to adapt themselves to their environment. Many of the biomolecules are in use for the well-being of mankind. While these biomolecules are mostly beneficial, their indiscriminate use can bring harmful effects not only to health but also to the environment and economy.

Sikkim Manipal Institute of Medical Sciences (SMIMS), Sikkim Manipal University (SMU), Gangtok, Sikkim, India, 737102.
* Corresponding author: bidita.k@smims.smu.edu.in

Public health remains to be at stake owing to numerous documented failures and the inefficiency of existing biomolecule-based therapies resulting in life-threatening conditions, primarily antibiotic resistance, microbial infections, and drug cytotoxicity. These challenges have led to undesirable consequences, like extended hospital stays, costlier alternative medical expenses, overall excess mortality, etc. When the final product builds up in the microbe, it can trigger global stress reactions which can lead to apoptosis. In addition, the formation of misfolded and physiologically inactive proteins might reduce the yield of recombinant proteins. Membrane peptides, high-molecular-weight molecules, and multi-domain polypeptides are commonly found in inclusion bodies. Furthermore, eukaryotic proteins expressed in a prokaryotic-based heterogeneous system may result in a compound that is not properly or incompletely modified by posttranslational kinases, which are essential for its functionality (Rosano and Ceccarelli 2014). This chapter aims to highlight the importance of biomolecules of microbial and natural origin and emphasizes the challenges associated with the use of such biomolecules.

Biomolecules

Biomolecules are defined as organic molecules comprising biomacromolecules (carbohydrate, protein, lipid, and nucleic acid) and biomicromolecules (simple sugar, peptide, amino acid, fatty acid, glycerol, and nucleotides), derived from plant or animal origin (Figure 1). Biomolecules are essential to all living organisms including bacteria, plants, and animals due to their distinct structural and functional properties (Rogers 2020). There are several major and few minor classes of biomolecules, which can be of microbial or natural origin as listed below:

i. Proteins and Amino acids
ii. Carbohydrates and starch/simple sugars
iii. Lipids and fatty acids/glycerol
iv. Nucleic acid (DNA/RNA) and nucleotides
v. Small organic molecules
vi. Inorganic ions (calcium, sodium, iron, magnesium, potassium, and chlorine)
vii. Combination of biomolecules (lipoproteins and glycoproteins)

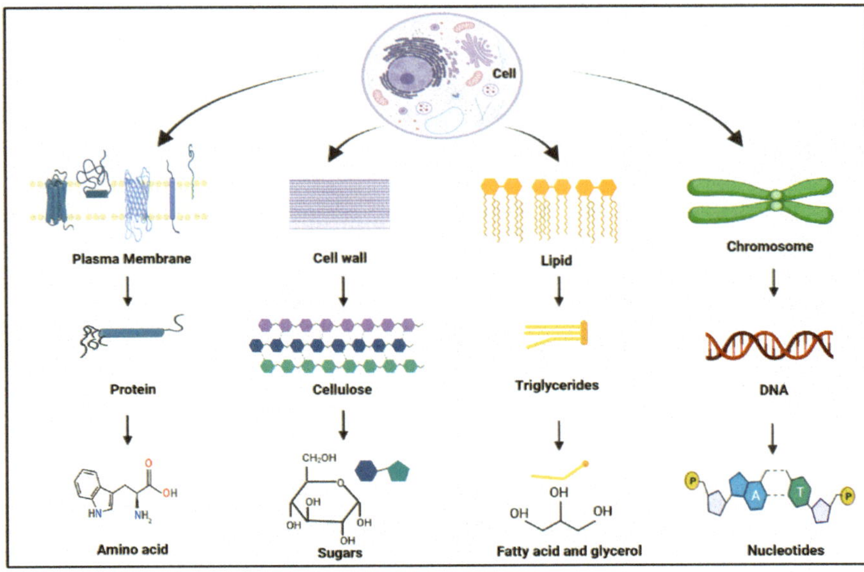

Figure 1. Four major classifications of biomolecules

Biomolecules of Different Origins and Associated Challenges

Microbial Natural Products and Challenges

Microbial natural products (NPs) are secondary metabolites derived from microbes including bacteria or fungi. They are valuable resources of therapeutics with varied chemical, structural diversity, and functions comprising antibiotics, antifungals, immunosuppressive, anti-inflammatory, and anti-cancer agents (Pham et al. 2019). Biomolecules of microbial and natural origin have had tremendous effects on modern drug discovery and imaging technologies through their extensive past and present use in the prevention, diagnosis, and treatment of human disease. Many natural molecules that can be used as new therapeutic compounds are found in low quantities in nature, making drug development difficult and practically unsustainable. As a result, expressing biosynthetic markers from the original sources in microbial hosts, particularly bacteria and fungus, is becoming a viable option (Song et al. 2014).

The microbial ecology is a vibrant system in which each species strives to outperform its competitors as intra-species competition is prevalent in such habitats. To combat competition, microbes use chemicals to triumph over one another (Nigam et al. 2014). Humans used these compounds to treat a variety of chronic disorders. Following the discovery of penicillin

(Bennett and Chung 2001), several antibiotics were found. For ages, antibiotics had proved to be useful in the treatment of dreadful infectious diseases. However, extensive use of antibiotics led to the emergence of multidrug resistance (MDR) in bacteria also known as "superbugs" which has presented the healthcare system with a new challenge (Dong et al. 2007).

Additionally, the microbial synthesis of natural and biological products faces several other obstacles. The challenges encountered by natural product-based therapeutic discoveries are minimal production titers, difficulties in molecule purification, and conformational characterization. Accumulation of the final by-product may elicit adverse effects, which can lead to downregulation and suppression of cell growth and proliferation. Furthermore, the inactive, misfolded, or altered peptides may reduce the output of the targeted protein. Eukaryotic proteins expressed in a prokaryotic system may result in a non-modified and non-translated product, essential for the functioning of a cell (Rosano and Ceccarelli 2014).

Microbial Enzymes

Enzymatic proteins can be found abundantly in nature. Pollen, dust mite, animal dander, and baking corn starch are all examples of polymers that can trigger allergies. Enzymes, like several other molecules, are foreign to the body that can cause allergic reactions on exposure to the allergen. Even the micro quantity of foreign antigen inhaled in the form of dust or aerosols can cause the immune system to generate corresponding antibodies. When re-exposed to the same antigen, the existence of these specific antibodies can induce the production of histamine in certain individuals. Any contact with such microbial enzymes may have a chance of producing sensitization, which could lead to hypersensitivity reactions. Watery eyes, a runny nose, and a scratchy throat are all symptoms that are manifested by a patient suffering from hay fever. The symptoms and effects are transient that may be mitigated by limiting/preventing the encounter of an allergen.

Enzymes are considered too complex to be produced using chemical methods and their pure form is extremely unstable due to which living beings are the only mode of production. The concern is that beneficial enzymatic compounds synthesized by microbes are frequently expressed on low scales and blended with a large number of other enzymes. These microbes can also be difficult to cultivate in industrial settings and can produce unwanted by-products. In addition to which, there is a risk of cross-contamination and the development of toxoids that might be fatal on use (Gurung et al. 2013).

Plant Biomolecules and Health Management

Plants produce a large pool of natural chemicals, which has contributed to a rich source of drugs and has led to the development of new drugs. An estimated 25% of prescribed drugs are derived from plants and many drugs are developed from synthetic analogues of plant-based compounds (Kala 2009). Other than its occupancy in modern drugs, plant-based extracts or their parts have been used in different traditional medical practices for ages, e.g., Ayurveda and Siddha in India, Unani medicine in the Middle East and South Asia and traditional medicines in China, Kampo Medicine in Japan, etc.

Plants can produce a wide range of phytochemicals, which can be broadly categorized into two groups, primary and secondary metabolites. While biomolecules like sugars and amino acids, involved in basic life functions are known as primary metabolites, in plants hundreds of other small molecules called secondary metabolites are found. Plants produce secondary metabolites through metabolic pathways to adapt themselves to their environment. More than 50,000 secondary metabolites have been reported in the plant kingdom (Teoh 2015).

Plant secondary metabolites are the molecules which bring out therapeutic effects in traditional herbal medicine as well as in modern medicine. They are classified on the basis of their chemical structures. The classes include phenolics, terpenoids, alkaloids, saponins, lipids, and carbohydrates (Hussein and El-Anssary 2018). Many of them are found to be pharmacologically active, e.g., flavonoids (a phenolic and antioxidant), polyphenolic catechins of tea (a phenolic, anti-oxidant, anti-metastatic, and preventive from epigenetic aberrations), vinblastine (alkaloid) and vinca alkaloids (anti-cancer), caffeine (an alkaloid, diuretic, stimulants of respiratory, cardiovascular, and central nervous system), essential oil (anti-microbial, analgesic, anti-inflammatory, etc. (Bag and Bag 2020, Hussein and El-Anssary 2018).

Endophytes

Endophytes are organisms that live within healthy plant cells, most common fungi, and bacteria. Endophytes play critical roles in plant growth and adaptability by producing protective chemicals that protect them from biotic and abiotic influences (Egan et al. 2016). The efficiency of the use of plant-based nutrients can be improved by endophytes through a variety of processes. This may be achieved in several ways, such as the development of extra-root hyphae for nutrient absorption, encouraging root growth, changing plant metabolism to boost nitrogen and phosphate intake, nitrogen fixation, and directly modifying soil or root exudates (Grodzinski et al. 2011).

Plant features such as pesticide and herbicide resistance can be developed using molecular approaches, such as gene editing and recombinant DNA technologies. Production of plant-specific endophyte inoculum on a large scale is required for commercial application. However, further research is needed to fully comprehend the dynamics of the host population. Reduced reliance on chemical fertilisers, insecticides, and fungicides will be aided by an improved and optimized generation of host-specific endophyte inoculum (Ogodo 2020).

Furthermore, bioactive secondary metabolites synthesized by endophytes such as alkaloids and lactones have antibacterial and anticancer effects. These bioactive molecules might be further investigated for medicinal, agricultural, and pharmaceutical applications (Pandey et al. 2017).

Herbal Medicines and Challenges

The past three decades have witnessed a tremendously growing popularity of the use of herbal medicines (Ekor 2014) due to their natural origin and affordable price. They have been used to treat a wide range of diseases, including asthma, tuberculosis, malaria, cholera, skin diseases, diabetes, hypertension, cancer, etc. (Mintah et al. 2019). It is largely assumed that, unlike synthetic drugs, herbal products are harmless for their natural origin which is untrue and misleading. Although many of these herbal drugs are promising enough, many of them are not tested for potential side effects or toxicity. Commonly used herbal medicines either include plant parts or crude extracts of them with more than one constituent, which are believed to work synergistically to bring a therapeutic effect (Ekor 2014). Many times herbal products are suggested as dietary supplements from which significant adverse effects may arise. Toxic effects have been reported from the indiscriminate use of many herbal products. To name a few, severe cardiotoxicity eventually leading to death has been found for aconitine-containing herbal preparation from Aconitum species (Dhesi et al. 2010, Tai et al. 1992). Similarly, serious toxicity for cardiovascular and central nervous systems have been reported for Ephedra sinica which is used to treat weight-loss dietary supplement (Hallas et al. 2008); liver cirrhosis for Tussilago farfara is used for acute and chronic cough and contact dermatitis and morbid spinal epidural hematoma for garlic extract used in the management of hypertension and hypercholesterolemia (Edgar et al. 2011). Therefore, rigorous testing, determination of dosages, good manufacturing procedure, toxicological evaluations, and quality monitoring should be strictly followed before marketing or suggesting herbal medicines.

However, there are many challenges in the implementation of pharmacovigilance due to several reasons (Ekor 2014). The same plant may be considered as food, supplementary food, or herbal medicine in different regions. Therefore, difficulty arises for the establishment of universal regulation on the usage of herbal products. Furthermore, it is difficult to single out toxic molecules in herbal medicines as they contain extracts from single or more plants or plant parts, which may contain mixtures of several phytochemicals. Due to a lack of internationally accepted quality control measurements, herbal products are often found contaminated by microorganisms, fungi, toxins, heavy metals, pesticides, dust pollen, etc., which can cause serious health effects (Zhou et al. 2019). Lack of knowledge for taxonomic identification of correct species, good collection practices, right plant selection, and cultivation are other challenges in quality control of traditional medicine.

Apart from pharmacovigilance, a major concern remains with the extensive exploitation of plant resources that can lead to the extinction of many valuable plants. It is estimated that around 15,000 medicinal plants may face extinction due to overuse, and at this rate of extinction, the world is losing one major drug every two years. Therefore, awareness of its sustainable use should be created among people and new policies are to be implemented to conserve the species (Roberson 2008).

Nutraceuticals and Challenges

Nutraceuticals, which combine 'nutrition' and 'pharmaceutical', include natural bioactive molecules with health benefits as well as medicinal effects. They may range from dietary supplements, processed food like cereals, beverages, soups, and isolated products of herbal origin to genetically modified foods (Rajam et al. 2010). They have non-specific biological effects that work by promoting well-being and preventing diseases, cosmetic effects (weight loss), etc. Nutraceuticals may be vitamins, minerals, fatty acids, amino acids, or plant extracts.

Unlike pharmaceuticals, nutraceuticals usually are not patented and do not require government sanction (Nasri et al. 2014) nor require a prescription to purchase them. While these make the consumers happy, unregulated use of nutraceuticals may impose serious health risks. They may cause side effects, e.g., allergic reactions, cardiac diseases, and/or interact with other nutraceuticals and drugs (Singh 2016). Therefore, the need for regulated use is important. Tea polyphenols have received special attention as nutraceuticals as consumption of tea is associated with several health benefits including protection from diabetes, microbial diseases, neurodegenerative diseases, cardiovascular diseases, etc. (Khan and Mukhtar 2018). Green tea, as well as black tea polyphenols, are also known to reduce the risk of different types of cancers. Research shows that these molecules can act as primary (inhibit mutagenesis) and tertiary (inhibits invasion and metastasis) chemopreventive agents (Arcone et al. 2016, Landis-Piwowar 2014). However, green tea polyphenols are known to have poor bioavailability which can be overcome by increasing the doses. This, on the other hand, may cause hepatotoxicity and can increase the risk of childhood carcinoma, neural tube birth defects, etc. (Bag and Bag 2019). However, research findings on nutraceuticals are frequently misinterpreted and the compounds are described as harmless and their beneficial effects are over projected for commercial interests due to high consumer demands (Siddiqui and Moghadasian 2020). Therefore, strict authoritative control should be introduced to avoid adverse effects on consumers. Quality control should also be introduced at the production level as stated previously for the herbal drugs.

Challenges in Nanomedicine

Other than synthetic polymers both plant- or animal-based proteins or polysaccharides are used in nanomedicines as carriers in drug-delivery systems (Gagliardi et al. 2021). For example, in nab-paclitaxel albumin nanoparticles are used to make a solvent-free drug delivery system, which makes it more efficient and less toxic in comparison to the solvent-based formulation of paclitaxel (Gradishar 2006). Gelatin, hyaluronic acid, and chitosan are other animal-based polymers and plant-based biopolymers, like cellulose, starch, soy-, proteins, etc., which are used for this purpose. Biopolymers of microbial or marine organism origin are also used. Challenges with using natural polymers include immune reactions. Usually, plant-based biopolymers are less immunogenic than animal-based biopolymers. The use of natural polymers in nanomedicine is also limited by the variation of the molecules with the source, complicated extraction process, and high production cost (Selvakumaran et al. 2015).

Surfactants and Challenges

Surfactants are amphiphilic (both hydrophobic and hydrophilic) substances, derived chemically (synthetic surfactants) or naturally produced by a range of microorganisms, including fungi and bacteria (biosurfactants). It can disperse itself across two immiscible fluids, lowering surface/interfacial tensions and causing the solubility of polar molecules in non-polar solvents (Otzen 2017). Biosurfactants have been in high demand as emulsifying agents, de-emulsifiers, wetting agents, dispersing additives, foaming agents, functional food components, and detergents. Biosurfactants have recently become the focus of many studies and implementations from the food industry to the oil industry due to many significant benefits over synthetic surfactants (Liu et al.

2015). A few examples are rhamnolipids and lipopeptides, secreted by bacteria (Pseudomonas, Burkholderia, and Bacillus), which are employed in a variety of industrial applications, crop protection, and biocontrol techniques (Fenibo et al. 2019). The use of biosurfactants as biopesticides in disease control has gained attention due to their multiple advantages; they are non-toxic, biodegradable in nature, compatible with the environment, and has the ability to be synthesized from renewable sources. Other biosurfactants, such as sophorolipids, mannosylerythritol lipids, and cellobiose lipids, have only been studied for their antibacterial capabilities against plant infections (Crouzet et al. 2020).

To enable the expanded use of biosurfactants in plant protection, their prices, field efficacy, and purity of compounds must be improved. To fully understand the benefits of biosurfactants in terms of environmental sustainability, manufacturing procedures that make them cost-effective and yield competitive results with synthetic equivalents must be developed and extended, assuring their economic viability. Trials on mixtures as well as highly purified molecules will be required in the future to better understand the mechanisms of action of biosurfactants. Biosurfactant usage in large-scale and commercial environmental applications has huge potential in the use of free to low-cost waste materials as substrates (Malik et al. 2015, Nikolva and Gutierrez 2021). Despite the present enthusiasm for these compounds, there are still a few issues to be resolved. The capacity to reliably detect and quantify the numerous contaminants of biosurfactants, which are normally generated by wild strains as well as effective separation and purification is critical. Additionally, molecular biology approaches to successfully adapt the biosurfactant generated to the specific demands of the product formulation will also be essential in future (Banat 2016).

Challenges Associated with the Use of Biomolecules in Medicine

Antibiotic Resistance

Antibiotics are defined as chemical compounds of microbial or synthetic origin, also termed antibacterial or antimicrobial drugs, that are used to treat or prevent certain forms of bacterial infections, which is done either by killing or blocking the proliferation of the bacteria (CDC 2020, Grennan et al. 2020). Antibiotic activity is largely attributed to the suppression of protein biosynthesis, DNA damage, and cell wall biosynthesis (Walsh 2000). Penicillin, which was discovered over 70 years ago, is still used with several other antibiotics as the same prominent antibacterial weapon since its invention (Lawrie 1985). The mechanism underlying the only existing treatment strategy is the use of antimicrobial agents in an attempt to eliminate the targeted pathogen. Despite its enormous success, focusing on the microorganism has two drawbacks:

i. Despite good prophylactic antibiotics, mortality is prevalent in communicable diseases. The host immune response causes tissue damage in addition to the direct effects of microbial virulence factors. Even after the eradication of the infection, immune-mediated injury leading to pulmonary, circulatory, and renal failure (sepsis) persists (Angus and van der Poll 2013). There are currently no treatments available to alter these harmful facets of the host immune response.
ii. The emergence of drug resistance potentially causes microorganisms to escape the reach of our only defence system against them. Unless something changes, infection-related fatalities are expected to rise dramatically, surpassing malignant disease in developed nations by 2050 (O'Neill 2016).

Antibiotic resistance can be transmitted in bacteria by a range of mechanisms, such as mutagenesis or genomic transfer, which can result in the selection of pre-existing forms, species, and variants. A mutation could change the configuration of the antimicrobial receptor, lowering its permeation, enhancing efflux, increasing the expression of an antibiotic-inactivating agent or bypassing an enzymatic route. Resistance can be transmitted horizontally by gene delivery via plasmids and transposons (Livermore 2003, Van Bambeke et al. 2003). The ampicillin-resistant factor blaTEM, which encodes TEM-1-lactamase, is the most prevalent and has expanded globally by this method. Few organisms integrate DNA from related species' dead cells, causing changes to their genes. Penicillin resistance in pneumococci has mostly proliferated through this process, which involves incorporating DNA produced from dead tissue of similar species and modifying their native genes (Livermore 2004, Spratt 1994).

Factors causing Antibiotic Resistance

Antibiotic resistance (AR) develops as a result of a two-fold effect in the context of diverse environmental and clinical contexts. For the summative impacts on overall resistance, each possible contributor including environmental habitat (soil and water) and clinical circumstances causing AR must be recognized to monitor the evolution of resistance. Following are the factors responsible for AR:

(i) Environmental: The soil resistome, rabs, and amr by aquatic microbiota are the prime contributor to Antimicrobial resistance (AMR). Certain environmental habitats are known to stimulate the genesis of AR as they include pools of genes comparable to antibiotic resistance genes (ARGs) in the hospital. Several antimicrobials synthesizing bacteria/fungi with a history of resistant strains have intrinsic potency or genes, which may be lost by one and acquired by the other microbes residing in the same niches of soil microbiota. This may be due to external factors like environmental stress, organic fertilizers, manure containing residual antibiotics (RAbs), ARGs, etc. AR in the aquatic ecosystem is a result of microbial exposure to RABs and is also dependent on the time of exposure (Adegoke et al. 2016).

(ii) Non-Environmental: Non-environmental factors have been identified to predict the emergence of AMR in Clinical and Sub-Clinical Settings. Some of the causative factors have gone unnoticed in spite of the growing rate of AMR outbreaks in hospitals. Therapeutic failure, a longer hospital stay, a higher expense of alternative treatment options, and possibly a higher risk of death are the possible outcomes. Self-medication is still a major barrier to the establishment of AMR. Self-medication is not just a phenomenon in developing nations. It includes countries where drug sales and administration are strictly regulated (Figueira et al. 2011, Pagán et al. 2006). Because orally delivered antimicrobial drugs (AMDs) are self-administered, they are more susceptible to self-medication than parenteral AMDs, resulting in multiple contraindications followed by the onset of AMR (Pan et al. 2012).

In many nations, a poor prescription auditing system has promoted antibiotic prescription without consideration of laboratory results and physician's consultation. Inappropriate prescription is as worse as self-medication because both induce AMR. These are also linked to patients' non-adherence to a physician-prescribed treatment schedule; the two anomalies resulting in the inability to attain and regulate the required plasma concentrations (inhibitory or lethal dose) during the drug's delivery time (Pan et al. 2012, Luyt et al. 2014).

Biosecurity

Biological warfare agents (BWAs) of microbial origin (viral, fungal, bacterial, and toxins) are among the most lethal biological weapons of mass destruction. BWAs such as Bacillus anthracis, variola major virus, Clostridium botulinum, and Yersinia pestis (plague) are troublesome and can be a major biological threat to mankind. This is because of their high transmission rate at low concentration, cryptic, asymptomatic nature (for several days) and affecting communities on a large scale. For example, 10 g of anthrax spores have the destructive potential of a tonne of nerve toxin sarin (Ivnitski et al. 2003). BWA aerosols are odorless, tasteless, and often undetectable after initial release. A BWA exposure, unlike a chemical weapon attack, usually does not result in an instant response.

Before a sufferer exhibits severe symptoms associated with exposure to a biological infection, an incubation time of several days is often expected. As a result, the impact of such a strike on affected individuals as well as the rate at which it spreads is obscured and amplified before it is detected (Beeching et al. 2002, Greenfield et al. 2002). The development of a real time based biodetection system is a need of the hour for the preparedness and respondents against a BWA attack. Biodetectors have the capacity to precisely anticipate the dispersion, concentration, and final target site of BWAs discharged into the environment in real time is essential to preparing and responding to a BWA release. Because of huge advancements in PCR and microarray technologies, precise and efficient screening of BWAs is being tackled with largely beneficial results. Advances in microarray technology indicate that hybridization-based techniques for detecting BWAs will be used in the near future (Connelly et al. 2012).

Drug Toxicity and Adverse Drug Reaction

Drug toxicity refers to side effects that occur when a drug's dose or plasma levels exceed the acceptable range, whether purposefully or negligently (drug overdose). Traditional chemotherapy fails to distinguish cancer cells from normal cells, killing both at the same time, and resulting in significant side effects. An adverse drug reaction (ADR) is an unfavorable or fatal reaction that occurs after taking a drug or a combination therapy in normal circumstances and is thought to be caused by the drug. An ADR usually necessitates the drug's discontinuation or dosage reduction. Drug abuse is defined as the use of recreational or pharmaceutical drugs that can result in addiction or dependency, major organ impairment (such as kidney, liver, or heart), psychological damage (aberrant behaviour patterns, psychosis, and impaired memory), or fatality (IUPHAR 2022).

Vaccines and Challenges

CDC defines vaccines as a substance that is used to boost the immune system's ability to fight disease. Vaccines are typically administered intradermal, intramuscularly or subcutaneously; however, some can also be given orally or through a nasal spray. Vaccines are now available to assist individuals of all ages to live longer and healthier by preventing more than 20 life-threatening illnesses. Every year, vaccines save 2–3 million fatalities due to illnesses, such as diphtheria, tetanus, pertussis, influenza, and measles. Vaccines are also important for preventing and controlling infectious disease outbreaks. They are essential in the fight against antimicrobial resistance and support global health security. For example, current influenza (Flu) vaccine is trivalent or quadrivalent but that can incorporate seasonal prevalent influenza (IAV) A (H1N1 and H3N2) and influenza B strains (NIH 2020).

COVID vaccination has helped in reducing the severity of the disease and its associated mortality. Frequent viral antigenic alterations and mutations pose a significant challenge for the IAV vaccine, resulting in vaccine incompatibility and ineffectiveness with a protection rate of 10 to 60% (Grohskopf 2020). Due to the biological complexity of the predominant microbes and the hurdles inherent in vaccine production and development for individuals of high risk is tough. The most well-known obstacles to failure or delay in vaccine development are twofold: the complexity of developing the production process, composition, and analytical assays as well as the difficulties of clinical assay adjustment (Heaton 2020).

Vaccine research and development (R&D) has challenges at all levels, including monitoring and reducing the time between vaccine discovery, manufacturing, and clinical development. It is important to assure and monitor the native antigen produced in response to the replication of the vaccine, the strength of elicited immune response for efficient protection and forecast the safety and efficacy of vaccines at an early stage. Adjuvants and immunopotentiators are used to tailor the appropriate protective immune responses, that is systems biology and other computational methods. This is essential to predict the safety and efficiency of vaccines, genomics and proteomics for vaccine antigen discovery, structural biology to broadly redesign protective antigens, and synthetic technologies to accelerate vaccine production (Oyston and Robinson 2012, Tameris et al. 2013)

Interference and Silencing of RNA and DNA

RNA, RNA interference (RNAi), and RNA antisense are genome-based technologies that have gained the most attention so far. The RNA molecules used in this method block the target gene(s) from being expressed. The disadvantages of this method are that RNA-based therapies are fragile, difficult to distribute, and can cause toxicity. However, solutions to address these issues are being explored, and many of these medicines are already in the early stages of development (Biospectrum 2022).

Fusion Proteins

In biotherapeutics, fusion proteins are typically used to combine a human and mouse antibody. The Fc region of a human is linked with a Fab region of the mouse to retain specificity and make the antibody 'humanized' so that the body does not recognize it as a foreign molecule causing rejection. This occurs when two or more genes coding for different proteins are combined to form a single protein. This process can happen naturally as in the case of the bcr-abl fusion protein in CML (chronic myeloid leukaemia) or it can be genetically modified to develop biotherapeutics (Biospectrum 2022).

Future Prospects, Safety Concerns and a Potential Approach to Resolve the Issue

(i) Despite the numerous hurdles encountered by the microbial synthesis of natural products and biologics, there can be a tremendous scope of development in terms of therapeutic substances of microbial origin. With standard rDNA technologies, a number of proposed changes such as genetic modification, ribosomal engineering, mutagenesis, and conformational gene upregulation can be utilized to enhance the optimum manufacture of natural substances and pharmaceuticals in microbial systems. Current techniques and advancements, like CRISPR/Cas (Clustered Regularly Interspaced Short Palindromic Repeats/CRISPR associated protein), must be explored as gene-editing tools for further development and increased production. Non-model native organisms can also be modified to be recombinant expression hosts using genetic engineering tools, allowing them to be used as platforms for multimodal biosynthesis to make synthetic natural compounds and their derivatives. (Rosano and Ceccarelli 2014).

(ii) Genome mining is yet another method for exploring secondary metabolites that involve retrieving data from genotyping. The use of genome mining to evaluate cryptic bacterial biosynthetic gene clusters (BGCs) has resulted in the discovery of new compounds. By introducing a predicted precursor to a culture of Pseudomonas fluorescens, a genome mining technique coupled with bioinformatics predictions was employed to identify the newer natural chemical orfamide A. A genome mining-based multimodal biosynthesis strategy was recently used to find novel representatives of the leinamycin family of natural compounds in a recent study. Because of its powerful anticancer properties, distinct genetic architecture, and intriguing modes of action, leinamycin has been regarded as a potential anticancer therapeutic candidate (Zerikly and Challis 2009, Pan et al. 2017).

(iii) Gene shuffling and ribosomal remodeling for improved secondary metabolite production are two ways that can contribute to advancements in the field of microbial natural products. Furthermore, the incorporation of 'omics' data holds enormous potential in the discovery of natural substance therapies, e.g., the use of metabolics to accurately assess biochemical changes/alteration as well as metabolic processes. The field of metagenomics has progressed in recent years by aiding in the interpretation of a wide range of unique and complex microbiological sources, such as soil and water microbiota as examples of harsh environments (Mohana et al. 2018).

(iv) X-ray crystallography and cryo-electron microscopy are other innovative techniques that can solve structural issues by conformational characterization studies and analysis of natural microbial products. Cryo-electron microscopy is being explored as a promising tool for assessing complex molecular architectures at a resolution near the level of an atom (Shoemaker and Ando 2018).

Conclusion

Biomolecules have promising prospects. Anything which has opportunities is inevitably associated with challenges and so is true with the varied biomolecules available for use. An insight into the challenges has led to opportunities for improvement and growth in terms of technique leading to higher yield, decreased adverse effects, cost-effectiveness, and rapid detection of resistance. Easy availability of the counter products needs to be regulated at every step from its manufacturing to its availability to the consumer. As there is no one size fits all solution for the challenges in life, the hurdles similarly faced in natural versus microbial biomolecules are different and so the solutions to them also need to be customized. Consumer awareness and change in the perspective toward natural biomolecules with a clear understanding of the facts is the need of the hour. There are many unexplored aspects of biomolecules and as the scientific community unravels the unknown, newer challenges will emerge which would be an opportunity to rediscover rather than be an obstacle.

References

Adegoke, A. A., Awolusi, O. O., and Stenström, T. A. (2016). Organic fertilizers: public health intricacies. Organic Fertilizers–From Basic Concepts to Applied Outcomes, 343–374.

Angus, D. C., and van der Poll, T. (2013). Severe sepsis and septic shock. New England Journal of Medicine, 369(9): 840–851. https://doi.org/10.1056/NEJMra1208623.

Arcone, R., Palma, M., Pagliara, V., Graziani, G., Masullo, M., and Nardone, G. (2016). Green tea polyphenols affect invasiveness of human gastric MKN-28 cells by inhibition of LPS or TNF-α induced Matrix Metalloproteinase-9/2. Biochimie open, 3: 56–63. https://doi.org/10.1016/j.biopen.2016.10.002.

Bag, N., and Bag A. (2020). Antimetastatic property of tea polyphenols. Nutr. Cancer, 72(3): 365–376. doi: 10.1080/01635581.2019.1638426.

Banat I. M. (2016). Biosurfactants: Challenges, successes and future opportunities 7th World Congress on Microbiology. https://microbiology.conferenceseries.com/europe/abstract/2016/biosurfactants-challenges-successes-and-future-opportunities.

Beeching, N. J., Dance, D. A., Miller, A. R., and Spencer, R. C. (2002). Biological warfare and bioterrorism. BMJ (Clinical research ed.), 324(7333): 336–339. https://doi.org/10.1136/bmj.324.7333.336.

Bennett, J. W., and Chung, K.-T. (2001). Alexander Fleming and the discovery of penicillin. In Advances in Applied Microbiology (Vol. 49, pp. 163–184). Academic Press. https://doi.org/10.1016/S0065-2164(01)49013-7.

Bhushan, B., Mahato, D. K., Verma, D. K., Kapri, M., and Srivastav, P. P. (2017). Potential health benefits of tea polyphenols—A review. In Engineering Interventions in Agricultural Processing. Apple Academic Press.

Biospectrum (March, 2022). Biomolecules and future biotherapeutics. https://www.biospectrumindia.com/features/70/8699/biomolecules-and-future-biotherapeutics.html.

CDC fact sheet. (2020). All antibiotic classes. https://arpsp.cdc.gov/profile/antibiotic-use/all-classes.

Connelly, J. T., Kondapalli, S., Skoupi, M., Parker, J. S., Kirby, B. J., and Baeumner, A. J. (2012). Micro-total analysis system for virus detection: microfluidic pre-concentration coupled to liposome-based detection. Analytical and Bioanalytical Chemistry, 402(1): 315–323. https://doi.org/10.1007/s00216-011-5381-9.

Crouzet, J., Arguelles-Arias, A., Dhondt-Cordelier, S., Cordelier, S., Pršić, J., Hoff, G., and Dorey, S. (2020). Biosurfactants in plant protection against diseases: Rhamnolipids and lipopeptides case study. Frontiers in Bioengineering and Biotechnology, 1014.

De, S., Malik, S., Ghosh, A., Saha, R., and Saha, B. (2015). A review on natural surfactants. RSC advances, 5(81): 65757–65767.

Dhesi, P., Ng, R., Shehata, M. M., and Shah, P. K. (2010). Ventricular tachycardia after ingestion of Ayurveda herbal antidiarrheal medication containing aconitum. Archives of Internal Medicine, 170(3): 303–305. https://doi.org/10.1001/archinternmed.2009.518.

Dong, Y.-H., Wang, L.-H., and Zhang, L.-H. (2007). Quorum-quenching microbial infections: Mechanisms and implications. Philosophical Transactions of the Royal Society B: Biological Sciences, 362(1483): 1201–1211. https://doi.org/10.1098/rstb.2007.2045.

Edgar, J. A., Colegate, S. M., Boppré, M., and Molyneux, R. J. (2011). Pyrrolizidine alkaloids in food: a spectrum of potential health consequences. Food additives and contaminants. Part A, Chemistry, Analysis, Control, Exposure and Risk Assessment, 28(3): 308–324. https://doi.org/10.1080/19440049.2010.547520.

Egan, J. M., Kaur, A., Raja, H. A., Kellogg, J. J., Oberlies, N. H., and Cech, N. B. (2016). Antimicrobial fungal endophytes from the botanical medicine goldenseal (Hydrastis canadensis). Phytochemistry Letters, 17: 219–225. https://doi.org/10.1016/j.phytol.2016.07.031.

Ekor, M. (2014). The growing use of herbal medicines: issues relating to adverse reactions and challenges in monitoring safety. Frontiers in pharmacology, 4, 177. https://doi.org/10.3389/fphar.2013.00177.

Fenibo, E. O., Ijoma, G. N., Selvarajan, R., and Chikere, C. B. (2019). Microbial surfactants: the next generation multifunctional biomolecules for applications in the petroleum industry and its associated environmental remediation. Microorganisms, 7(11): 581. https://doi.org/10.3390/microorganisms7110581.

Figueira, V., Vaz-Moreira, I., Silva, M., and Manaia, C. M. (2011). Diversity and antibiotic resistance of *Aeromonas* spp. in drinking and wastewater treatment plants. Water Research, 45(17): 5599–5611. https://doi.org/10.1016/j.watres.2011.08.021.

Gagliardi, A., Giuliano, E., Venkateswararao, E., Fresta, M., Bulotta, S., Awasthi, V., and Cosco, D. (2021). Biodegradable Polymeric Nanoparticles for Drug Delivery to Solid Tumors. Frontiers in Pharmacology, 12:601626. https://doi.org/10.3389/fphar.2021.601626.

Gradishar, W. J. (2006). Albumin-bound paclitaxel: a next-generation taxane. Expert Oopinion on Pharmacotherapy, 7(8): 1041–1053. https://doi.org/10.1517/14656566.7.8.1041.

Greenfield, R. A., Lutz, B. D., Huycke, M. M., and Gilmore, M. S. (2002). Unconventional biological threats and the molecular biological response to biological threats. The American Journal of the Medical Sciences, 323(6): 350–357. https://doi.org/10.1097/00000441-200206000-00007.

Grennan, D., Varughese, C., and Moore, N. M. (2020). Medications for treating infection. JAMA, 323(1): 100–100. https://doi.org/10.1001/jama.2019.17387.

Grodzinski, B., King, W. A., and Yada, R. (2011). 4.01—Introduction. pp. 1–7. *In*: Moo-Young, M. (Ed.). Comprehensive Biotechnology (Second Edition). Academic Press. https://doi.org/10.1016/B978-0-08-088504-9.00240-3.

Grohskopf, L. A. (2020). Prevention and Control of Seasonal Influenza with Vaccines: Recommendations of the Advisory Committee on Immunization Practices — the United States, 2020–21 Influenza Season. MMWR. Recommendations and Reports, 69. https://doi.org/10.15585/mmwr.rr6908a1.

Gurung, N., Ray, S., Bose, S., and Rai, V. (2013). A broader view: microbial enzymes and their relevance in industries, medicine, and beyond. BioMed Research International, 2013, e329121. https://doi.org/10.1155/2013/329121.

Hallas, J., Bjerrum, L., Stovring, H., and Andersen, M. (2008). Use of a prescribed ephedrine/caffeine combination and the risk of serious cardiovascular events: a registry-based case-crossover study. Am. J. Epidemiol. 168: 966–973. doi: 10.1093/aje/kwn191.

Heaton, P. M. (2020). Challenges of developing novel vaccines with particular global health importance. Frontiers in Immunology, 11. https://www.frontiersin.org/article/10.3389/fimmu.2020.517290.

Hussein, R. A., and El-Anssary, A. A. (2018). Plants secondary metabolites: the key drivers of the pharmacological actions of medicinal plants. *In*: Philip, F. Builders (Ed.). Herbal Medicine. IntechOpen. https://doi.org/10.5772/intechopen.76139.

IUPHAR. (2022). The Pharmacology Education Project (PEP), The International Union of Basic and Clinical Pharmacology (IUPHAR), Adverse Drug Resistance. https://www.pharmacologyeducation.org/clinical-pharmacology/adverse-drug-reactions.

Ivnitski, D., O'Neil, D. J., Gattuso, A., Schlicht, R., Calidonna, M., and Fisher, R. (2003). Nucleic acid approaches for detection and identification of biological warfare and infectious disease agents. BioTechniques, 35(4): 862–869. https://doi.org/10.2144/03354ss03.

Kala, C. P. (2009). Medicinal plants conservation and enterprise development. Medicinal Plants—International Journal of Phytomedicines and Related Industries, 1(2): 79. https://doi.org/10.5958/j.0975-4261.1.2.011.

Khan, N., and Mukhtar, H. (2018). Tea polyphenols in promotion of human health. Nutrients, 11(1): 39. https://doi.org/10.3390/nu11010039.

Landis-Piwowar, K. R., and Iyer, N. R. (2014). Cancer chemoprevention: current state of the art. Cancer Growth and Metastasis, 7: 19–25. https://doi.org/10.4137/CGM.S11288.

Lawrie, R. (1985). First clinical use of penicillin. British Medical Journal (Clinical Research Ed.), 290(6465): 397. https://www.ncbi.nlm.nih.gov/pmc/articles/PMC1417391/.

Liu, J. F., Mbadinga, S. M., Yang, S. Z., Gu, J. D., and Mu, B. Z. (2015). Chemical structure, property and potential applications of biosurfactants produced by Bacillus subtilis in petroleum recovery and spill mitigation. International Journal of Molecular Sciences, 16(3): 4814–4837. https://doi.org/10.3390/ijms16034814.

Livermore, D. (2004). Can better prescribing turn the tide of resistance? Nature Reviews Microbiology, 2(1): 73–78. https://doi.org/10.1038/nrmicro798.

Livermore, D. M. (2003). Bacterial resistance: origins, epidemiology, and impact. Clinical Infectious Diseases, 36(Supplement_1): S11–S23. https://doi.org/10.1086/344654.

Luyt, C. E., Bréchot, N., Trouillet, J. L., and Chastre, J. (2014). Antibiotic stewardship in the intensive care unit. Critical care (London, England), 18(5): 480. https://doi.org/10.1186/s13054-014-0480-6.

Mintah, S. O. , Asafo-Agyei, T., Archer, M., Junior, P. A., Boamah, D., Kumadoh, D., Appiah, A., Ocloo, A., Boakye, Y. D., and Agyare, C. (2019). Medicinal plants for treatment of prevalent diseases. *In*: Perveen, S., and Al-Taweel, A. (Eds.). Pharmacognosy - Medicinal Plants. IntechOpen. https://doi.org/10.5772/intechopen.82049.

Mohana, N. C., Rao, H. Y., Rakshith, D., Mithun, P. R., Nuthan, B. R., and Satish, S. (2018). Omics based approach for biodiscovery of microbial natural products in antibiotic resistance era. Journal of Genetic Engineering and Biotechnology, 16(1): 1–8.

Muhamad, I. I., Selvakumaran, S., and Lazim, N. A. (2015). Designing Polymeric Nanoparticles for Targeted Drug Delivery System.

Nasri, H., Baradaran, A., Shirzad, H., and Rafieian-Kopaei, M. (2014). New concepts in nutraceuticals as alternative for pharmaceuticals. International Journal of Preventive Medicine, 5(12): 1487–1499.

Nigam, A., Gupta, D., and Sharma, A. (2014). Treatment of infectious disease: Beyond antibiotics. Microbiological Research, 169(9): 643–651. https://doi.org/10.1016/j.micres.2014.02.009.

NIH (August 13, 2020). National Institite of allergy and infectious disease. https://www.niaid.nih.gov/research/vaccines.

Nikolova, C., and Gutierrez, T. (2021). Biosurfactants and their applications in the oil and gas industry: current state of knowledge and future perspectives. Frontiers in Bioengineering and Biotechnology, 46.

O'Neill, J. (2016). Review on antimicrobial resistance: Tackling drug-resistant infections globally: final report and recommendations. Review on Antimicrobial Resistance: Tackling Drug-Resistant Infections Globally: Final Report and Recommendations. https://www.cabdirect.org/globalhealth/abstract/20163354200.

Ogodo, A. C. (2020). Chapter 11—Biological control of plant pests by endophytic microorganisms. pp. 127–134. *In*: Egbuna, C., and Sawicka, B. (Eds.). Natural Remedies for Pest, Disease and Weed Control. Academic Press. https://doi.org/10.1016/B978-0-12-819304-4.00011-7.

Otzen, D. E. (2017). Biosurfactants and surfactants interacting with membranes and proteins: same but different? Biochim. Biophys. Acta., 1859: 639–649.

Oyston, P., and Robinson, K. (2012). The current challenges for vaccine development. Journal of Medical Microbiology, 61(7): 889–894.

Pagán, J. A., Ross, S., Yau, J., and Polsky, D. (2006). Self-medication and health insurance coverage in Mexico. Health policy (Amsterdam, Netherlands), 75(2): 170–177. https://doi.org/10.1016/j.healthpol.2005.03.007.

Pan, G., Xu, Z., Guo, Z., Hindra, Ma, M., Yang, D., Zhou, H., Gansemans, Y., Zhu, X., Huang, Y., Zhao, L. X., Jiang, Y., Cheng, J., Van Nieuwerburgh, F., Suh, J. W., Duan, Y., and Shen, B. (2017). Discovery of the leinamycin family of natural products by mining actinobacterial genomes. Proceedings of the National Academy of Sciences of the United States of America, 114(52): E11131–E11140. https://doi.org/10.1073/pnas.1716245115.

Pan, H., Cui, B., Zhang, D., Farrar, J., Law, F., and Ba-Thein, W. (2012). Prior knowledge, older age, and higher allowance are risk factors for self-medication with antibiotics among university students in southern China. PloS one, 7(7): e41314. https://doi.org/10.1371/journal.pone.0041314.

Pandey, P. K., Singh, M. C., Singh, S., Singh, A. K., Kumar, M., Pathak, M., Shakywar, R. C., and Pandey, A. K. (2017). Inside the plants: endophytic bacteria and their functional attributes for plant growth promotion. International Journal of Current Microbiology and Applied Sciences, 6(2): 11–21. https://doi.org/10.20546/ijcmas.2017.602.002.

Pham, J. V., Yilma, M. A., Feliz, A., Majid, M. T., Maffetone, N., Walker, J. R., Kim, E., Cho, H. J., Reynolds, J. M., Song, M. C., Park, S. R., and Yoon, Y. J. (2019). A review of the microbial production of bioactive natural products and biologics. Frontiers in Microbiology, 10: 1404. https://doi.org/10.3389/fmicb.2019.01404.

Rajam, S.A., Nagasamy, V., and Vishnu, V. (2010). Nutraceutical as medicine. An International Journal of Advances In Pharmaceutical Sciences, 1(2): 132–145.

Roberson, E. (2008). Nature's Pharmacy, Our Treasure Chest: Why We Must Conserve Our Natural Heritage. Center for Biological Diversity. www. biologicaldiversity. org.

Rogers, K. (2020, March 18). biomolecule. Encyclopedia Britannica. https://www.britannica.com/science/biomolecule.

Rosano, G. L., and Ceccarelli, E. A. (2014). Recombinant protein expression in *Escherichia coli*: Advances and challenges. Frontiers in Microbiology, 5. https://www.frontiersin.org/article/10.3389/fmicb.2014.00172.

Shoemaker, S. C., and Ando, N. (2018). X-rays in the Cryo-Electron Microscopy Era: structural biology's dynamic future. Biochemistry, 57(3): 277–285. https://doi.org/10.1021/acs.biochem.7b01031.

Siddiqui, R. A., and Moghadasian, M. H. (2020). Nutraceuticals and nutrition supplements: challenges and opportunities. Nutrients, 12(6): 1593. https://doi.org/10.3390/nu12061593.

Singh, R. K. (2016). Nutraceuticals in Reproductive and Developmental Disorders.

Song, M. C., Kim, E. J., Kim, E., Rathwell, K., Nam, S.-J., and Yoon, Y. J. (2014). Microbial biosynthesis of medicinally important plant secondary metabolites. Natural Product Reports, 31(11): 1497–1509. https://doi.org/10.1039/C4NP00057A.

Spratt, B. G. (1994). Resistance to antibiotics mediated by target alterations. Science, 264(5157): 388–393. https://doi.org/10.1126/science.8153626.

Tai, Y. T., But, P. P., Young, K., and Lau, C. P. (1992). Cardiotoxicity after accidental herb-induced aconite poisoning. Lancet (London, England), 340(8830): 1254–1256. https://doi.org/10.1016/0140-6736(92)92951-b.

Tameris, M. D., Hatherill, M., Landry, B. S., Scriba, T. J., Snowden, M. A., Lockhart, S. et al. Trial Study Team. (2013). Safety and efficacy of MVA85A, a new tuberculosis vaccine, in infants previously vaccinated with BCG: a randomised, placebo-controlled phase 2b trial. The Lancet, 381(9871): 1021–1028.

Teoh, E. S. (2015). Genus: Calanthe to Cyrtosia. Medicinal Orchids of Asia, 171–250. https://doi.org/10.1007/978-3-319-24274-3_9.

Van Bambeke, F., Glupczynski, Y., Plésiat, P., Pechère, J. C., and Tulkens, P. M. (2003). Antibiotic efflux pumps in prokaryotic cells: Occurrence, impact on resistance and strategies for the future of antimicrobial therapy. Journal of Antimicrobial Chemotherapy, 51(5): 1055–1065. https://doi.org/10.1093/jac/dkg224.

Walsh, C. (2000). Molecular mechanisms that confer antibacterial drug resistance. Nature, 406(6797): 775–781. https://doi.org/10.1038/35021219.

Zerikly, M., and Challis, G. L. (2009). Strategies for the discovery of new natural products by genome mining. Chembiochem : a European Journal Oof Chemical Biology, 10(4): 625–633. https://doi.org/10.1002/cbic.200800389.

Zhou, X., Li, C. G., Chang, D., and Bensoussan, A. (2019). Current status and major challenges to the safety and efficacy presented by chinese herbal medicine. Medicines (Basel, Switzerland), 6(1): 14. https://doi.org/10.3390/medicines6010014.

Index

A

Actinobacteria 4, 7
Antibiotics 2–4, 7, 8
Antimicrobial peptides 15, 19, 23, 25, 112, 113, 117
antimicrobial resistance 15, 22, 29, 35
Anti-Tuberculosis 44–50

B

bacteriocins 23, 25–28
bacteriophages 15, 23, 28, 30, 31, 35
Bioactive compounds 76–78
bioavailability 59, 155, 163, 166, 173, 211
Biomolecule 43–45, 57, 60, 68, 178, 179, 181, 183, 186, 189, 199, 204–206, 208, 212, 218
Biosensors 178, 179, 181–184, 186–192, 194–199

C

Chagas disease 157, 159, 160
clinical diagnosis 187
Communicable Disease 56
Cyclodextrins (CDs) 154

D

direct killing 121, 128
DNA Technology 82
Drugs 75–82

E

enveloped and non-enveloped viruses 130
Epidemics 75, 80

G

Gram-positive and Gram-negative bacteria 113, 119, 120, 124

H

host cell 130, 131, 134, 135, 137, 138
Human African Trypanosomiasis (HAT) 159, 161

I

immune modulation 121
Infection 56, 57, 60–65, 67, 69
Infectious Disease 74, 179, 190, 191, 194, 199

L

lab-on-a-chip 197
Leishmaniasis 154, 156–158, 162–165

M

Marine Environment 75, 78, 79, 81
Marine Organisms 74, 76, 78
mechanism of action 126, 132, 134, 137, 139
Microbes 75, 76, 78
microbial molecules 48
Mycobacterium 6, 8
Mycobacterium tuberculosis 86, 89–91, 101, 103

N

nanoparticles 23, 32, 33
nanotechnology 195, 199
natural macromolecules 43, 44
neglected tropical diseases (NTDs) 157

P

pathogens 14–17, 22, 24–27, 35
Pharmaceutical industry 78, 81
Plant products 5
primary metabolites 204, 208

S

secondary metabolites 204, 206, 208, 209, 217

T

Treatment 56, 57, 61, 62, 69
Tuberculosis 43–50, 86–103

V

viral infections 130, 131, 134, 136–138

W

WHO global priority pathogens list 9